MW00592793

CRACKING THE REGENTS

SEQUENTIAL MATH III

479

THE PRINCETON REVIEW

1999–2000

CRACKING THE REGENTS

SEQUENTIAL MATH III

DAVID S. KAHN

1999–2000 Edition

Random House, Inc.
New York

Princeton Review Publishing
2315 Broadway
New York, NY 10024
E-mail: info@review.com

Published in the United States by Random House, Inc., New York, and
simultaneously in Canada by Random House of Canada Limited, Toronto.

ISBN 0-375-75273-0

ISSN 1097-1130

Editor: Gretchen Feder
Design: Chris J. Thomas
Production Editor: Kristen Azzara
Production Coordinator: Scott Harris

Manufactured in the United States of America.

9 8 7 6 5 4 3 2 1

1999–2000 Edition

ACKNOWLEDGMENTS

First of all, I would like to thank Arnold Feingold and Peter B. Kahn for once again doing every problem, reading every word, and otherwise lending their invaluable assistance. I also want to thank my editor, Lesly Atlas, and my producer, Amy Bryant, and the rest of The Princeton Review for their assistance. Thanks also to Gary King, Nancy Schneider, and Blase Caruana, for reviewing all of the test problems. Thanks to Frank, without whose advice I probably wouldn't be doing this altogether, and to Carl, for reasons beyond counting. Thanks to Jeffrey and Miriam for moral support. Thanks Mom.

Finally, I would like to thank the people who really made all of this effort worthwhile—my students. Your support truly helped make this book possible, so thank you:

Aaron G., Aaron R., Abby B., Abby F., Aidan, Alex, Alex and Gabe, Alexandra, Alexes, Alexis, Ali, Alicia, Allison and Andrew S., Ally, Alyssa, Amanda B., Amanda C., Amanda M., Amelia, Andrea T., Andrea V., Andrew B., Andrew S., Andy and Allison, Anna L-W., Anna F., Anna M., April, Arthur, Arya, Asheley and Freddy, Ashley, Ashley and Lauren, Bethany and Lesley, Betsy and Jon, Blythe, Brett, Brian, Brooke, Butch, Caitlin, Caroline, Chad, Christine, Christine W., Chrissy, Claudia, Corinne, Courtney, Craig, Daniel, Danielle, Dara P., Dara M., Deborah, Devon, Dora, Eairinn, Elisa, Emily B., Emily C., Emily G., Emily L., Emily S., Emily T. and Erica H., Eric and Lauren, Erin, Frank, Gabby, Geoffrey, George, Gloria, Heather, Hilary, Holli, Holly G., Holly K., Ingrid, Jackson, Jaclyn, Jacob, Jan, Jason, Jason Q., Jay, Jenna, Jen, Jennifer B., Jennifer W., Jesse, Jessica, Jessica L., Jocelyn, Johanna, John, John and Dan, Jon, Jordan, Josh, Judie, Julia, Julie H., Julie and Dana, Kat, Kate D., Kate L., Kate S., Katie, Katrina, Kimberly, Kitty, Laura F., Laura G., Laura Z., Lauren R., Lauren T., Lauren W., Lee R., Lee S., Leigh, Lila, Lily, Lily Hayes, Lindsay F., Lindsay R., Lindsay, Lisa, Lizzy, Magnolia, Mara, Mariel, Marielle, Marietta, Marisa, Marsha, Mary M., Matt, Matt S., Matthew B., Matt B., Matthew F., Matt V., Maya and Rohit, Melissa, Meredith, Milton, Morgan, Nadia, Nathania, Nicole, Nikki, Nora, Oliver, Omar, Oren, Paige, Pam, Peter, Rachel A., Rachel B., Rachel F., Rachel M., Rachel R., Ramit, Rayna, Rebecca, Ricky, Ruthie, Sally, Sam C., Sam L., Sam W., Samantha W., Samar, Samuel C., Sara, Sarah L., Sarah M-D., Sarah S., Sascha, Sasha, Saya, Sonja, Sonja and Talya, Sophia, Sophie, Stacy, Stacey, Stephanie, Stefanie, Tammy and Hayley, Taylor, Tenley, Terrence, Tracy, Tripp H., Tripp W., Waleed, and Zach.

If I forgot anyone, I apologize. I'll get you next time.

CONTENTS

PART I

HOW TO CRACK
THE SYSTEM

INTRODUCTION

WHAT IS THE PRINCETON REVIEW?

The Princeton Review is an international test preparation company with branches in all major U.S. cities and several cities abroad. In 1981, John Katzman started teaching an SAT prep course in his parents' living room. Within five years Katzman (in partnership with fellow test guru Adam Robinson) had established the largest SAT coaching program in the country.

The Princeton Review's phenomenal success in improving students' scores on standardized tests was (and continues to be) the result of a simple, innovative, and radically effective approach: study the test, not what the test *claims* to test. This approach has led to the development of techniques for taking standardized tests based on the principles the test writers themselves use to write the tests.

The Princeton Review has found that its methods work not just for cracking the SAT, but for any standardized test. We've successfully applied our system to the SAT II, AP, ACT, GMAT, and LSAT, to name just a few. As a result of hundreds of hours of exhaustive study, we are now applying that system to the New York State Regents exam. This book uses our time-tested principle: figure out what the test-givers want, then teach that to the test-takers in a comprehensive *and* fun way.

We also publish books and CD-ROMs on an enormous variety of education- and career-related topics. If you're interested, check out our Web site at www.review.com.

WHAT IS THE REGENTS EXAMINATION?

The Regents are annual exams given in a variety of subjects—math, English, history, sciences, and foreign languages. To get a Regents Diploma, you must pass a certain number of these exams in required subjects. Among other things, you must pass the math exams. Sequential III is the last in this series of math exams; you have probably already taken the Regents examinations in Sequential Math I and Sequential Math II and passed them. Thus, in this book we will assume that you are familiar with a few basic aspects of taking the Regents.

You usually take the Regents in June, in lieu of, or in addition to, finals, but they are also offered in January and August. Most of you will take the Sequential III exam in eleventh grade, but some of you take it earlier or later.

The test is administered in school by your teachers and graded by them as well. The New York State Department of Education sets guidelines for grading these exams. These guidelines maintain uniformity in the grading process and give your teacher specific rules to follow, particularly when it comes to awarding partial credit for answers.

THE FUTURE OF THE REGENTS

In the next few years the Regents are going to change the exams. Fortunately, you probably will have graduated before then so you won't have to worry about it. But you can tell your younger friends and siblings that big changes are on the horizon—this will make you sound well-informed and will invite people to seek your advice in many other matters. In 1999 the mathematics and English exams will be made mandatory, and eventually there will be new versions of the math exams (Regents Math A and Math B, instead of I, II, and III). In 2001 the Regents Board plans to release two new exams—the Integrated Science Exam and the Social Studies Exam—which will also be mandatory. We would not be surprised if there were delays in these release dates, but these changes are definitely coming, so be ready!

HOW TO USE THIS BOOK

This book contains twelve past Regents exams and explanations of how to arrive at the correct answer for each question. Taking these practice exams can help you dramatically improve your performance on the exam, and along with The Princeton Review test-taking tips will help you become a better standardized test taker in general. This book also contains an overview of the material that is on the exam. You should study this material before you take any of the exams and review it again when you discover which areas are your weak points. If you find that you want a more in-depth review of the material, we suggest that you consult The Princeton Review's *High School Review Math III*. This book covers all of these topics in more detail and depth and includes lots of practice problems.

The primary value of *Cracking the Regents Sequential Math III* is that it contains the most recent exams with their explanations. Pay careful attention to the explanations to your wrong answers. We attack each question type the same way so you can develop a uniform approach to cracking each type of problem found on the test. We have also included a section we call Target Practice. If you want to work on a particular weak area, refer to this section to find the questions that specifically target that area.

CHAPTER 1
ABOUT THE EXAM

WHAT'S ON THE TEST

The point of the Regents exam is to test what you learned throughout the year in your math class. Specifically, the Regents Math III test covers the following areas:

Trigonometry

The bulk of the exam tests your knowledge of trigonometry. This includes general trigonometry involving the various trig ratios; converting from radians to degrees and the other way around; the laws of sines and cosines; finding the area of a triangle using trig; the trig formulas (which are provided for you on a formula sheet at the beginning of each exam); the graphs of the sine and cosine functions; and trig identities. In Part II of each exam, you will find at least one question on these topics.

This is a lot of material and should have involved several months of class time this past year. If you don't know *any* of this stuff, you have a lot of preparation time to tackle.

Algebra

The exam will contain a few questions covering advanced algebra including: equations containing radicals; equations containing rational functions; and equations containing exponents. You will be expected either to solve the equations or to simplify expressions containing these terms and to graph a very simple exponential equation. You need to know inverse variations, quadratic equations and inequalities, absolute value equations and inequalities, and the binomial expansion. In Part II of each exam, you will find at least one question on these topics.

Complex Numbers

The exam will cover a few questions covering basic complex numbers including the following: adding, subtracting, or multiplying complex numbers; graphing complex numbers; and powers of i. Most of these questions are very simple. You will *not* be expected to know DeMoivre's theorem. Sometimes there is a question in Part II of the exam on these topics.

Geometry

The only geometry covered on the test are circle rules and transformations. Questions on the circle rules include the following: secants and tangents; chords; arc lengths; and degree measures of arcs and angles. Questions on transformations include the follwoing: translations; rotations; reflections; and dilations. You will also be expected to know that all of these transformations preserve area except for dilations. There is virtually always a question in Part II of the exam on circle rules, and part of a question or an entire question on transformations.

Functions

The exam will have a couple of questions involving composite functions; whether a graph is of a function; logarithms—graphing and solving equations involving them; and some very simple algebra theory. There is always a question in Part II of the exam on at least one of these topics.

Probability and Statistics

There are always a few questions involving slightly more difficult probability (specifically binomial probability and summations) and statistics (namely standard deviation of a set of data and some basic questions involving a normal distribution) problems than on the Sequential I and II exams. There is always a question in Part II of the exam on at least one of these topics.

CHAPTER 2
PREPARING FOR THE EXAM

The exam is generally given at the end of your course, so the best preparation is to keep up with class assignments. You should have covered all of these topics in school and had a chance to practice many problems. The idea of the Regents exam is to give you a cumulative final exam where you can prove your knowledge of all of the subjects. This means that if you don't already know your stuff you have a lot of reviewing to do. One of the best ways to review is to use this book.

First, read through the material in the "Princeton Review Techniques" section (page 12). There you will learn how to do two very important things: *plug in* and *backsolve*. These are integral techniques for knowing how to solve problems when you don't know how to get the right answer. You will find that once you get good at plugging in and backsolving you will get more questions correct and you will move through the test faster.

Next read the section called "Math Review: What You Should Know," a quick review of the most important parts of the Sequential III curriculum (page 17). If you don't already know all of this stuff, this is the time to go back and learn it. We recommend that you review your textbook to find good practice problems. Or, as we pointed out before, you can refer to the companion book to this one—*High School Review Math III*—which explains all of the Regents topics in depth. No matter which method you choose, you must make sure that you know most or all of the material for the exams before you attempt the tests. Otherwise you are just going to be frustrated.

Next, read the sections called "How to Remember the Special Angles" and "Rules to Remember." These will give you a summary of the rules that you need to plant firmly in your brain so that you can do the problems easily. One of the most important aspects to getting a good score in trigonometry is knowing all of the trig values of the special angles. Think of them as your "times tables." Make flash cards and memorize the values, or use the method that we show you. Whichever you choose, knowing these values and rules is crucial to getting a good score.

Now you are ready to take the exams. Before you do, make yourself a study chart that will enable you to keep track of what kinds of questions you are getting wrong. Along the left-hand margin of a piece of paper, write the name (by corresponding date) for each of the practice exams. Going across the top of the paper, write the following column headings: *trig, geometry, algebra, prob/stat, complex, functions*. Now it's time to take a diagnostic test. After you are done, go through the answers and note in the appropriate column of your grid the problem numbers that you answered wrong. The answers each have headings to help you with this. For example, suppose that you take the January 1996 exam and that you get question 4 wrong. In the explanations section it tells you that this is a question on law of sines, which is a trig topic, so under the *trig* column put the number 4. Or suppose that you got question 27 wrong. In the explanations section it says that this is a question on statistics, so under the *prob/stat* column write the number 27. Got the idea? This will help you keep track of what your weak areas are. Do you see a pattern in what you are getting wrong? If so, you can use the "Target Practice" section to zero in on a particular area. If your difficulties are spread out, then keep drilling—take more tests, follow more explanations.

Practice a couple more tests and keep using the grid. By about the fourth test, you should be getting most of the problems in your weak areas correct. You should find that the questions are similar from one test to the next (this is why the Regents is a *standardized* test), and thus you should know what to do when you get to each question type. Now you can focus on speed. Do you have time left at the end of the test? If so, then you're fine. If not, start to push yourself. You can't get a high score if you don't finish. You will have several practice tests to time yourself and focus on your speed. You have three hours to complete the test. Although Part I is worth more than Part II, you should be able to complete it faster. You should be aiming to finish Part I in about seventy-five minutes. This will leave you plenty of time to go back and check your answers at the end.

Be realistic with your practice schedule. You are not going to do four of these tests in a day. In fact, if you can do four in a day and do well, you don't need our help! A good way to review these is to do one test every other day, if you have the time. If that means that you don't end up doing all of the tests in the book, that's okay. After you have done about eight practice tests, you should be in pretty good shape. Naturally, your goal is to take *all* of the tests in the book, but you may not have the time. Try not to start these tests late at night. This is not a realistic simulation of the exam, and you will make mistakes because you are tired, which is discouraging and counterproductive. Try to do the tests over the weekend or right after school. Make sure that you will not be disturbed while you are practicing. Remember, you are trying to simulate the real thing. Don't answer the phone during practice and **NEVER** look at the answers until you have finished. You won't be able to look at them during the real thing and looking gives you a mental crutch that is very difficult to throw away. That's it. Good luck. Get to work!

CHAPTER 3
TAKING THE REGENTS EXAM

If you are taking the Regents Sequential III exam, you have probably already taken the Sequential I and II exams, so this will be déjà vu for you. But just in case you have a bad memory, here is what you need to know:

1. Not only are you allowed to use a scientific calculator for the exam, you MUST use one. The calculator must have logarithmic and trigonometric features. It MAY NOT have infrared communication ability (for obvious reasons), but it may have statistical functions, π, and other cool features. It MAY NOT be a graphing calculator.

2. You must have a straight edge and should probably bring a compass.

3. You are supposed to be provided with blank paper for writing Part II answers, and graph paper (for graphing), but we wouldn't take any chances if we were you! In other words, come prepared!

4. You have to take the test in blue or black pen. You may only use pencil for graphs and diagrams.

SCORING

There is no partial credit for Part I of the test. There are 35 questions and you must answer 30 of them.

Which Ones To Skip

There doesn't seem to be a particular type of question in Part I that is harder than the others, but you should probably skip the question that we label "algebra theory." This usually asks about "fields," and if you don't know what a field is, you will have no idea how to answer it.

There is usually a "probability" question that can be time-consuming. If you don't know how to do it, skip it.

Remember that about two-thirds of Part I is multiple choice, which means that there are Princeton Review techniques that you can use to get the right answer (we'll get to them in a little bit). So, you will probably prefer to skip a question that is not multiple choice and stick to the ones that are.

The test is NOT in order of difficulty. However, you will probably find that Part I stuff is much easier than Part II stuff. Most of the difficulty will come from how well you know the material. If you are weak in trigonometry, you are going to have a *very* difficult time with this test.

Each of the Part I questions is worth two points, for a total of sixty points for the section. Remember that no *one* question is worth more than another, so don't spend too long on any question. Move on!

Part II usually contains seven questions, four of which you are supposed to answer. They are worth ten points each, for a total of forty points for the section. Most of the questions consist of more than one part, and the value of each part is given along with the question. Pay attention to the values! They will tell you what is hard and what is easy.

For example, there is usually a question on "circle rules" (chords, secants, arcs, etc.) that has five parts, each one worth two points. You usually need the answer to part (a) to figure out part (b), and so on. So, if you know your circle rules, this will be an easy question. If you don't, skip it.

OFFICIAL REGENTS GUIDELINES

There are no "fractional points" on the test. In other words, you can't get an 87.5 on this test. An arithmetic error that does not change the nature of the question should receive a deduction of one point, while an error due to a violation of some cardinal principle should receive a deduction ranging from 30 to 50 percent, depending on the relative importance of the principle in the solution of the problem. The maximum deduction for arithmetic errors in *each part* of a question is one point.

This means that if you know what you are doing on a question, but you are sloppy, the penalty is small. If, on the other hand, you have *no idea* what you are doing, then why are you answering the question?!

Here are some easy ways to lose partial credit:

- Failure to reduce fractions to lowest terms.
- Failure to round correctly.
- Leaving a perfect square inside the radical.
- Not identifying the unknown variable (usually x).
- Not labeling correctly.

If a question asks for $a + bi$ form, then an answer such as $\dfrac{5 + 3i}{2}$ is incorrect. The correct way to write the answer is $\dfrac{5}{2} + \dfrac{3i}{2}$.

Make sure that your graphs are correct, neat, legible, and labeled accurately. *Scale* is not that critical.

Some steps of a proof are more important than others. Don't skip a step, but don't worry that you might have missed one.

Label the axes!

If the question says to draw graphs on the same set of axes, don't make separate graphs.

There are lots of other things that the Regents exam is picky about, but they are usually *not* as picky as your teachers. What does this mean? It means that you should do all of those annoying little things that the teacher has been telling you to do all year!

USING THE CALCULATOR

As we're sure you know, the Regents require you to use a calculator on the exam. This gives them the freedom to ask you to find bigger, messier numbers, and means that you no longer have to use log and trig tables. It also gives you a big advantage because now you don't have to worry about messing up calculations. There are many types of calculators out there, so we couldn't possibly address all of the different kinds. You are **NOT** allowed to use a graphing calculator on the exam.

Make sure that you know how to switch back and forth from degrees to radians.

The sine, cosine, and tangent keys should be clearly marked. Remember, if you want to find $\sin^2 x$, you first have to find $\sin x$ and then square it. So, for example, if you wanted to find $\sin^2 10°$, you would enter: (**sin** 10)^2 or (**sin** 10)x^y2 (it depends on the calculator).

To find the root of a number, if you don't have a root key (which will look like $\sqrt[x]{y}$) raise the number to the corresponding fractional power. For example, to find $\sqrt[4]{16}$, you would enter: 16^(1/4).

The log key is marked **log**. Obviously! You won't need to use **ln** or e^x.

We're sure that there are lots of other goodies in your calculator, but we will leave it to you to discover them. The main thing is to be comfortable using a calculator so that it doesn't slow you down.

CHAPTER 4
PRINCETON REVIEW TECHNIQUES

There are two techniques that will be very helpful for multiple-choice questions—*plugging in* and *backsolving*. Of course, these are not a substitute for knowing the math, but if you can become adept with these techniques, you will be able to get right answers to questions where you are not sure of the correct mathematical method. They aren't necessarily faster, and if you can do the tests using the math that you already know, then you won't need them. But, if you aren't able to answer a question using your math skills, then you should use *plugging in* and *backsolving*.

PLUGGING IN

If you are given a problem that has variables in all the answer choices, you can convert the problem from an algebra problem to an arithmetic problem and make your life much easier.

How do you do this?

Step (1): Make up a number for the variable or variables.

Step (2): Plug the number into the problem and get an answer.

Step (3): Plug the number into the answer choices and pick the choice that gives you the same answer.

This stuff is always easier to understand if we do an example.

<u>Example:</u> (From the January 1992 Regents)

When combined, $\dfrac{2}{x+3} + \dfrac{1}{x}$ is equivalent to

(1) $\dfrac{3x+3}{x^2+3x}$ (3) $\dfrac{3x+3}{2x+3}$

(2) $\dfrac{3x+1}{x^2+3x}$ (4) $\dfrac{3}{2x+3}$

Now, you could certainly do the algebra to get this right, but why bother? Just plug in!

Step (1): Make up a value for x. Let's let $x = 4$.

Step (2): Plug $x = 4$ into the problem. We get:

$$\frac{2}{x+3} + \frac{1}{x} = \frac{2}{4+3} + \frac{1}{4} = \frac{2}{7} + \frac{1}{4} = \frac{15}{28}$$

Step (3): Plug $x = 4$ into the answer choices and see which one gives us $\frac{15}{28}$.

Choice (1): $\frac{3x+3}{x^2+3x} = \frac{3(4)+3}{(4)^2+3(4)} = \frac{15}{28}$. There it is!

Wasn't that easy? There is one thing that you should watch out for. Sometimes, because of the particular numbers that you picked, more than one choice will give the answer you are looking for. If so, you should repeat the plugging in using different numbers.

So, just in case, let's check the other three choices.

Choice (2): $\frac{3x+1}{x^2+3x} = \frac{3(4)+1}{(4)^2+3(4)} = \frac{13}{28}$. Nope. Wrong answer.

Choice (3): $\frac{3x+3}{2x+3} = \frac{3(4)+3}{2(4)+3} = \frac{15}{11}$. Wrong answer.

Choice (4): $\frac{3}{2x+3} = \frac{3}{2(4)+3} = \frac{3}{11}$. Wrong answer.

Those of you who are more algebra savvy will have spotted that the other three choices couldn't possibly match the first choice. But, if you aren't that confident in your abilities, *always check all of the answer choices*.

Let's do another example.

<u>Example:</u> (From the June 1992 Regents)

The expression $\cos 2A - \cos^2 A$ is equivalent to
(1) $\cos^2 A + 1$ (3) $-\sin^2 A$
(2) $\sin^2 A - 1$ (4) $\cos^2 A$

This is testing your knowledge of the trig identities. Why not plug in?

Step (1): Make up a value for A. Let's let $A = 30°$.

Step (2): Plug $A = 30°$ into the problem. We get:

$$\cos 60° - \cos^2 30° = \frac{1}{2} - \frac{3}{4} = -\frac{1}{4}$$

You could do this on your calculator, or you could do this knowing the special angles. Just make sure that you are in **degree** mode if use your calculator.

Step (3): Plug $A = 30°$ into the answer choices and see which one gives us $-\dfrac{1}{4}$.

Choice (1): $\cos^2 A + 1 = \cos^2 30° + 1 = \dfrac{7}{4}$. Wrong answer.

Choice (2): $\sin^2 A - 1 = \sin^2 30° - 1 = -\dfrac{3}{4}$. Wrong answer.

Choice (3): $-\sin^2 A = -\sin^2 30° = -\dfrac{1}{4}$ Bingo! Just in case, check choice (4).

Choice (4): $\cos^2 A = \cos^2 30° = \dfrac{3}{4}$ Wrong!

You will find that, once you get the hang of plugging in, it will bail you out when you are in a jam. It is one of the best techniques ever developed for solving math problems, and even real mathematicians use it sometimes!

Now let's learn the other technique.

BACKSOLVING

Sometimes you will be given a problem where all you will have to do is check the answer choices to see which one gives the answer that you are looking for. Then, rather than work out the algebra, you can just use your calculator.

The technique is quite simple:

Step (1): Work out the problem on your calculator.

Step (2): Work out each of the choices on your calculator and pick the one that matches.

Let's do an example.

Example: (from the June 1992 Regents)

If $\cos x = 0.8$, what is a value of $\tan \frac{x}{2}$?

(1) $\dfrac{1}{3}$ (3) 3

(2) $\dfrac{1}{9}$ (4) 9

This question is testing whether you know your half-angle formulas, but why not just have the calculator do the work for you?

Step (1): Use the calculator to find x.

$x = \cos^{-1} 0.8 \approx 37°$. You use the **2nd cos** button, if you didn't know.

Now, find $\tan \dfrac{37°}{2} \approx 0.3346$.

Step (2): Look at the answers. The only one that's close is choice (1), which just happens to be the right answer!

Let's do another one.

Example: (from the June 1991 Regents)

The value of $\cos 64° \cos 26° - \sin 64° \sin 26°$ is

(1) 1

(3) $\dfrac{\sqrt{3}}{2}$

(2) $\dfrac{1}{2}$

(4) 0

This question is testing whether you know your angle addition formulas, but why not just have the calculator do the work for you?

Step (1): Use the calculator to find $\cos 64° \cos 26° - \sin 64° \sin 26° = 0$. Make sure that you are in **degree** mode.

Step (2): Look at the answers. Choice (4) is the winner.

The Regents expect you to work with the formulas, which, by the way, are provided for you. But you don't have to! Usually, all that you have to do is to plug in or backsolve and you can get to the right answer.

When we go over the answers to test questions, if there is a technique that can be used we will show it to you. First, we will go through the problem the traditional way. Then, we will show you "another way to get the right answer." This is the magic phrase that tells you that you can use a technique or your calculator to get the right answer.

CHAPTER 5
MATH REVIEW:
WHAT YOU SHOULD KNOW

THIS IS A VERY IMPORTANT SECTION. DON'T SKIP IT!

In this book we are going to use the phrase "you should know" about a bunch of things. This is a signal that the Regents expect you to know this stuff, and that your calculator may not help you. Whenever you are reading over a solution and you see the magic phrase "you should know," and you *don't know it*, refer back to this section and review the material.

Most of the stuff that you are expected to have in your memory has to do with trigonometry.

(1) Do you know the definitions of the trig functions?

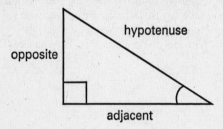

Given the right triangle above: the sine of an angle is $\dfrac{opposite}{hypotenuse}$. The cosine of an angle is $\dfrac{adjacent}{hypotenuse}$.

The tangent of an angle is $\dfrac{opposite}{adjacent}$.

$$\csc \theta = \frac{1}{\sin \theta}; \ \sec \theta = \frac{1}{\cos \theta}; \ \tan \theta = \frac{\sin \theta}{\cos \theta}; \ \cot \theta = \frac{\cos \theta}{\sin \theta} = \frac{1}{\tan \theta}$$

(2) Do you know your special angles?

$$0°, 30°, 45°, 60°, 90°, 120°, 135°, 150°, 180°, 210°,$$
$$225°, 240°, 270°, 300°, 315°, 330°$$

You should also know the radian equivalents of all of these angles.

$$0, \frac{\pi}{6}, \frac{\pi}{4}, \frac{\pi}{3}, \frac{\pi}{2}, \frac{2\pi}{3}, \frac{3\pi}{4}, \frac{5\pi}{6}, \pi,$$

$$\frac{7\pi}{6}, \frac{5\pi}{4}, \frac{4\pi}{3}, \frac{3\pi}{2}, \frac{5\pi}{3}, \frac{7\pi}{4}, \frac{11\pi}{6}$$

These are the angles from the four parts of the coordinate axes, the two special triangles, the $30° - 60° - 90°$ triangle, and the $45° - 45° - 90°$ triangle.

On the next page is a table of all of the trig values for the special angles. You should know ALL of them. Naturally, it won't be easy to memorize all of them, but you should be able to derive the ones that you don't memorize. These are the most important things to know for trigonometry. You will need to know these if you take any math in the future (such as calculus), so if you are planning to take more math after Sequential III, then you MUST learn these. Look for the patterns, make flashcards, do whatever it takes.

TABLE OF SPECIAL ANGLES

Deg.	Rad.	sin	cos	tan	cot	sec	csc
0	0	0	1	0	und.	1	und.
30	$\frac{\pi}{6}$	$\frac{1}{2}$	$\frac{\sqrt{3}}{2}$	$\frac{\sqrt{3}}{3}$	$\sqrt{3}$	$\frac{2}{\sqrt{3}}$	2
45	$\frac{\pi}{4}$	$\frac{\sqrt{2}}{2}$	$\frac{\sqrt{2}}{2}$	1	1	$\sqrt{2}$	$\sqrt{2}$
60	$\frac{\pi}{3}$	$\frac{\sqrt{3}}{2}$	$\frac{1}{2}$	$\sqrt{3}$	$\frac{\sqrt{3}}{3}$	2	$\frac{2}{\sqrt{3}}$
90	$\frac{\pi}{2}$	1	0	und.	0	und.	1
120	$\frac{2\pi}{3}$	$\frac{\sqrt{3}}{2}$	$-\frac{1}{2}$	$-\sqrt{3}$	$-\frac{\sqrt{3}}{3}$	-2	$\frac{2}{\sqrt{3}}$
135	$\frac{3\pi}{4}$	$\frac{\sqrt{2}}{2}$	$-\frac{\sqrt{2}}{2}$	-1	-1	$-\sqrt{2}$	$\sqrt{2}$
150	$\frac{5\pi}{6}$	$\frac{1}{2}$	$-\frac{\sqrt{3}}{2}$	$-\frac{\sqrt{3}}{3}$	$-\sqrt{3}$	$-\frac{2}{\sqrt{3}}$	2
180	π	0	-1	0	und.	-1	und.
210	$\frac{7\pi}{6}$	$-\frac{1}{2}$	$-\frac{\sqrt{3}}{2}$	$\frac{\sqrt{3}}{3}$	$\sqrt{3}$	$-\frac{2}{\sqrt{3}}$	-2
225	$\frac{5\pi}{4}$	$-\frac{\sqrt{2}}{2}$	$-\frac{\sqrt{2}}{2}$	1	1	$-\sqrt{2}$	$-\sqrt{2}$
240	$\frac{4\pi}{3}$	$-\frac{\sqrt{3}}{2}$	$-\frac{1}{2}$	$\sqrt{3}$	$\frac{\sqrt{3}}{3}$	-2	$-\frac{2}{\sqrt{3}}$
270	$\frac{3\pi}{2}$	-1	0	und.	0	und.	-1
300	$\frac{5\pi}{3}$	$-\frac{\sqrt{3}}{2}$	$\frac{1}{2}$	$-\sqrt{3}$	$-\frac{\sqrt{3}}{3}$	2	$-\frac{2}{\sqrt{3}}$
315	$\frac{7\pi}{4}$	$-\frac{\sqrt{2}}{2}$	$\frac{\sqrt{2}}{2}$	-1	-1	$\sqrt{2}$	$-\sqrt{2}$
330	$\frac{11\pi}{6}$	$-\frac{1}{2}$	$\frac{\sqrt{3}}{2}$	$-\frac{\sqrt{3}}{3}$	$-\sqrt{3}$	$\frac{2}{\sqrt{3}}$	-2

As you can see, there are a lot of special angles to learn. *In the next chapter, we will show you how easily you can memorize or derive the trig ratios of the special angles.* Of course, you have gone through all of these many times in school, and should be aware of all the patterns.

Identities and Formulas

You will be given a sheet with all the trig formulas and identities that you are supposed to know. We suggest you study the sheet beforehand so that you are comfortable with them.

Quadrants

You should know where the various trig functions are positive and negative.

- All trig functions are positive in quadrant I.
- Sine and Cosecant are positive in quadrant II.
- Tangent and Cotangent are positive in quadrant III.
- Cosine and Secant are positive in quadrant IV.

Graphs

You should be familiar with the basic graphs of sine, cosine, and tangent.

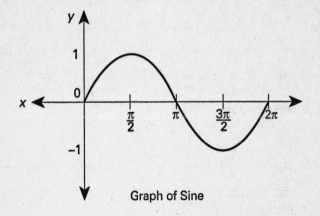

Graph of Sine

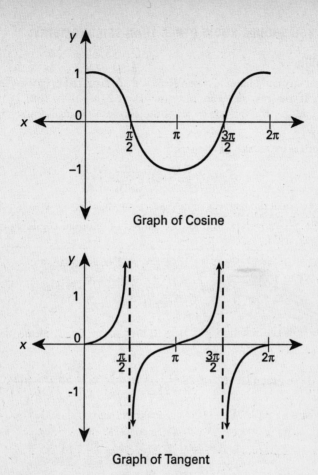

Graph of Cosine

Graph of Tangent

The Pythagorean Identity of Trigonometry

You should know that $\sin^2 \theta + \cos^2 \theta = 1$ for *any* angle θ

STUFF YOU SHOULD KNOW OTHER THAN TRIGONOMETRY

Probability

In probability and binomial expansions, you are expected to know what a factorial is. If you see $n!$, it means $n! = n(n-1)(n-2)(n-3) \ldots (3)(2)(1)$. In other words, multiply the integer by all of the successively smaller integers until you get to 1.

Factorials are *always* of integers.

$$0! = 1 \text{ and } 1! = 1.$$

You can't take the factorial of a negative number.

You should know how to find the number of different combinations of n objects taken r times.

$$\text{The formula is } {_nC_r} = \frac{n!}{(n-r)!\, r!}.$$

Sometimes the notation $\binom{n}{r}$ is used instead of ${_nC_r}$. It means the same formula.

Let's do an example: ${_5C_2} = \frac{5!}{(5-2)!\, 2!} = \frac{5!}{3!\, 2!} = \frac{5 \cdot 4 \cdot 3 \cdot 2 \cdot 1}{(3 \cdot 2 \cdot 1)(2 \cdot 1)} = 10$

There are a couple of special combinations that you should know.

$${_nC_n} = {_nC_0} = 1 \text{ and } {_nC_{n-1}} = {_nC_1} = n$$

For example:

$${_4C_4} = 1 \text{ and } {_4C_0} = 1$$
$${_4C_3} = 4 \text{ and } {_4C_1} = 4$$

When you are told that something has a normal distribution then you should know the following:

In a normal distribution, with a mean of $\bar{x}$ and a standard deviation of σ:

- approximately 68% of the outcomes will fall between $\bar{x} - \sigma$ and $\bar{x} + \sigma$
- approximately 95% of the outcomes will fall between $\bar{x} - 2\sigma$ and $\bar{x} + 2\sigma$
- approximately 99.5% of the outcomes will fall between $\bar{x} - 3\sigma$ and $\bar{x} + 3\sigma$.

For example, if you are told that the scores on a test are normally distributed, that the mean score is 100 and the standard deviation is 10, then you know that approximately 68% of the scores are between 90 and 110, 95% of the scores are between 80 and 120, and 99.5% of the scores are between 70 and 130.

You also should know that the closer a score is to the mean, the more likely it is to occur.

COMPLEX NUMBERS

In order to find a power of i, you divide the exponent by 4 and just use the remainder as the power. Then you use the following rule:

$$i^0 = 1$$
$$i^1 = i$$
$$i^2 = -1$$
$$i^3 = -i$$

If you have trouble memorizing the pattern, you can easily derive them.

Anything to the zero power is 1, so $i^0 = 1$.

Anything to the first power is itself, so $i^1 = i$.

By definition, $i^2 = -1$.

Finally, $i^3 = i^1 \cdot i^2$, and since $i^2 = -1$, $i^3 = i^1(-1) = -i$.

Then the cycle starts over.

For example, if you wanted to find i^{75}, you would divide 75 by 4. You get 18, remainder 3. Therefore, $i^{75} = i^3 = -i$.

LOGARITHMS

You should know the following log rules:

$$\text{(i)} \quad \log(AB) = \log A + \log B$$

$$\text{(ii)} \quad \log\left(\frac{A}{B}\right) = \log A - \log B$$

$$\text{(iii)} \quad \log A^B = B \log A$$

You should also know the *definition of a logarithm*: $\log_b x = a$ means that $b^a = x$, and you should know the *change of base rule*:

$$\log_b x = \frac{\log x}{\log b}$$

Let's do some examples of logarithms.

<u>Example:</u> If we know that $\log 4 = 0.602$ and $\log 3 = 0.477$, then we can find $\log 12$.

Because $12 = 3 \cdot 4$, we can rewrite $\log 12$ as $\log(3 \cdot 4)$. Using rule (i), we get $\log(3 \cdot 4) = \log 3 + \log 4 = 0.602 + 0.477 = 1.079$.

If you use your calculator, you will find that this matches the value of $\log 12$.

<u>Example:</u> If $\log 4 = 0.602$ and $\log 3 = 0.477$, then we can find $\log \dfrac{64}{\sqrt{27}}$

First, we use rule (ii) to rewrite $\log \dfrac{64}{\sqrt{27}}$ as $\log 64 - \log \sqrt{27}$.

Next, because $64 = 4^3$ and $\sqrt{27} = \sqrt{3^3} = 3^{\frac{3}{2}}$, we can rewrite this as $\log 4^3 - \log 3^{\frac{3}{2}}$.

Next, using rule (iii), we can rewrite this as $3 \log 4 - \dfrac{3}{2} \log 3$.

Substituting $\log 4 = 0.602$ and $\log 3 = 0.477$, we get:

$$3(0.602) - \frac{3}{2}(0.477) = 1.090.$$

<u>Example:</u> Use the change of base rule to find $\log_{100} 1000$

The rule says that $\log_{100} 1000 = \dfrac{\log 1000}{\log 100} = \dfrac{3}{2}$.

<u>Example:</u> Use the change of base rule to find $\log_3 100Y$.

The rule says that $\log_3 100 = \dfrac{\log 100}{\log 3} = \dfrac{2}{0.477} = 4.193$.

ALGEBRA THEORY

There are some basic properties of arithmetic that you should know:

Commutative Property: $a + b = b + a$ or $ab = ba$

Associative Property: $a + (b + c) = (a + b) + c$ or $a(bc) = (ab)c$

Inverse Property: $a + (-a) = 0$ or $a\left(\dfrac{1}{a}\right) = 1$

Distributive Property: $a(b + c) = ab + ac$

MISCELLANEOUS

You should know the quadratic formula. Given $ax^2 + bx + c = 0$, then

$$x = \frac{-b \pm \sqrt{b^2 - 4ac}}{2a}$$

You should know the Pythagorean theorem. Given a right triangle with legs of lengths a and b, and hypotenuse of length c, then $a^2 + b^2 = c^2$.

The converse is also true. In other words, if you have a triangle with sides of lengths $a, b,$ and c, where $a^2 + b^2 = c^2$, then the triangle is a right triangle.

You should know how to factor the difference of two squares:

$$a^2 - b^2 = (a - b)(a + b)$$

For example, $9x^2 - 64 = (3x - 8)(3x + 8)$.

You should know that $x^{\frac{a}{b}} = \sqrt[b]{x^a} = \left(\sqrt[b]{x}\right)^a$

You should know that $x^{-a} = \dfrac{1}{x^a}$.

Finally, you should know all of the stuff in Sequential I and II, all of which is considered fair game on the Sequential III test. Don't say that we didn't warn you!

CHAPTER 6
HOW TO REMEMBER
THE SPECIAL ANGLES

You're probably thinking, "You must be joking if you think that I can memorize all of these!" Well, the good news is that you don't have to know all of them. In fact, if you memorize just a few things, then you can easily figure out all of the rest. Here's what you need to memorize:

First, memorize these sines and cosines:

$$\sin 30° = \frac{1}{2} \qquad \sin 45° = \frac{\sqrt{2}}{2} \qquad \sin 60° = \frac{\sqrt{3}}{2}$$

$$\cos 30° = \frac{\sqrt{3}}{2} \qquad \cos 45° = \frac{\sqrt{2}}{2} \qquad \cos 60° = \frac{1}{2}$$

See how the two complement each other?

Now we'll learn how to derive the rest of the trig ratios from just these ratios.

Remember that $\tan \theta = \dfrac{\sin \theta}{\cos \theta}$. This means that we can find the tangents of these angles by dividing the appropriate sine by the appropriate cosine.

$$\tan 30° = \frac{\sin 30°}{\cos 30°} = \frac{\frac{1}{2}}{\frac{\sqrt{3}}{2}} = \frac{1}{\sqrt{3}}$$

$$\tan 45° = \frac{\sin 45°}{\cos 45°} = \frac{\frac{\sqrt{2}}{2}}{\frac{\sqrt{2}}{2}} = 1$$

$$\tan 60° = \frac{\sin 60°}{\cos 60°} = \frac{\frac{\sqrt{3}}{2}}{\frac{1}{2}} = \sqrt{3}$$

Now to find the other three ratios, you just turn the corresponding ratio upside down.

$$\cot\theta = \frac{1}{\tan\theta} \quad \text{so} \quad \cot 30° = \sqrt{3} \qquad \cot 45° = 1 \qquad \cot 60° = \frac{1}{\sqrt{3}}$$

$$\sec\theta = \frac{1}{\cos\theta} \quad \text{so} \quad \sec 30° = \frac{2}{\sqrt{3}} \qquad \sec 45° = \sqrt{2} \qquad \sec 60° = 2$$

$$\csc\theta = \frac{1}{\sin\theta} \quad \text{so} \quad \sin 30° = 2 \qquad \csc 45° = \sqrt{2} \qquad \csc 60° = \frac{2}{\sqrt{3}}$$

See how easy it is to derive the other ratios once you know the sines and cosines?

Now here is the second set to memorize:

$\sin 0° = 0$	$\sin 90° = 1$	$\sin 180° = 0$	$\sin 270° = -1$
$\cos 0° = 1$	$\cos 90° = 0$	$\cos 180° = -1$	$\cos 270° = 0$

Once again we can derive the rest of the trig ratios from just these.

Using the identities, we get:

$\tan 0° = 0$	$\tan 90° = und.$	$\tan 180° = 0$	$\tan 270° = und.$
$\cot 0° = und.$	$\cot 90° = 0$	$\cot 180° = und.$	$\cot 270° = 0$
$\sec 0° = 1$	$\sec 90° = und.$	$\sec 180° = -1$	$\sec 270° = und.$
$\csc 0° = und.$	$\csc 90° = 1$	$\csc 180° = und.$	$\csc 270° = -1$

Now you can see that just by knowing the sines and cosines of seven angles—0°, 30°, 45°, 60°, 90°, 180°, 270°—you can find all of the rest.

The last step to memorizing the special angles is to learn the mnemonic for the quadrants:

All Students Take Calculus

which stands for:

All trig ratios are positive in quadrant I.

Sine and cosecant are positive in quadrant II. (The rest are negative.)

Tangent and cotangent are positive in quadrant III. (The rest are negative.)

Cosine and secant are positive in quadrant IV. (The rest are negative.)

Now, once you are given an angle, just figure out the reference angle (that is, the acute angle that the line forms with the x-axis) and then use the appropriate trig ratio from above.

<u>Example</u>: Find csc 330°.

First, draw a little picture to help figure out the reference angle.

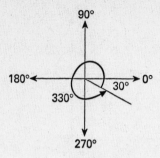

Now that we know that the reference angle is 30°, we remember that $\sin 30° = \dfrac{1}{2}$, which means that $\csc 30° = 2$. In quadrant IV, cosine and secant are positive, and the rest are negative, so $\csc 330° = -2$.

Let's do another one.

<u>Example</u>: Find cot 225°.

First, draw a little picture to help figure out the reference angle.

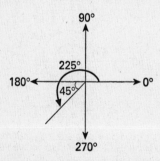

Now that we know that the reference angle is 45°, we remember that $\sin 45° = \cos 45° = \dfrac{\sqrt{2}}{2}$, which means that $\tan 45° = 1$ and thus $\cot 45° = 1$. In quadrant III, tangent and cotangent are positive, so $\cot 225° = 1$.

<u>Example</u>: Find sec 120°.

First, draw a little picture to help figure out the reference angle.

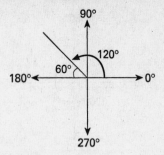

Now that we know that the reference angle is 60°, we remember that $\cos 60° = \dfrac{1}{2}$, which means that sec 60° = 2. In quadrant II, sine and cosecant are positive, and the rest are negative, so sec 120° = –2 .

Now for a slightly harder one.

<u>Example</u>: Find tan 495°.

First, draw a little picture to help figure out the reference angle.

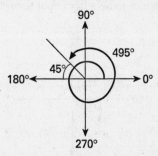

Notice that the angle is bigger than 360° so, as we go around the coordinate axes, we continue past 360° and go around a second time. This is allowed! In fact, you can keep going around as many times as you want!

Now that we know that the reference angle is 45°, we remember that $\sin 45° = \cos 45° = \dfrac{\sqrt{2}}{2}$, which means that tan 45° = 1. In quadrant II, sine and cosecant are positive, and the rest are negative, so tan 495° = –1.

CHAPTER 7
RULES TO REMEMBER

Here are some of the rules that are frequently tested on the Regents Exam:

CIRCLE RULES

1. *The measure of an inscribed angle is half of the arc it subtends (intercepts).*
2. *A central angle has the same measure as the arc it subtends.*
3. *The measure of an angle formed by a pair of secants or a secant and a tangent is equal to half of the difference between the larger and the smaller arcs that are formed by the secants or the secant and the tangent.*
4. *An angle formed by a tangent and a chord is equal to half of its intercepted arc.*
5. *If two chords are congruent, they cut off equal arcs.*
6. *Given a point exterior to a circle, the square of the tangent segment to the circle is equal to the product of the lengths of the secant and its external segment.*
7. *If two chords intersect within a circle, then the angle between the two chords is the average of their intercepted arcs.*
8. *The two tangents drawn to a circle from the same point are congruent.*
9. *A diameter of a circle that intersects a chord at right angles **bisects** the chord.*
10. *If two chords intersect within a circle, then the product of the lengths of the segments of one chord is equal to the product of the lengths of the other chord.*

ARC LENGTH

In a circle with a central angle, θ, measured in radians, and a radius, r, the arc length, s, is found by $s = r\theta$.

LOGARITHMS

1. $\log 1 = 0$

2. $\log_B B = 1$

3. $\log A + \log B = \log(AB)$

4. $\log A - \log B = \log\left(\dfrac{A}{B}\right)$

5. $\log A^B = B \log A$

6. $\log_B A = \dfrac{\log A}{\log B}$

BINOMIAL PROBABILITY

If the probability of a particular outcome is **p**, *then the probability of that outcome occurring* **r** *times out of a possible* **n** *times is* $_nC_r(p)^r(1-p)^{n-r}$.

STATISTICS

In a normal distribution, with a mean of $\bar{x}$ *and a standard deviation of* σ:
approximately 68% of the outcomes will fall between $\bar{x} - \sigma$ *and* $\bar{x} + \sigma$.
approximately 95% of the outcomes will fall between $\bar{x} - 2\sigma$ *and* $\bar{x} + 2\sigma$.
approximately 99.5% of the outcomes will fall between $\bar{x} - 3\sigma$ *and* $\bar{x} + 3\sigma$.

BINOMIAL THEOREM

If you expand $(a + b)^n$, *you get:*

$$_nC_0a^n + {}_nC_1a^{n-1}b^1 + {}_nC_2a^{n-2}b^2 + \ldots + {}_nC_{n-2}a^2b^{n-2} + {}_nC_{n-1}a^1b^{n-1} + {}_nC_nb^n$$

QUADRATIC EQUATIONS

1. Given an equation of the form $ax^2 + bx + c = 0$, the sum of the roots is $\dfrac{-b}{a}$ and the product of the roots is $\dfrac{c}{a}$.

2. Given an equation of the form $ax^2 + bx + c = 0$,

   If $b^2 - 4ac < 0$, the equation has two imaginary roots.

   If $b^2 - 4ac = 0$, the equation has one rational root.

   If $b^2 - 4ac > 0$, and $b^2 - 4ac$ is a perfect square, then the equation has two rational roots.

If $b^2 - 4ac > 0$, and $b^2 - 4ac$ is not a perfect square, then the equation has two irrational roots.

CONIC SECTIONS

A graph of the form $\dfrac{x^2}{a^2} + \dfrac{y^2}{b^2} = 1$, where a and b have the same value, is the equation of a circle. If a and b have different values, the graph is that of an ellipse.

EXPONENTS

1. A number raised to the power zero is equal to one.

2. A number raised to the power one is equal to itself.

3. When a number raised to a power is itself raised to a power, we multiply the exponents.

4. A number raised to a negative power is the same as the reciprocal of that number raised to the corresponding positive power.

5. A number raised to $\dfrac{1}{n}$ is the same as the nth root of that number.

TARGET PRACTICE

If you want to practice specific types of problems, find the topic below that you want to practice, then do the problems that are listed.

PART I PROBLEMS:

Absolute Value Equations

January 1995: 19

June 1995: 18

August 1995: 18

January 1996: 24

June 1996: 14

August 1996: 15

January 1997: 19

June 1997: 6

August 1997: 17

January 1998: 23

June 1998: 21

August 1998: 24

Algebra Theory

June 1995: 20

January 1997: 28

Arc Length

June 1995: 8

August 1995: 8

January 1996: 15

January 1997: 13

June 1997: 14

June 1998: 11

August 1998: 6, 10

Binomial Expansions

January 1995: 31

June 1995: 15

August 1995: 35

January 1996: 32

January 1997: 32

June 1997: 31

January 1998: 29

June 1998: 35

August 1998: 34

Circle Rules

January 1995: 1

June 1995: 3, 7

August 1995: 6, 10

January 1996: 9, 12

June 1996: 3, 12, 15

August 1996: 1, 16, 17

January 1997: 1, 4, 17

June 1997: 13

August 1997: 11, 15

January 1998: 4, 15

June 1998: 10, 15

August 1998: 11, 18

Complex Numbers

January 1995: 10, 22

June 1995: 2, 27

August 1995: 1, 19
January 1996: 7, 22
June 1996: 9, 16, 20
August 1996: 8, 12
January 1997: 2, 8
June 1997: 5, 12, 19, 22
August 1997: 1, 10, 26
January 1998: 1, 17
June 1998: 20
August 1998: 8, 30

Composite Functions

June 1996: 8
January 1997: 10
June 1997: 15
August 1997: 9
January 1998: 20
August 1998: 2

Conic Sections

January 1995: 25
June 1995: 35
January 1996: 34
June 1996: 22
August 1996: 35
January 1997: 24
June 1997: 32
August 1997: 23
January 1998: 21, 32
June 1998: 17
August 1998: 29

Converting from Degrees to Radians (or vice versa)

January 1995: 2
August 1995: 5
January 1996: 2
June 1996: 2
August 1996: 3
January 1997: 7
June 1997: 1
August 1997: 3
June 1998: 30
August 1998: 1

Exponential Equations

January 1995: 8
June 1995: 6
January 1996: 4
June 1996: 7
August 1996: 9
January 1997: 6
June 1997: 4
August 1997: 2
January 1998: 10, 24
August 1998: 19

Exponents

January 1995: 15, 20
January 1996: 16
August 1996: 13
January 1997: 15
June 1997: 18, 24, 28
June 1998: 2
August 1998: 10

Functions

January 1995: 17, 35
August 1995: 24, 25
June 1997: 11, 25, 34
August 1997: 5, 31
January 1998: 7, 25
June 1998: 1, 29

Inverse Variation

January 1995: 32
August 1995: 9
August 1996: 21
January 1997: 14
August 1997: 16
January 1998: 14
June 1998: 14
August 1998: 15

Inverses

June 1995: 33
August 1995: 32
January 1996: 26
June 1996: 23
August 1996: 22

Law of Cosines

June 1995: 26
January 1996: 21
June 1996: 32
August 1996: 10
January 1997: 11
August 1997: 29
June 1998: 28
August 1998: 27

Law of Sines

January 1995: 4
June 1995: 34
August 1995: 4, 29
June 1996: 27, 33
August 1996: 14
January 1997: 3
June 1997: 2
August 1997: 30
January 1998: 2, 34
June 1998: 7, 34
August 1998: 4

Logarithms

January 1995: 29, 34
June 1995: 19
August 1995: 14, 15
January 1996: 19
June 1996: 4, 26
August 1996: 20
January 1997: 33
June 1997: 17
August 1997: 18
January 1998: 18
August 1998: 13

Probability

January 1995: 33
June 1995: 13
August 1995: 23
January 1996: 23
June 1996: 29
August 1996: 29

January 1997: 30
June 1997: 23
August 1997: 35
January 1998: 1
June 1998: 9
August 1998: 22

Quadratic Equations
January 1995: 23
June 1995: 28
August 1995: 33
January 1996: 28
June 1996: 19, 35
August 1996: 28, 32
January 1997: 26, 35
June 1997: 30, 33
August 1997: 32, 34
June 1998: 22, 33
August 1998: 28, 35

Quadratic Inequalities
June 1995: 24
August 1995: 26
January 1996: 33
June 1996: 30
August 1996: 31
January 1997: 31
January 1998: 33
June 1998: 26
August 1998: 23

Radical Equations
June 1995: 9
August 1995: 3

January 1996: 1
June 1996: 1, 24, 25
August 1996: 4
January 1997: 22
June 1997: 8
August 1997: 13
January 1998: 5
June 1998: 4
August 1998: 26

Rational Expressions
June 1995: 5, 12
August 1995: 27, 31
January 1996: 10, 20
June 1996: 6, 18
August 1996: 2, 30
June 1997: 5, 10, 16
August 1997: 7, 27
January 1998: 8
June 1998: 3, 18
August 1998: 9

Statistics
January 1995: 27
June 1995: 17
August 1995: 20
January 1996: 18
June 1996: 31
August 1996: 6
January 1997: 12
June 1997: 26
August 1997: 24
January 1998: 30

Trigonometry Formulas

August 1995: 21
August 1996: 19
January 1997: 21
June 1997: 27
August 1997: 8
January 1998: 19
June 1998: 16
August 1998: 21

Trigonometry Graphs

January 1995: 12, 24
June 1995: 1
August 1995: 13, 22
January 1996: 17, 27
June 1996: 34
August 1996: 24
January 1997: 25
June 1997: 21
August 1997: 19, 25
January 1998: 22
August 1998: 20

Trigonometry Identities

January 1995: 30
June 1995: 23
August 1995: 34
January 1996: 30
August 1996: 25
January 1997: 16
June 1997: 20
August 1997: 20
June 1998: 24
August 1998: 25

PART II PROBLEMS:

Binomial Expansions
June 1996: 41(a)

Circle Rules
January 1995: 37

June 1995: 37

August 1995: 37

January 1996: 39

June 1996: 36

August 1996: 36

January 1997: 41

June 1997: 37

August 1997: 38

January 1998: 39

June 1998: 37

August 1998: 42

Complex Numbers
June 1995: 40(a)

Composite Functions
January 1996: 41

Exponential Equations
January 1995: 41(a)

June 1995: 42(a)

June 1996: 41(b)

August 1997: 37, 39(a)

August 1998: 39

Law of Cosines
January 1995: 38(a)

June 1995: 39(a)

August 1995: 41(a)

June 1996: 42(a)

August 1996: 40(a)

June 1997: 41(a), 41(b)

August 1997: 40(a)

January 1998: 41(a)

June 1998: 42(b)

August 1998: 40(a)

Law of Sines
August 1996: 40(b)

January 1997: 41(c)

June 1997: 41(b)

August 1997: 40(b)

August 1998: 40(b)

Logarithms
June 1995: 42(c)

August 1995: 42(b)

January 1996: 37(b)

June 1996: 40(b)

August 1996: 41

January 1997: 37(b), 42

June 1997: 42

January 1998: 40(b)

June 1998: 38(a)

Probability
August 1995: 40

January 1996: 39

August 1996: 42(a)

January 1997: 39(b)

June 1997: 39(b)

August 1997: 41

June 1998: 39(a)

August 1998: 41

Quadratic Equations

January 1995: 42(b)

August 1995: 42(a)

January 1996: 38(a)

January 1997: 36(b)

June 1997: 40(b)

January 1998: 38(b), 40(a)

June 1998: 41

August 1998: 38(b)

Radical Equations

August 1996: 37(a)

Rational Expressions

January 1995: 39(b)

June 1995: 40(b)

January 1996: 36(b)

August 1996: 37(b)

January 1997: 36(a), 40(a)

June 1997: 40(a)

January 1998: 40(a)

June 1998: 41(a)

August 1998: 38(a)

Statistics

January 1995: 42(b)

August 1995: 42(a)

January 1996: 36(a)

August 1996: 42(b)

January 1997: 39(a)

June 1997: 39(a)

August 1997: 42(b)

January 1998: 42

June 1998: 39(b)

August 1998: 41

Transformations

January 1995: 36(b), 36(c), 41(b)

June 1995: 42(b)

August 1995: 38

January 1996: 42(c), 42(d)

August 1997: 36(a), 37(b) and (d)

January 1998: 38(a)

August 1998: 39(a)

Trigonometric Area

January 1995: 38(b)

June 1995: 39(b)

August 1995: 41(b)

June 1996: 42(b)

January 1998: 41(b)

Trigonometric Equations

January 1995: 40

June 1995: 38

August 1995: 39

January 1996: 38(b)

June 1996: 38

August 1996: 38(a)

January 1997: 37(a)

June 1997: 38

August 1997: 39(b)

January 1998: 37

June 1998: 40

August 1998: 37

Trigonometry (General)

June 1995: 41(b)

January 1996: 42

Trigonometry Graphs

January 1995: 36(a)

June 1995: 36(b)

August 1995: 36

January 1996: 40

June 1996: 37

August 1996: 39

January 1997: 38

June 1997: 36

August 1997: 36

January 1998: 36

June 1998: 36

August 1998: 36

Trigonometry Identities

January 1995: 42(a)

June 1995: 41(a)

January 1996: 37(a)

June 1996: 40(a)

August 1996: 38(b)

January 1997: 40(b)

August 1997: 42(a)

PART II

EXAMS AND EXPLANATIONS

Formulas

Pythagorean and Quotient Identities

$$\sin^2 A + \cos^2 A = 1 \qquad \tan A = \frac{\sin A}{\cos A}$$

$$\tan^2 A + 1 = \sec^2 A \qquad \cot A = \frac{\cos A}{\sin A}$$

$$\cot^2 A + 1 = \csc^2 A$$

Functions of the Sum of Two Angles

$$\sin (A + B) = \sin A \cos B + \cos A \sin B$$

$$\cos (A + B) = \cos A \cos B - \sin A \sin B$$

$$\tan (A + B) = \frac{\tan A + \tan B}{1 - \tan A \tan B}$$

Functions of the Difference of Two Angles

$$\sin (A - B) = \sin A \cos B - \cos A \sin B$$

$$\cos (A - B) = \cos A \cos B + \sin A \sin B$$

$$\tan (A - B) = \frac{\tan A - \tan B}{1 + \tan A \tan B}$$

Law of Sines

$$\frac{a}{\sin A} = \frac{b}{\sin B} = \frac{c}{\sin C}$$

Law of Cosines

$$a^2 = b^2 + c^2 - 2bc \cos A$$

Functions of the Double Angle

$$\sin 2A = 2 \sin A \cos A$$

$$\cos 2A = \cos^2 A - \sin^2 A$$

$$\cos 2A = 2 \cos^2 A - 1$$

$$\cos 2A = 1 - 2 \sin^2 A$$

$$\tan 2A = \frac{2 \tan A}{1 - \tan^2 A}$$

Functions of the Half Angle

$$\sin \frac{1}{2} A = \pm\sqrt{\frac{1 - \cos A}{2}}$$

$$\cos \frac{1}{2} A = \pm\sqrt{\frac{1 + \cos A}{2}}$$

$$\tan \frac{1}{2} A = \pm\sqrt{\frac{1 - \cos A}{1 + \cos A}}$$

Area of Triangle

$$K = \frac{1}{2} ab \sin C$$

Standard Deviation

$$\text{S.D.} = \sqrt{\frac{1}{n} \sum_{i=1}^{n} \left(x_i - \bar{x}\right)^2}$$

EXAMINATION
JANUARY 1995

Part I

Answer 30 questions from this part. Each correct answer will receive 2 credits. No partial credit will be allowed. Write your answers in the spaces provided on the separate answer sheet. Where applicable, answers may be left in terms of π or in radical form. [60]

1 In the accompanying diagram of circle O, chords $\overline{AC}$ and $\overline{WF}$ are drawn, $\overline{AOF}$ is a diameter, $\overline{AC} \parallel \overline{WF}$ and m$\angle AFW = 60$. Find m$\overarc{AC}$.

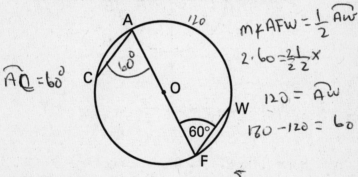

(handwritten notes)
120

$AC = 60°$

m$\angle AFW = \frac{1}{2}\overarc{AW}$

$2 \cdot 60 = \frac{2}{2}\frac{1}{2}x$

$120 = \overarc{AW}$

$180 - 120 = 60$

2 Express 450° in radian measure.

(handwritten) $\cancel{450} \cdot \frac{\pi}{\cancel{180}} = \frac{5\pi}{2}$

3 Factor completely: $2x^3 - 98x$

(handwritten) $2x(x^2 - 49)$ $2x(x-7)(x+7)$

4 In $\triangle ABC$, sin $A = 0.4293$, sin $C = 0.4827$, and $a = 34.5$ centimeters. Find, to the *nearest tenth* of a centimeter, the measure of c.

(handwritten)
$\dfrac{34.5}{0.4293} = \dfrac{c}{0.4827}$

$0.4293c = 16.65315$

$c = 38.8$

[OVER]

5 Evaluate: $\displaystyle\sum_{k=2}^{4} k^2 - k$

[handwritten] $2^2 - 2 + 3^2 - 3 + 4^2 - 4$
$4 - 2 + 9 - 3 + 16 - 4$
$2 + 6 + 12 = 20$

6 In the accompanying diagram, unit circle O has radii $\overline{OB}$, $\overline{OC}$, and $\overline{OD}$. Central angle θ is drawn and $\overline{CA} \perp \overline{OB}$. The length of which line segment represents $\sin \theta$?

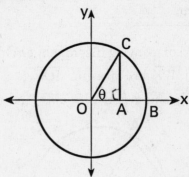

7 Evaluate: $\cos \dfrac{\pi}{2} + \sin \dfrac{3\pi}{2}$

[handwritten] $\dfrac{\pi}{2}$ · $\dfrac{180}{\pi}$ $\dfrac{3\pi}{2}$ $\dfrac{90}{180}{\pi}$
90
$\cos 90 + \sin 270 = -1$

8 Solve for x: $27^{x+2} = 9^{2x}$

[handwritten] $3^{3(x+2)} = 3^{2 \cdot 2x}$
$9^{\frac{3}{2}(x+2)} = 9^{2x}$
$3^{4(x+2)} = 3^{4x}$
$3(x+2) = 12x - 1$
$3x + 6 = 12x - 1$
$6 = 9x$
$2 = x$

9 In $\triangle ABC$, m$\angle A = 60$, $b = 4$, and $c = 4$. What is the area of $\triangle ABC$?

[handwritten] $A = \frac{1}{2} bc \cdot \sin A$
$A = \frac{1}{2}(4)(4) \cdot \sin 60$
$A = 8 \cdot \sin 60$

10 If $4 + 2i - (a + 4i) = 9 - 2i$, find the value of a.

[handwritten] $4 + 2i - a - 4i = 9 - 2i$ $-2i + 2i - 9 + 4 = a$
$4 - 2i - a = 9 - 2i$ $-5 = a$

11 Find the value of $\sin \left(\text{Arc} \tan \dfrac{\sqrt{3}}{3} \right)$

Directions (12–35): For *each* question chosen, write the *numeral* preceding the word or expression that best completes the statement or answers the question.

12 What is the amplitude of the graph of the equation $y = 3 \sin 2x$?

(1) $\dfrac{1}{2}$

(3) 3

(2) 2

(4) $\dfrac{1}{3}$

13 If $\sin A < 0$ and $\cos A < 0$, in which quadrant does $\angle A$ terminate?

(1) I

(3) III

(2) II

(4) IV

14 What are the coordinates of the point of intersection of the graphs of the equations $y = x^2$ and $xy = 8$?

(1) (4,2)

(3) (1,8)

(2) (2,4)

(4) (8,64)

15 Which equation is equivalent to $y = 10^x$?

(1) $y = -10^{-x}$

(3) $y = \left(\dfrac{1}{10}\right)^{-x}$

(2) $y = -10^{-x}$

(4) $y = \left(\dfrac{1}{10}\right)^{x}$

[OVER]

16 What is the graph of the solution set of
    $15 < 3x + 5 < 21$?

(1)

(2)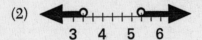

(3)

(4)

17 If $f(x) = \dfrac{x-4}{x+4}$, then $f(4a)$ equals

(1) $\dfrac{a-1}{a+1}$ (3) $\dfrac{4a-1}{4a+1}$

(2) $\dfrac{a+1}{a-1}$ (4) $\dfrac{4a+1}{4a-1}$

18 Which graph illustrates a quadratic relation whose domain is all real numbers?

(1)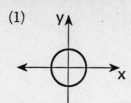

(3)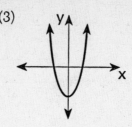

(2)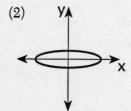

(4)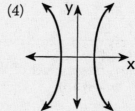

19 The graph below represents the solution to which inequality?

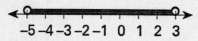

$$-5 -4 -3 -2 -1 \quad 0 \quad 1 \quad 2 \quad 3$$

(1) $|x + 8| \le 3$

(3) $|x + 1| \le 4$

(2) $|x + 1| < 4$

(4) $|x + 6| > 1$

20 The value of $(-64)^{\frac{2}{3}}$ is

(1) 16

(3) $-\dfrac{1}{16}$

(2) −16

(4) 512

[OVER]

21 Which expression is equivalent to:
cos 100° cos 80° − sin 100° sin 80°?

(1) 1   (3) −1
(2) 0   (4) cos 20

22 The expression $\dfrac{1}{5 + 2i}$ is the equivalent to

(1) $\dfrac{5 + 2i}{21}$ (3) $\dfrac{5 - 2i}{21}$

(2) $\dfrac{5 + 2i}{29}$ (4) $\dfrac{5 - 2i}{29}$

23 What is the product of the roots of the equation
$2x^2 - 9x + 6 = 0$?

(1) $\dfrac{9}{2}$ (3) 3

(2) $-\dfrac{9}{2}$ (4) $\dfrac{1}{3}$

24 Between −2π and 2π the graph of the equation
$y = \cos x$ is symmetric with respect to

(1) the y-axis   (3) the origin
(2) the x-axis   (4) $y = x$

25 Which equation represents an ellipse?

(1) $x^2 + y^2 = 400$
(2) $25x^2 + 16y^2 = 400$
(3) $x^2 - y^2 = 400$
(4) $xy = 400$

26 Which figure has 120° rotational symmetry?
  (1) rhombus
  (2) regular pentagon
  (3) square
  (4) equilateral triangle

27 On a standardized test, the mean is 48 and the standard deviation is 4. Approximately what percent of the scores will fall in the range from 36 to 60?
  (1) 34%   (3) 95%
  (2) 68%   (4) 99%

28 In the interval $0 \leq x < 2\pi$, the solutions of the equation $\sin^2 x = \sin x$ are

  (1) $0, \dfrac{\pi}{2}, \pi$ (3) $0, \dfrac{\pi}{2}, \dfrac{3\pi}{2}$

  (2) $\dfrac{\pi}{2}, \dfrac{3\pi}{2}$ (4) $\dfrac{\pi}{2}, \pi, \dfrac{3\pi}{2}$

29 The expression $\dfrac{1}{3} \log m - 2 \log n$ is equivalent to

  (1) $\log \left(\dfrac{1}{3} m - 2n \right)$ (3) $\log \left(\sqrt[3]{m} - n^2 \right)$

  (2) $\log \left(\dfrac{m^3}{\sqrt{n}} \right)$ (4) $\log \left(\dfrac{\sqrt[3]{m}}{n^2} \right)$

[OVER]

30 The expression $\dfrac{\sin^2 B}{\cos B} + \cos B$ is equivalent to

(1) 1

(3) $\dfrac{1}{\sec B}$

(2) $\dfrac{1}{\cos B}$

(4) $\sin^2 B$

31 The fifth term in the expansion of $(3a - b)^6$ is

(1) $135a^2b^4$

(3) $-18ab^5$

(2) $540a^3b^3$

(4) $-135a^2b^4$

32 If x varies inversely as y, which statement is true?

(1) When x is multiplied by 2, y is multiplied by 2.
(2) When x is multiplied by 2, y is divided by 2.
(3) When x is divided by 2, y is divided by 2.
(4) When x is increased by 2, y is decreased by 2.

33 If the probability of winning a game is $\dfrac{3}{5}$, then the probability of winning exactly 3 games out of 4 played is

(1) $\dfrac{27}{125}$

(3) $\dfrac{216}{625}$

(2) $\dfrac{54}{625}$

(4) $\dfrac{532}{625}$

34 Which equation is represented by the graph in the accompanying diagram?

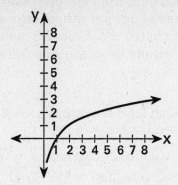

(1) $y = \log x$ (3) $y = 2^x$
(2) $y = \log_2 x$ (4) $y = 10^x$

35 Which fraction is defined for all real numbers?

(1) $\dfrac{x^2 - 1}{(x - 1)^2}$ (3) $\dfrac{x^2 - 1}{x^2}$

(2) $\dfrac{x^2 - 1}{x + 1}$ (4) $\dfrac{x^2 - 1}{x^2 + 1}$

[OVER]

Part II

Answer *four* questions from this part. Clearly indicate the necessary steps, including appropriate formula substitutions, diagrams, graphs, charts, etc. Calculations that may be obtained by mental arithmetic or the calculator do not need to be shown. [40]

36   *a*   Sketch and label the function $y = 2 \sin \frac{1}{2} x$ in the interval $-2\pi \leq x \leq 2\pi$. [4]

     b   On the same set of axes, sketch the function drawn in part *a* after a dilation $D_{\frac{1}{2}}$. Label the graph *b*. [4]

     c   Write an equation of the function graphed in part *b*. [2]

37   In circle O, $\overrightarrow{FA}$ is the tangent, $\overline{FEDB}$ is a secant, $\overline{ADC}$ and $\overline{AB}$ are chords, $\text{m}\widehat{CE} = 40$, $\text{m}\widehat{AB} = 130$, and $\text{m}\angle CAB = 60$.

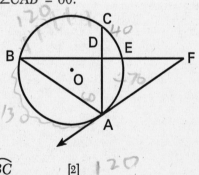

Find:

     a $\text{m}\widehat{BC}$ [2]

     b $\text{m}\angle EBA$ [2]

     c $\text{m}\angle ADE$ [2]

     d $\text{m}\angle F$ [2]

     e $\text{m}\angle FAC$ [2]

38 In $\triangle ABC$, m$\angle A$ = 42°20′, AC = 2.0 feet, and AB = 18 inches.

  a Find BC to the *nearest tenth*. [*Indicate the unit of the measure.*]   [7]

  b Find the area of $\triangle ABC$ to the *nearest tenth*. [*Indicate the unit of the measure.*]   [3]

39 a Using the accompanying set of data, find the standard deviation to the *nearest tenth*.   [6]

Measure (x_i)	Frequency (f_i)
80	5
85	7
90	9
95	4

  b Simplify: $\dfrac{1 - \dfrac{3}{\cos x}}{\dfrac{9}{\cos^2 x} - 1}$ [4]

40 Find, to the *nearest degree*, all values of x in the interval $0° \leq x < 360°$ that satisfy the equation $3 + \tan^2 x = 5 \tan x$.   [10]

[OVER]

41 **a** Solve the equation $9^{(x^2+x)} = 3^4$ for all values of x. [*Only an algebraic solution will be accepted.*] [4]

b Triangle ABC has coordinates $A(-1, 2)$, $B(6, 2)$, and $C(3, 4)$.

(1) On graph paper, draw and label $\triangle ABC$. [1]

(2) Graph and state the coordinates of $\triangle A'B'C'$, the image of $\triangle ABC$ after the composition $R_{90°} \circ r_{x\text{-axis}}$. [3]

(3) Write a transformation equivalent to $R_{90°} \circ r_{x\text{-axis}}$. [2]

42 **a** For all values of x for which the expressions are defined, prove the following is an identity:

$$\sec^2 x + \csc^2 x = \left(\tan x + \cot x\right)^2 \quad [5]$$

b Solve for x and express the roots in the terms of i:

$$-3x^2 + 2x = 2 \quad [5]$$

ANSWER KEY

Part I

1. 60	13. (3)	25. (2)
2. $\dfrac{5\pi}{2}$	14. (2)	26. (4)
3. $2x(x + 7)(x - 7)$	15. (3)	27. (4)
4. 38.8	16. (1)	28. (1)
5. 20	17. (1)	29. (4)
6. $\overline{CA}$	18. (3)	30. (2)
7. -1	19. (2)	31. (1)
8. 8	20. (1)	32. (2)
9. $4\sqrt{3}$	21. (3)	33. (3)
10. -5	22. (4)	34. (2)
11. $\dfrac{1}{2}$	23. (3)	35. (4)
12. (3)	24. (1)	

ANSWERS AND EXPLANATIONS
JANUARY 1995

Part I

In problems 1–11, you must supply the answer. Because of this, you will not be able to **plug-in** or **backsolve**. But, you can use your calculator to simplify calculations. In addition, you will find that these problems tend to test your basic knowledge of Sequential Math III and are not tricky. Many can be solved just by knowing the right formula and by plugging in the numbers.

PROBLEM 1: CIRCLE RULES

Given parallel lines cut by a transversal, the alternate interior angles are congruent. Here, $m\angle AFW = 60$; therefore, because $AC\|FW$, $m\angle CAF = 60$.

The measure of an inscribed angle is always half of the measure of the arc that it subtends (intercepts). Therefore, because $m\angle CAF = 60$, arc $\overset{\frown}{CF}$ has measure 120.

Finally, we are given that AOF is a diameter, so arc $\overset{\frown}{ACF}$ has measure 180.

Therefore, arc $\overset{\frown}{AC}$ has measure $180 - 120 = 60$.

PROBLEM 2: CONVERTING FROM DEGREES TO RADIANS

All we have to do is multiply $450°$ by $\dfrac{\pi}{180°}$.

$$450°\left(\frac{\pi}{180°}\right) = \frac{5\pi}{2}.$$

PROBLEM 3: FACTORING

First, factor $2x$ out of both terms: $2x^3 - 98x = 2x(x^2 - 49)$

Next, the expression $(x^2 - 49)$ is the difference of two squares, so we can factor it into $(x - 7)(x + 7)$. Therefore, we get:

$$2x^3 - 98x = 2x(x - 7)(x + 7).$$

PROBLEM 4: LAW OF SINES

The Law of Sines says that, in any triangle, with angles $A, B,$ and $C,$ and opposite sides of $a, b,$ and $c,$ respectively, $\dfrac{a}{\sin A} = \dfrac{b}{\sin B} = \dfrac{c}{\sin C}$. (This is on the formula sheet.)

Here, $\dfrac{34.5}{0.4293} = \dfrac{c}{0.4827}$, so $c = \dfrac{(34.5)(0.4827)}{0.4293} \approx 38.7914$. Rounded to the nearest tenth, we get $c \approx 38.8$.

PROBLEM 5: SUMMATIONS

To evaluate a sum, we plug each of the consecutive values for k into the expression, starting at the bottom value and ending at the top value, and sum the results. We get:

$$\sum_{k=2}^{4}\left(k^2 - k\right) = \left[\left(2^2 - 2\right)\right] + \left[\left(3^2 - 3\right)\right] + \left[\left(4^2 - 4\right)\right] = 2 + 6 + 12 = 20$$

Another way to do this is to use your calculator. Push **2nd STAT** (**LIST**) and under the **MATH** menu, choose **SUM**. Then, again push **2nd STAT** and under the **OPS** menu, choose **SEQ**.

We put in the following: (*expression, variable, start, finish, step*)

For *expression*, we put in the formula, using x as the variable

For *variable*, we ALWAYS put in x.

For *start*, we put in the bottom value.

For *finish*, we put in the top value.

For *step*, we ALWAYS put in 1.

Therefore, we put in the calculator: **SUM SEQ** $(x^2 - x, x, 2, 4, 1)$. The calculator should return the value 20.

PROBLEM 6: TRIGONOMETRY

You should know that the sine of an angle in a right triangle is $\dfrac{\text{opposite}}{\text{hypotenuse}}$. Thus, $\sin\theta = \dfrac{\overline{AC}}{\overline{OC}}$. Because this is a unit circle, $mOC = 1$ and thus $\sin\theta = \overline{AC}$.

PROBLEM 7: TRIGONOMETRY

You should know the value of both of these because they are special angles. $\cos\dfrac{\pi}{2} = 0$ and $\sin\dfrac{3\pi}{2} = -1$, therefore:

$$\cos\frac{\pi}{2} + \sin\frac{3\pi}{2} = 0 + -1 = -$$

Of course, you can always find the value on your calculator. Make sure that you are in **radian** mode!

PROBLEM 8: EXPONENTIAL EQUATIONS

In order to solve this equation, we need to have a common base for the two sides of the equation. Because $3^3 = 27$ and $3^2 = 9$, we can rewrite the equation as $\left(3^3\right)^{x+2} = \left(3^2\right)^{2x-1}$.

The rules of exponents say that *when a number raised to a power is itself raised to a power, we multiply the exponents.* Thus, we get:

$$3^{3x+6} = 3^{4x-2}$$

Now, we can set the powers equal to each other and we get $3x + 6 = 4x - 2$ and therefore $x = 8$.

Problem 9: Trigonometric Area

The area of a triangle, if we are given the lengths of the two sides a and b, and their included angle θ, is $A = \frac{1}{2}ab\sin\theta$. In other words, if we are given SAS (side, angle, side), we find the area by multiplying $\frac{1}{2}$ by the product of the two sides by the sine of the included angle. Here, the two sides are b and c and the included angle is A. Plugging in, we get: $A = \frac{1}{2}(4)(4)\sin 60° = 4\sqrt{3}$.

Problem 10: Complex Numbers

First, to find the difference of $4 + 2i$ and $a + 4i$ we subtract the real components and the imaginary components separately. This gives us:
$(4 - a) + (2 - 4)i = 9 - 2i$. Therefore, $4 - a = 9$ and $a = -5$.

Problem 11: Trigonometry

$Arc\tan\frac{\sqrt{3}}{3}$ means "what angle θ has $\tan\theta = \frac{\sqrt{3}}{3}$?" Although there are an infinite number of answers, when the A in Arctangent is capitalized, you only use the principal angle. This means that, if the Arctangent is of a positive number, use the **first** quadrant answer; and, if the Arctangent is of a negative number, use the **fourth** quadrant answer. This is a special angle and you should know that $\tan 30° = \frac{\sqrt{3}}{3}$, so $Arc\tan\frac{\sqrt{3}}{3} = 30°$.

Now, we just have to find $\sin 30°$. This is a special angle, so you should know that $\sin 30° = \frac{1}{2}$.

Multiple-Choice Problems

For problems 12–35, you will find that you can sometimes **plug-in** or **backsolve** to get the right answer. You will also find that a calculator will simplify the arithmetic. In addition, most of the problems only require knowing which formula to use.

PROBLEM 12: TRIGONOMETRY GRAPHS

A graph of the form $y = a \sin bx$ or $y = a \cos bx$ has an amplitude of $|a|$ and a period of $\frac{2\pi}{b}$.

Therefore, the graph has an amplitude of 3.

The answer is (3).

PROBLEM 13: TRIGONOMETRY

The cosine of an angle is negative in quadrants II and III.

The sine of an angle is negative in quadrants III and IV.

The answer is (3).

PROBLEM 14: GRAPHS

First, we rearrange the second equation to obtain $y = \frac{8}{x}$. Now, because both equations are solved for y, we can set them equal to each other and solve for x. The solution will be the x-coordinate of the intersection. We get: $\frac{8}{x} = x^2$.

Now multiply both sides by x: $8 = x^3$.

Therefore, $x = 2$.

Now, plug $x = 2$ into either equation to find the value of y: $y = \frac{8}{2} = 4$

Therefore, the coordinates are $(2, 4)$.

The answer is (2).

PROBLEM 15: EXPONENTS

A number raised to a negative power is the same as the reciprocal of that number raised to the corresponding positive power.

Let's apply this rule to the answer choices:

Choice (1): $y = -10^{-x} = -\dfrac{1}{10^x}$. This is not the same as $y = 10^x$.

Choice (2): $y = 10^{-x} = \dfrac{1}{10^x}$. This is not the same as $y = 10^x$.

Choice (3): $y = \left(\dfrac{1}{10}\right)^{-x} = 10^x$. This is what we are looking for.

The answer is (3).

Another way to get the right answer is to **plug-in**. Make up a number for x. For example, let $x = 2$. Then $y = 10^2 = 100$. Now plug $x = 2$ into each answer choice and see which one matcher.

Choice (1): $y = -10^{-2} = -0.01$. This is not the answer.

Choice (2): $y = 10^{-2} = 0.01$. This is not the answer.

Choice (3): $y = \left(\dfrac{1}{10}\right)^{-2} = 100$. This is what we are looking for

PROBLEM 16: INEQUALITIES

We break this inequality apart into two inequalities:
$$15 < 3x + 5 \text{ and } 3x + 5 < 21.$$
We can solve each of these separately and we get:
$$\frac{10}{3} < x \text{ and } x < \frac{16}{3}.$$

The answer is (1).

PROBLEM 17: FUNCTIONS

All we do is plug in $4a$ for x and we get: $f(x) = \dfrac{4a - 4}{4a + 4}$. We can simplify this by factoring 4 out of the numerator and denominator.

We get:

$$f(x) = \frac{4(a - 1)}{4(a + 1)} = \frac{(a - 1)}{(a + 1)}$$

The answer is (1).

Another way to get this right is to **plug-in**. Make up a value for a. For example, let's let $a = 2$. Then $4a = 8$. If we plug 8 into $f(x)$, we get $f(8) = \dfrac{8 - 4}{8 + 4} = \dfrac{1}{3}$. Now plug $a = 2$ into the answers and see which one matches.

Choice (1): $\dfrac{2 - 1}{2 + 1} = \dfrac{1}{3}$. This is the right answer, but just in case . . .

Choice (2): $\dfrac{2 + 1}{2 - 1} = 3$. Wrong!

Choice (3): $\dfrac{8 - 1}{8 + 1} = \dfrac{7}{9}$. Wrong!

Choice (4): $\dfrac{8 + 1}{8 - 1} = \dfrac{9}{7}$. Wrong!

PROBLEM 18: GRAPHS

Choice (1): The domain is only the numbers on the x-axis inside the circle.

Choice (2): The domain is only the numbers on the x-axis inside the ellipse.

Choice (3) The domain is all real numbers.

Choice (4): The domain is only the numbers on the x-axis from the vertices of the hyperbolas outwards.

The answer is (3).

PROBLEM 19: ABSOLUTE VALUE INEQUALITIES

When we have an absolute value equation, we break it into two equations.

First, rewrite the left side without the absolute value and leave the right side alone. Second, rewrite the left side without the absolute value, and make the right side negative.

Here, we are going to have to figure out each answer choice to find the one that matches the solution, which, by the way, is: $-5 < x < 3$.

Choice (1): $|x + 8| \leq 3$. This becomes $x + 8 \leq 3$ or $x + 8 \geq -3$. Solving each of these, we get: $x \leq -5$ or $x \geq -11$. This doesn't match what we are looking for.

Choice (2): $|x + 1| < 4$. This becomes $x + 1 < 4$ and $x + 1 > -4$. Solving each of these, we get: $x < 3$ and $x > -5$. This is the right answer.

The answer is (2).

We don't have to check the other two, because they can't give the same solution set.

By the way, you should have noticed that the answer couldn't be choices 1 or 3 because they would give answers with filled circles on a number line, not empty circles. Do you know why?

PROBLEM 20: EXPONENTS

When you raise a number to a fractional power, you follow the rule $x^{\frac{a}{b}} = \sqrt[b]{x^a}$.

Therefore, $(-64)^{\frac{2}{3}} = \sqrt[3]{(-64)^2}$. We can find the root first, or the power. Let's find the cube root first. We get:

$$\sqrt[3]{-64} = -4$$

Now, we square the result and we get: $(-4)^2 = 16$.

The answer is (1).

Of course, you could have done this on your calculator. Just be sure to put the power and the -64 in parentheses!

PROBLEM 21: TRIGONOMETRY

Notice that this problem has the form $\cos A \cos B - \sin A \sin B$. Each Regents exam comes with a formula sheet that contains all of the trig formulas that you will need to know. If you look under <u>Functions of the Sum of Two Angles</u>, you will find the formula $\cos (A + B) = \cos A \cos B - \sin A \sin B$. Thus we can rewrite the problem as $\cos 100° \cos 80° - \sin 100° \sin 80° = \cos(100° + 80°) = \cos 180°$.

This is a special angle that you should know. $\cos 180° = -1$.

The answer is (3).

Another way to answer this question was to use your calculator to find the value of $\cos 100° \cos 80° - \sin 100° \sin 80°$. **Make sure that your calculator is in degree mode.** You should get:

$$\cos 100° \cos 80° - \sin 100° \sin 80° = -1.$$

PROBLEM 22: COMPLEX NUMBERS

Whenever we have a fraction where the denominator has a complex number in it, we can simplify the expression by multiplying the top and bottom by the *complex conjugate* of the denominator. The complex conjugate of $a + bi$ is $a - bi$. Thus, the complex conjugate of $5 + 2i$ is $5 - 2i$.

When you multiply two complex numbers, remember to FOIL.

We get:

$$\frac{1}{5 + 2i} \cdot \frac{5 - 2i}{5 - 2i} = \frac{5 - 2i}{25 - 10i + 10i - 4i^2} = \frac{5 - 2i}{25 - 4(-1)} = \frac{5 - 2i}{29}$$

The answer is (4).

PROBLEM 23: QUADRATIC EQUATIONS

Given an equation of the form $ax^2 + bx + c = 0$, *the product of the roots is* $\frac{c}{a}$, *and the sum of the roots is* $-\frac{b}{a}$.

Here, we have the equation $2x^2 - 9x + 6 = 0$, so the product of the roots is $\frac{c}{a} = \frac{6}{2} = 3$.

The answer is (3).

Another way to get the answer is actually to find the roots and multiply them. The equation doesn't factor, so we use the quadratic formula

$$x = \frac{-b \pm \sqrt{b^2 - 4ac}}{2a}$$

Plugging in, we get $x = \frac{-(-9) \pm \sqrt{(-9)^2 - 4(2)(6)}}{2(2)} = \frac{9 \pm \sqrt{33}}{4}$. The

roots are thus $\frac{9 + \sqrt{33}}{4} \approx 3.686$ and $\frac{9 - \sqrt{33}}{4} \approx 0.814$. If we multiply them, we get 3.000, which is closest to answer (3).

PROBLEM 24: TRIGONOMETRY GRAPHS

The graph of $y = \cos x$ looks like this:

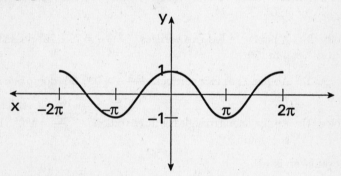

This is obviously symmetric with respect to the y-axis.

The answer is (1).

PROBLEM 25: CONIC SECTIONS

A graph of the form $\frac{x^2}{a^2} + \frac{y^2}{b^2} = 1$, *where a and b have the same value, is the equation of a circle. If a and b have different values, the graph is that of an ellipse.*

If we divide answer choice (1) through by 400, we get: $\frac{x^2}{400} + \frac{y^2}{400} = 1$.

a and b have the same value, so this is a circle.

If we divide answer choice (2) through by 400, we get: $\dfrac{x^2}{16} + \dfrac{y^2}{25} = 1$.

a and b have different values, so this is an ellipse.

Choices (3) and (4) are hyperbolas.

The answer is (2).

PROBLEM 26: TRANSFORMATIONS

If a figure has 120° rotational symmetry, then one can turn it through a third of a circle and it will look the same. An easy test is to divide the number of vertices into 360° and see if the answer goes into 120°. If yes, then, it has 120° rotational symmetry. If the quotient doesn't divide into 120° then it doesn't have 120° rotational symmetry.

Choice (1): A rhombus has four vertices. $\dfrac{360°}{4} = 90°$. 90° doesn't go into 120°.

Choice (2): A pentagon has five vertices. $\dfrac{360°}{5} = 72°$. 72° doesn't go into 120°.

Choice (3): A square has four vertices. $\dfrac{360°}{4} = 90°$. 90° doesn't go into 120°.

Choice (4): An equilateral triangle has three vertices. $\dfrac{360°}{3} = 120°$. 120° goes into 120°.

The answer is (4).

PROBLEM 27: STATISTICS

You should know the following rule about normal distributions.

> In a normal distribution, with a mean of $\bar{x}$ and a standard deviation of σ:
>
> - approximately 68% of the outcomes will fall between $\bar{x} - \sigma$ and $\bar{x} + \sigma$
> - approximately 95% of the outcomes will fall between $\bar{x} - 2\sigma$ and $\bar{x} + 2\sigma$
> - approximately 99.5% of the outcomes will fall between $\bar{x} - 3\sigma$ and $\bar{x} + 3\sigma$.

Here, we are told that $\overline{x} = 48$ and $\sigma = 4$. Thus $\overline{x} - 3\sigma = 48 - 12 = 36$ and $\overline{x} + 3\sigma = 48 + 12 = 60$. Therefore, 99% of the scores lie between 36 and 60.

The answer is (4).

PROBLEM 28: TRIGONOMETRIC EQUATIONS

First, subtract $\sin x$ from both sides. We get: $\sin^2 x - \sin x = 0$.

Note that the trig equation $\sin^2 x - \sin x = 0$ has the same form as a quadratic equation $x^2 - x = 0$. Just as we could factor the quadratic equation, so too can we factor the trig equation. We get: $\sin x \, (\sin x - 1) = 0$.

This gives us $\sin x = 0$ and $\sin x - 1 = 0$. If we solve each of these equations, we get: $\sin x = 0$ and $\sin x = 1$. Therefore $x = \sin^{-1}(0)$ and $x = \sin^{-1}(1)$. The solutions to both of these are easy because they are special angles.

$\sin^{-1}(0) = 0, \pi$ and $\sin^{-1}(1) = \dfrac{\pi}{2}$.

The answer is (1).

Another way to get the right answer is to **backsolve**. Try each answer choice to see which ones work. Eliminate the answer choices that don't.

Does $\sin^2(0) = \sin(0)$? Yes. Therefore, the answer can only be (1) or (3).

Both choices (1) and (3) contain $\dfrac{\pi}{2}$ so we don't have to check that answer.

Does $\sin^2(\pi) = \sin(\pi)$? Yes. Therefore, choice (1) is the answer.

PROBLEM 29: LOGARITHMS

You should know the following log rules:

(i) $\log A + \log B = \log(AB)$

(ii) $\log A - \log B = \log\left(\dfrac{A}{B}\right)$

(iii) $\log A^B = B \log A$

First, using rule (iii), we can rewrite $\dfrac{1}{3} \log m - 2 \log n$ as:

$$\log \sqrt[3]{m} - \log n^2$$

Next, using rule (ii), we can rewrite $\log \sqrt[3]{m} - \log n^2$ as $\log \dfrac{\sqrt[3]{m}}{n^2}$.

The answer is (4).

Another way to get the right answer is to **plug-in**. First, let's make up values for m and n. Let $m = 8$ and $n = 3$.

Then we use our calculator to find $\dfrac{1}{3} \log 8 - 2 \log 3 = -0.653$.

Next, we plug $m = 8$ and $n = 3$ into each of the answer choices to see which one matches.

Choice (1): Does $\log\left(\dfrac{8}{3} - 6\right) \approx -0.653$? If you plug it into the calculator, you get an error because you can't take the log of a negative number, so choice (1) is not the answer.

Choice (2): Does $\log\left(\dfrac{8^3}{3}\right) \approx -0.653$? If you plug it into the calculator, you get ≈ 2.232, so choice (2) is not the answer.

Choice (3): Does $\log(2 - 9) \approx -0.653$? If you plug it into the calculator, you get an error because you can't take the log of a negative number, so choice (3) is not the answer.

Choice (4): Does $\log \dfrac{2}{9} \approx -0.653$? If you plug it into the calculator, you get ≈ -0.653, so choice (4) **is** the answer.

PROBLEM 30: TRIGONOMETRIC IDENTITIES

First, let's get a common denominator of $\cos B$. This gives us:

$$\frac{\sin^2 B}{\cos B} + \frac{\cos^2 B}{\cos B} = \frac{\sin^2 B + \cos^2 B}{\cos B}.$$

Next, use the identity $\sin^2 B + \cos^2 B = 1$ (which is on the formula sheet) to obtain $\dfrac{\sin^2 B + \cos^2 B}{\cos B} = \dfrac{1}{\cos B}$.

The answer is (2).

Another way to get the right answer is to **plug-in** a value for B. For example, let $B = 30°$.

Now we plug $B = 30°$ into the problem and we get:

$$\frac{\sin^2 30°}{\cos 30°} + \cos 30° \approx 1.155.$$

Next, plug $B = 30°$ into each of the answer choices and find the one that matches.

Choice (1) is clearly wrong.

For choice (2) we get $\dfrac{1}{\cos 30°} \approx 1.155$. This is the right answer.

For choice (3) we get $\dfrac{1}{\sec 30°} \approx 0.866$. This is the wrong answer.

For choice (4) we get $\sin^2 30° = \dfrac{1}{4}$. This is the wrong answer.

PROBLEM 31: BINOMIAL EXPANSIONS

The binomial theorem says that if you expand $(a + b)^n$, you get the following terms:

$$_nC_0 a^n + {_nC_1} a^{n-1} b^1 + {_nC_2} a^{n-2} b^2 + \dots + {_nC_{n-2}} a^2 b^{n-2} + {_nC_{n-1}} a^1 b^{n-1} + {_nC_n} b^n$$

Therefore, if we expand $(3a - b)^6$, we get:

$$_6C_0(3a)^6 + {_6C_1}(3a)^5(-b)^1 + {_6C_2}(3a)^4(-b)^2 + {_6C_3}(3a)^3(-b)^3 + {_6C_4}(3a)^2(-b)^4 +$$
$$_6C_5(3a)^1(-b)^5 + {_6C_6}(-b)^6$$

Next, we use the rule that $_nC_r = \dfrac{n!}{(n-r)!\,r!}$. This gives us:

$$(3a)^6 + 6(3a)^5(-b)^1 + 15(3a)^4(-b)^2 + 20(3a)^3(-b)^3 +$$
$$15(3a)^2(-b)^4 + 6(3a)(-b)^5 + (-b)^6$$

which simplifies to:

$$(3a - b)^6 = 729a^6 - 1458a^5 b + 1215a^4 b^2 - 540a^3 b^3 + 135a^2 b^4 - 18ab^5 + b^6.$$

The fifth term is $135a^2b^4$.

The answer is (1).

A shortcut is to know the rule that the *rth* term of the binomial expansion of $(a - b)^n$ is $_nC_{r-1}(a)^{n-r+1}(b)^{r-1}$. Thus the fifth term is:

$_6C_{5-1}(3a)^{6-5+1}(-b)^{5-1}$, which can be simplified to $_6C_4(3a)^2(-b)^4 = 135a^2b^4$.

PROBLEM 32: INVERSE VARIATION

When two variables, x, and y, are said to vary inversely, their product is always a constant. In other words, $xy = k$, where k is a constant.

Therefore, if x is multiplied by 2, y has to be divided by 2 to keep the product the same.

The answer is (2).

PROBLEM 33: PROBABILITY

*If the probability of a particular outcome is **p**, then the probability of that outcome occurring **r** times out of a possible **n** times is* $_nC_r(p)^r(1 - p)^{n-r}$. This is known as *binomial probability*.

Here, the probability of winning is $\dfrac{3}{5}$. The probability of winning 3 out of a possible 4 games is $_4C_3\left(\dfrac{3}{5}\right)^3\left(1 - \dfrac{3}{5}\right)^{4-3}$. This can be simplified to $_4C_3\left(\dfrac{3}{5}\right)^3\left(\dfrac{2}{5}\right)^1$.

We can then evaluate $_4C_3$ according to the formula $_nC_r = \dfrac{n!}{(n - r)!\,r!}$.

If we plug in, we get $_4C_3 = \dfrac{4!}{(4 - 3)!\,3!} = \dfrac{4!}{1!\,3!} = \dfrac{4 \cdot 3 \cdot 2 \cdot 1}{(1)(3 \cdot 2 \cdot 1)} = 4$.

Therefore, the probability of winning 3 out of 4 games is:

$$4\left(\dfrac{3}{5}\right)^3\left(\dfrac{2}{5}\right) = \dfrac{216}{625}$$

The answer is (3).

PROBLEM 34: LOGARITHMS

The graph of the equation $y = \log_b x$ looks like this:

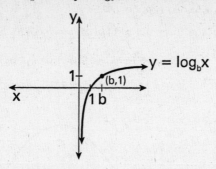

This graph goes through the point $(2, 1)$ so it is the graph of the equation $y = \log_2 x$.

The answer is (2).

PROBLEM 35: FUNCTIONS

A function is undefined when the denominator equals zero. Thus, if we want a function to be defined for all real numbers, the denominator can never be zero. Let's look at each of the choices.

Choice (1): $\dfrac{x^2 - 1}{(x-1)^2}$ has a denominator equal to zero when $x = 1$.

Choice (2): $\dfrac{x^2 - 1}{x + 1}$ has a denominator equal to zero when $x = -1$.

Choice (3): $\dfrac{x^2 - 1}{x^2}$ has a denominator equal to zero when $x = 0$.

Choice (4): $\dfrac{x^2 - 1}{x^2 + 1}$ has a denominator that is never equal to zero because x^2 is always greater than or equal to zero, so $x^2 + 1$ is always positive.

The answer is (4).

Part II

PROBLEM 36: TRIGONOMETRIC GRAPHS

(a) A graph of the form $y = a \sin bx$ or $y = a \cos bx$ has an amplitude of $|a|$ and a period of $\dfrac{2\pi}{b}$. Therefore, the equation $y = 2 \sin \dfrac{1}{2} x$ has an amplitude of 2 and a period of 4π. The graph looks like this:

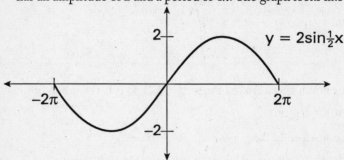

Remember, in graphing sines and cosines, the shape of the graph doesn't change, but it can be stretched or shrunk depending on the amplitude or period.

(b) A *dilation* D_k means that you should multiply the function by k. Thus, a dilation of $\dfrac{1}{2}$ means to multiply the x and y coordinates of the function by a scale factor of $\dfrac{1}{2}$, or shrink it by a scale factor of $\dfrac{1}{2}$.

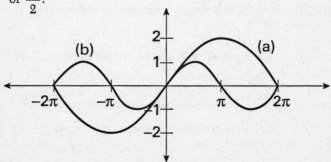

(c) When we multiply the function $y = 2\sin\dfrac{1}{2}x$ by $\dfrac{1}{2}$ we get: $y = \sin x$

PROBLEM 37: CIRCLE RULES

(a) *The measure of an inscribed angle is half of the arc it subtends (intercepts).*

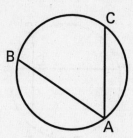

Which makes $\dfrac{1}{2}\,m\widehat{BC} = m\angle CAB = 60$ and therefore $m\widehat{BC} = 120$.

(b) Here we can use the same rule as in part (a). If we can find the measure of arc $\widehat{EA}$, then the measure of angle EBA is half of that number

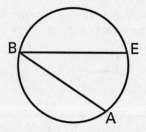

We can find the measure of arc $\widehat{EA}$ because we know that the sum of the degrees of the arcs of a circle is $360°$. In other words, $m\widehat{BC} + m\widehat{CE} + m\widehat{AB} + m\widehat{EA} = 360$.

Plugging in the measures of the arcs that we already know, we get:

$$120 + 40 + 130 + m\overset{\frown}{EA} = 360$$

$$m\overset{\frown}{EA} = 360 - 120 - 40 - 130 = 70$$

Therefore, $m\angle EBA = 35$.

(c) *If two chords intersect within a circle, then the angle between the two chords is the average of their intercepted arcs.*

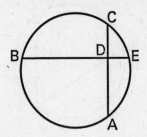

Therefore, $m\angle CDE = \dfrac{m\overset{\frown}{CE} + m\overset{\frown}{AB}}{2} = \dfrac{40 + 130}{2} = 85$.

Since $\angle CDE$ and $\angle ADE$ are supplementary, we know that:

$$m\angle ADE = 180 - 85 = 95 \cdot$$

(d) *The measure of an angle formed by a pair of secants, or a secant and a tangent, is equal to half of the difference between the larger and the smaller arcs that are formed by the secants, or the secant and the tangent.*

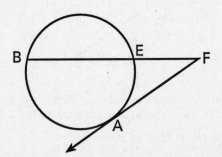

Here, the larger and smaller arcs formed by angle F are $\overset{\frown}{AB}$ and $\overset{\frown}{AE}$ respectively.

$$m\angle F = \frac{m\overset{\frown}{AB} - m\overset{\frown}{AE}}{2} = \frac{130 - 70}{2} = 30.$$

(e) *An angle formed by a tangent and a chord is equal to half of its intercepted arc.*

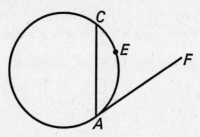

We can find the measure of arc $\overset{\frown}{AC}$ by adding the measures of arcs $\overset{\frown}{AE}$ and $\overset{\frown}{CE}$. We get: $m\overset{\frown}{AC} = 70 + 40 = 110$.

Therefore, $m\angle FAC = \frac{1}{2} m\overset{\frown}{AC} = \frac{1}{2} 110 = 55$.

Problem 38:

(a) Law of Cosines

First, let's draw a picture of the situation.

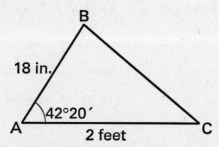

We can use the Law of Cosines. The Law of Cosines says that, in any triangle, with angles $A, B,$ and C, and opposite sides of a, b, and c, respectively, $c^2 = a^2 + b^2 - 2ab\cos C$. (This is on the formula sheet.) If we have a triangle and we are given SAS (side, angle, side), then the side opposite the angle we seek is c, and the angle is C.

Here, we are seeking BC and the angle opposite it is A.

We also need to do a couple of conversions. First of all, 2 feet is 24 inches.

Second, we convert an angle from a number of minutes to a decimal by dividing the number of minutes by 60. That is, $20' = \frac{20}{60}^\circ = \frac{1}{3}^\circ$ and thus $42°20' = 42\frac{1}{3}^\circ$.

We have:

$$(BC)^2 = 24^2 + 18^2 - 2(24)(18)\cos 42\frac{1}{3}^\circ$$

$$(BC)^2 = 576 + 324 - 864(0.7392)$$

$$(BC)^2 = 900 - 638.7 = 261.3$$

$BC \approx 16.2$ inches. (Make sure that you include the inches. You could also have done this in feet, in which case you would have gotten 1.3 feet.)

(b) Trigonometric Area

The area of a triangle, if we are given the lengths of the two sides a and b, and their included angle θ, is $A = \frac{1}{2}ab\sin\theta$. In other words, if we are given SAS (side, angle, side), we find the area by multiplying $\frac{1}{2}$ by the product of the two sides by the sine of the included angle. Here, the two sides are AC and AB and the included angle is $42\frac{1}{3}^\circ$. Plugging in, we get:

$$A = \frac{1}{2}(24)(18)\sin 42\frac{1}{3}^\circ \approx 145.5 \text{ in}^2 \text{ (or 1.0 ft}^2)$$

PROBLEM 39:

(a) Statistics

The method for finding a standard deviation is simple, but time-consuming.

First, find the average of the scores. We do this by multiplying each score by the number of people who obtained that score, and adding them up.

$(80)(5) + (85)(7) + (90)(9) + (95)(4) = 2,185$. Next, we divide by the total number of people $(5 + 7 + 9 + 4) = 25$, to obtain the average

$$\bar{x} = \frac{2185}{25} = 87.4.$$

Second, subtract the average from each actual score.

$$80 - 87.4 = -7.4$$
$$85 - 87.4 = -2.4$$
$$90 - 87.4 = 2.6$$
$$95 - 87.4 = 7.6$$

Third, square each difference.

$$(-7.4)^2 = 54.76$$

$$(-2.4)^2 = 5.76$$

$$(2.6)^2 = 6.76$$

$$(-7.6)^2 = 57.76$$

Fourth, multiply the square of the difference by the corresponding number of people and sum.

$$(54.76)(5) = 273.8$$

$$(5.76)(7) = 40.32$$

$$(6.76)(9) = 60.84$$

$$(57.76)(4) = 231.04$$

$$273.8 + 40.32 + 60.84 + 231.04 = 606$$

Fifth, divide this sum by the total number of people.

$$\frac{606}{25} = 24.24.$$

Last, take the square root of this number. This is the standard deviation. $\sigma = \sqrt{24.24} \approx 4.923$

Rounded to the nearest tenth, we get $\sigma \approx 4.9$

(b) Rational Expressions

First, we need to get a common denominator for the numerator and the denominator separately.

$$\frac{1 - \dfrac{3}{\cos x}}{\dfrac{9}{\cos^2 x} - 1} = \frac{\dfrac{\cos x}{\cos x} - \dfrac{3}{\cos x}}{\dfrac{9}{\cos^2 x} - \dfrac{\cos^2 x}{\cos^2 x}} = \frac{\dfrac{\cos x - 3}{\cos x}}{\dfrac{9 - \cos^2 x}{\cos^2 x}}$$

Now, invert the bottom and multiply instead of divide.

$$\frac{\dfrac{\cos x - 3}{\cos x}}{\dfrac{9 - \cos^2 x}{\cos^2 x}} = \frac{\cos x - 3}{\cos x} \cdot \frac{\cos^2 x}{9 - \cos^2 x}$$

Next, factor the denominator of the second expression. We get:

$$\frac{\cos x - 3}{\cos x} \cdot \frac{\cos^2 x}{9 - \cos^2 x} = \frac{\cos x - 3}{\cos x} \cdot \frac{\cos^2 x}{(3 - \cos x)(3 + \cos x)}$$

Finally, cancel like terms.

$$\frac{-\cos x}{(3 + \cos x)}$$

PROBLEM 40: TRIGONOMETRIC EQUATIONS

(a) First, subtract 5 tan x from both sides.

$$\tan^2 x - 5 \tan x + 3 = 0$$

Note that the trig equation $\tan^2 x - 5 \tan x + 3 = 0$ has the same form as a quadratic equation $x^2 - 5x + 3 = 0$. We would use the quadratic formula to solve for x, so here we will use the formula to solve for tan x.

We get:

$$\tan x = \frac{-(-5) \pm \sqrt{(-5) - 4(1)(3)}}{2} = \frac{5 \pm \sqrt{13}}{2}.$$

This gives us

$$\tan x = \frac{5 + \sqrt{13}}{2} \approx 4.303 \text{ and } \tan x = \frac{5 - \sqrt{13}}{2} \approx 0.697.$$

Therefore $x = \tan^{-1}(4.303)$ and $x = \tan^{-1}(0.697)$.

There is not a special angle that gives either of these values for tangent, so we need to use the calculator to find the value. Using the calculator, push **2nd tan (4.303) ENTER**. (**MAKE SURE THAT THE CALCULATOR IS IN DEGREE MODE!**) This gives us $76.9° \approx 77°$ as an answer. For the other angle, we get $34.9° \approx 35°$.

But, we are asked to find all of the values of x between $0°$ and $360°$, so these are not the only answers.

We can find the required values of x by making a small graph and plotting the reference angle of $x = 77°$

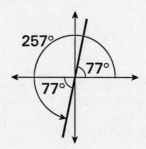

The other place where tangent is positive and has a reference angle of $77°$ is in quadrant III. The other angle then is $180° + 77° = 257°$. Similarly, we also get another angle of $180° + 35° = 215°$.

Therefore, the answers are $x = 35°$, $77°$, $215°$, and $257°$.

PROBLEM 41:

(a) Exponential Equations

$$9^{(x^2+x)} = 3^4.$$

In order to solve this equation, we need to have a common base for the two sides of the equation. Because $3^2 = 9$, we can rewrite the left side as $\left(3^2\right)^{x^2+x} = 3^4.$

The rules of exponents say that *when a number raised to a power is itself raised to a power, we multiply the exponents.* Thus, we get:

$$3^{2x^2+2x} = 3^4.$$

Now, we can set the powers equal to each other:

$$2x^2 + 2x = 4$$

Subtract 4 from both sides:

$$2x^2 + 2x - 4 = 0$$

Divide through by 2:

$$x^2 + x - 2 = 0$$

Factor:

$$(x + 2)(x - 1) = 0$$

Therefore, $x = -2$ or $x = 1$.

Whenever you are doing equations involving *logarithms, radicals, rational expressions, or exponents*, you should always check the answers.

Does $9^{(-2)^2-2} = 3^4$? $9^2 = 81 = 3^4$. Yes.

Does $9^{(1)^2+1} = 3^4$? $9^2 = 81 = 3^4$. Yes.

Therefore $x = -2$ or $x = 1$ are the answers.

(b) Transformations

(1) First, we plot the coordinates.

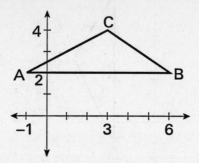

(2) Here we are asked to find the results of a composite transformation $\left(R_{90°} \circ r_{x\text{-}axis}\right)$. *Whenever we have a composite transformation, always do the **right** one first.*

$r_{x\text{-}axis}$ means that we reflect the graph in the *x-axis*; that is, we turn it upside down. When a graph is reflected in the *x*-axis, we change the sign of the *y*-coordinates and leave the *x*-coordinates alone. This gives us:

$$r_{x\text{-}axis}A(-1, 2) \rightarrow (-1, -2)$$

$$r_{x\text{-}axis}B(6, 2) \rightarrow (6, -2)$$

$$r_{x\text{-}axis}C(3, 4) \rightarrow (3, -4)$$

Now we do the other transformation to *the graph that we just reflected*.

$R_{90°}$ means that we rotate the graph by 90° counterclockwise. This means that we transform the coordinates as follows: $R_{90°}(x, y) \rightarrow (-y, x)$. This gives us:

$$R_{90°}(-1, -2) \rightarrow A'(2, -1)$$

$$R_{90°}(6, -2) \rightarrow B'(2, 6)$$

$$R_{90°}(3, -4) \rightarrow C'(4, 3)$$

Now we can plot our transformed triangle.

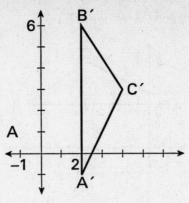

(c) Notice what our composite transformation did.

$$A(-1, 2) \rightarrow A'(2, -1)$$
$$B(6, 2) \rightarrow B'(2, 6)$$
$$C(3, 4) \rightarrow C'(4, 3)$$

In each case, we simply switched the x and y coordinates. This is equivalent to a reflection in the line $y = x$. We write this as $r_{y=x}$.

PROBLEM 42:

(a) Trigonometric Identities

All of the identities that you will need to know are contained on the formula sheet.

First, let's FOIL the right-hand side.

$$\sec^2 x + \csc^2 x = \tan^2 x + 2 \tan x \cot x + \cot^2 x$$

Next, notice that $\tan x \cot x = \dfrac{\sin x}{\cos x} \cdot \dfrac{\cos x}{\sin x} = 1$, which gives us:

$$\sec^2 x + \csc^2 x = \tan^2 x + 2 + \cot^2 x$$

Next, using the identities $1 + \tan^2 x = \sec^2 x$ and $1 + \cot^2 x = \csc^2 x$ we can rewrite the left side as:

$$\left(1 + \tan^2 x\right) + \left(1 + \cot^2 x\right) = \tan^2 x + 2 + \cot^2 x$$

Now, simplify the left side

$$\tan^2 x + 2 + \cot^2 x = \tan^2 x + 2 + \cot^2 x$$

and we have proved the identity.

(b) Quadratic Equations

If we subtract 2 from both sides, we get $-3x^2 + 2x - 2 = 0$.

We can now use the quadratic formula to solve for x.

The formula says that, given a quadratic equation of the form $ax^2 + bx + c = 0$, the roots of the equation are:

$$x = \frac{-b \pm \sqrt{b^2 - 4ac}}{2a}$$

Plugging in to the formula we get:

$$x = \frac{-2 \pm \sqrt{(2)^2 - 4(-3)(-2)}}{2(-3)}$$

$$= \frac{-2 \pm \sqrt{4 - 24}}{-6}$$

$$= \frac{-2 \pm \sqrt{-20}}{-6}$$

$$= \frac{-2 \pm i\sqrt{20}}{-6}$$

$$= \frac{-2 \pm 2i\sqrt{5}}{-6}$$

$$= \frac{1 \pm i\sqrt{5}}{3}$$

Formulas

Pythagorean and Quotient Identities

$$\sin^2 A + \cos^2 A = 1 \qquad \tan A = \frac{\sin A}{\cos A}$$

$$\tan^2 A + 1 = \sec^2 A \qquad \cot A = \frac{\cos A}{\sin A}$$

$$\cot^2 A + 1 = \csc^2 A$$

Functions of the Sum of Two Angles

$$\sin (A + B) = \sin A \cos B + \cos A \sin B$$

$$\cos (A + B) = \cos A \cos B - \sin A \sin B$$

$$\tan (A + B) = \frac{\tan A + \tan B}{1 - \tan A \tan B}$$

Functions of the Difference of Two Angles

$$\sin (A - B) = \sin A \cos B - \cos A \sin B$$

$$\cos (A - B) = \cos A \cos B + \sin A \sin B$$

$$\tan (A - B) = \frac{\tan A - \tan B}{1 + \tan A \tan B}$$

Law of Sines

$$\frac{a}{\sin A} = \frac{b}{\sin B} = \frac{c}{\sin C}$$

Law of Cosines

$$a^2 = b^2 + c^2 - 2bc \cos A$$

Functions of the Double Angle

$$\sin 2A = 2 \sin A \cos A$$

$$\cos 2A = \cos^2 A - \sin^2 A$$

$$\cos 2A = 2 \cos^2 A - 1$$

$$\cos 2A = 1 - 2 \sin^2 A$$

$$\tan 2A = \frac{2 \tan A}{1 - \tan^2 A}$$

Functions of the Half Angle

$$\sin \frac{1}{2} A = \pm\sqrt{\frac{1 - \cos A}{2}}$$

$$\cos \frac{1}{2} A = \pm\sqrt{\frac{1 + \cos A}{2}}$$

$$\tan \frac{1}{2} A = \pm\sqrt{\frac{1 - \cos A}{1 + \cos A}}$$

Area of Triangle

$$K = \frac{1}{2} ab \sin C$$

Standard Deviation

$$\text{S.D.} = \sqrt{\frac{1}{n} \sum_{i=1}^{n} \left(x_i - \bar{x}\right)^2}$$

EXAMINATION
JUNE 1995

Part I

Answer 30 questions from this part. Each correct answer will receive 2 credits. No partial credit will be allowed. Write your answers in the spaces provided. Where applicable, answers may be left in terms of π or in radical form. [60]

1   What is the amplitude of the graph of the equation $y = 2 \sin \frac{1}{3}x$? $|a| = 2$

2   Express $4\sqrt{-49} + 3\sqrt{-16}$ as a monomial in terms of i.
$4 \cdot 7i + 3 \cdot 4i$
$28i + 12i = 40i$

3   In a circle, chords $\overline{AB}$ and $\overline{CD}$ intersect at E. If $AE = 21$, $EB = 5$, and $ED = 7$, find CE.
$CE = 15$

4   If $\csc \theta = -\frac{4}{3}$ and $\cos \theta > 0$, in which quadrant does θ terminate? IV

5   Express in the simplest form: $\frac{2}{x} \cdot \frac{2x}{1} = 4$

6   Solve for x: $4^x = 8^{x-1}$.

$2^{2x} = 2^{3(x-1)}$

$2x = 3x - 3$

$3 = x$

7 In a circle, an inscribed angle intercepts an arc whose measure is $(14x - 2)°$. Express in terms of x, the number of degrees in the measure of the inscribed angle.

$7x - 1$

8 In a circle, a central angle of 3.5 radians intercepts an arc of 24.5 centimeters. Find the number of centimeters in the radius of the circle.

9 Solve for x: $\sqrt{2x - 3} - 2 = 5$.

26

10 If $f(x) = 4 \sin \dfrac{x}{3}$, find $f(\pi)$.

$2\sqrt{3}$

11 Evaluate: $\dfrac{2}{3} \displaystyle\sum_{a=1}^{4} (a + 1)^2$. 34

12 Express $\dfrac{5}{4 - \sqrt{13}}$ as an equivalent fraction with a rational denominator.

$\dfrac{25 - \sqrt{13}}{3}$

13 When Nick plays cards with Lisa, the probability that Nick will win is $\dfrac{6}{10}$. If they play three games of cards and there are no ties, what is the probability that Lisa will win all three games?

$4/625$

14 What is the image that results from this composition of transformations?

$$r_{x-axis} \circ R_{0,90°}(-3,0) \quad (0, -3)$$

15 What is the third term in the expansion of $(x + 1)^5$?

$10x^3$

Directions (16–35): For *each* question chosen, write the *numeral* preceding the word or expression that best completes the statement or answers the question.

16 Which mapping represents a dilation?
  (1) $(x,y) \rightarrow (y,x)$
  (2) $(x,y) \rightarrow (x + 2y + 2)$
  (3) $(x,y) \rightarrow (-y,-x)$
  (4) $(x,y) \rightarrow (2x,2y)$

17 On a standardized test, Phyllis scored 84, exactly one standard deviation above the mean. If the standard deviation for the test is 6, what is the mean score for the test?
  (1) 72   (3) 84
  (2) 78   (4) 90

18 What is the solution set of the equation $|3x + 2| = 5$?

  (1) $\{1\}$ (3) $\left\{1, \dfrac{7}{3}\right\}$

  (2) $\left\{\dfrac{7}{3}\right\}$ (4) $\left\{-1, \dfrac{7}{3}\right\}$

[OVER]

19 Log $\sqrt{\dfrac{a}{b}}$ is equivalent to

(1) $\dfrac{1}{2} \log a - \log b$

(2) $\dfrac{1}{2} (\log a - \log b)$

(3) $\dfrac{1}{2} (\log a + \log b)$

(4) $\dfrac{1}{2} \log a + \log b$

20 For all values of a and b, what is the additive inverse of $a + bi$?

(1) $a - bi$ (3) $a + bi$
(2) $-a + bi$ (4) $-a - bi$

21 The expression Arc $\cos\left(\dfrac{1}{2}\right)$ is equal to

(1) 30°   (3) 60°
(2) 45°   (4) 90°

22 In the interval $0 \le x \le 2\pi$, in how many points will the graphs of the equation $y = \sin x$ and $y = \dfrac{1}{2}$ intersect?

(1) 1   (3) 3
(2) 2   (4) 4

23 The expression $\dfrac{\sin 2A}{2\cos^2 A}$ is the equivalent to

   (1) $\sin A$ (3) $\cot A$

   (2) $\tan A$ (4) $2\tan A$

24 What is the solution set for the inequality $x^2 - 4x - 5 < 0$?

   (1) $\{x \mid -1 < x < 5\}$

   (2) $\{x \mid -5 < x < 1\}$

   (3) $\{x \mid x > 5 \text{ or } x < -1\}$

   (4) $\{x \mid x > 1 \text{ or } x < -5\}$

25 If $\tan A = 8$ and $\tan B = \dfrac{1}{2}$, what is the value of $\tan (A + B)$?

   (1) $\dfrac{4}{3}$ (3) $-\dfrac{15}{6}$

   (2) $\dfrac{17}{10}$ (4) $-\dfrac{17}{6}$

26 The sides of a triangle measure 6, 7, and 9. What is the cosine of the largest angle?

   (1) $-\dfrac{4}{84}$ (3) $\dfrac{4}{84}$

   (2) 81 (4) $-\dfrac{1}{81}$

27 The expression i^{10} is equivalent to

   (1) 1   (3) -1

   (2) i (4) $-i$

[OVER]

28 The roots of the quadratic equation $5x^2 - 2x = -3$ are

(1) imaginary
(2) real and irrational
(3) real, rational, and unequal
(4) real, rational, and equal

29 In the accompanying figure, $\overline{OP} = 1$.

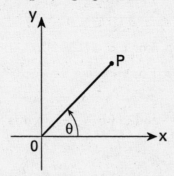

What are the coodinates of point P?

(1) $(\sin \theta, \cos \theta)$ (3) $(\cos \theta, \sin \theta)$
(2) $(-\sin \theta, -\cos \theta)$ (4) $(-\cos \theta, -\sin \theta)$

30 The transformation $R_{90°}$ maps the point $(5,3)$ onto the point whose coordinates are

(1) $(-3, 5)$ (3) $(3, -5)$
(2) $(3, 5)$ (4) $(5, -3)$

31 The value of $\cos \dfrac{\pi}{3} - \sin \dfrac{3\pi}{2}$ is

(1) $1\dfrac{1}{2}$ (3) $-\dfrac{1}{2}$

(2) $\dfrac{1}{2}$ (4) $-1\dfrac{1}{2}$

32 For the interval $-\pi \leq x \leq \pi$, which graph represents the image of the equation $y = \cos x$ after a reflection in the y-axis?

(1)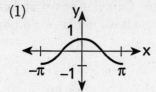

(3)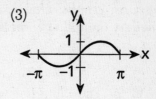

(2)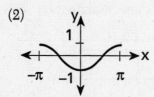

(4)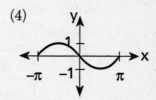

33 What is the inverse relation of the function whose equation is $y = 3x - 2$?

(1) $y = x$

(3) $y = 2x - 3$

(2) $y = 3x + 2$

(4) $y = \dfrac{x + 2}{3}$

34 If $a = 5$, $b = 7$, and m $\angle A = 30$, how many distinct triangles can be constructed?

(1) 1

(3) 3

(2) 2

(4) 0

35 An ellipse is formed by the graph of the equation

(1) $xy = 36$

(3) $9x^2 = 36 + 4y^2$

(2) $4x^2 - 9y^2 = 36$

(4) $9x^2 = 36 - 4y^2$

[OVER]

Part II

Answer four questions from this part. Clearly indicate the necessary steps, including appropriate formula substitutions, diagrams, graphs, charts, etc. Calculations that may be obtained by mental arithmetic or the calculator do not need to be shown. [40]

36 *a* The table below shows raw scores on an 80-question entrance examination. Find the standard deviation of these examination scores to the *nearest tenth*. [6]

x_i	f_i
40	5
50	4
60	6
70	3
80	2

b In the interval $0 \leq x \leq 2\pi$, sketch the graph of the equation $y = 2 \sin \frac{1}{2}x$. [4]

37 In the accompanying diagram of circle O, the ratio $m\widehat{BC}:m\widehat{CA}:m\widehat{AN}:m\widehat{NB}$ is 5:4:1:2. Chord $\overline{CB}$ is extended to external point M, chords $\overline{AB}$ and $\overleftrightarrow{CN}$ intersect at D, and tangent $\overrightarrow{MN}$ is drawn.

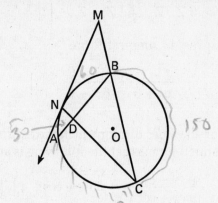

Find:

a $m\widehat{BC}$ [2]

b $m\angle ABC$ [2]

c $m\angle NMC$ [2]

d $m\angle NDA$ [2]

e $m\angle MND$ [2]

38 Find, to the *nearest degree,* all values of x in the interval $0° \le x < 360°$ that satisfy the equation $6\cos^2 x - 7\cos x + 2 = 0$.   [10]

39 a   Find, to the *nearest degree,* the measure of the largest angle of a triangle whose sides measure 22, 34, and 50.   [7]

[OVER]

b Find, to the *nearest integer*, the area of the triangle described in part *a*. [3]

40 *a* If $Z_1 = -1 + 6i$ and $Z_2 = 4 + 2i$, graphically represent Z_1, Z_2, and $Z_1 + Z_2$. [3]

b Express in simplest form:

$$\frac{1 - x^2}{6x + 6} \div \frac{x^4 - 1}{6x^2 + 6} \quad [7]$$

41 *a* For all values of *x* for which the expressions are defined, prove that the following is an identity:

$$\frac{\sec x + \csc x}{\tan x + \cot x} = \sin x + \cos x \quad [7]$$

b Find all values of *x* in the interval $0 \le x \le \pi$ that make the following fraction undefined:

$$\frac{1}{\sin 2x} \quad [3]$$

42 *a* Sketch the graph of the equation $y = 2^x$ and label the graph *a*. [3]

b On the same set of axes, graph the reflection of $y = 2^x$ in the *y*-axis and label the graph *b*. [3]

c Using logarithms, find *x*, to the *nearest hundredth*: $2^x = 5$. [4]

ANSWER KEY

Part I

1. 2

2. $40i$

3. 15

4. IV

5. 4

6. 3

7. $7x - 1$

8. 7   *Omit*

9. 26

10. $2\sqrt{3}$

11. 36

12. $\dfrac{5\left(4 + \sqrt{13}\right)}{3}$

13. $\dfrac{64}{1000}$

14. (0,3)

15. $10x^3$

16. (4)

17. (2)

18. (3)

19. (2)

20. (4)

21. (3)

22. (2)   *Omit*

23. (2)

24. (1)

25. (4)

26. (3)

27. (3)

28. (1)   *Omit*

29. (3)   *Omit*

30. (1)

31. (1)

32. (1)

33. (4)

34. (2)

35. (4)

ANSWERS AND EXPLANATIONS
JUNE 1995

Part I

In problems 1–15, you must supply the answer. Because of this, you will not be able to **plug-in** or **backsolve**. But, you can use your calculator to simplify calculations. In addition, you will find that these problems tend to test your basic knowledge of Sequential Math III and are not tricky. Many can be solved just by knowing the right formula and by plugging in the numbers.

PROBLEM 1: TRIGONOMETRY GRAPHS

A graph of the form $y = a \sin bx$ or $y = a \cos bx$ has an amplitude of $|a|$ and a period of $\frac{2\pi}{b}$.

Therefore, the graph has an amplitude of 2.

PROBLEM 2: COMPLEX NUMBERS

We can write $4\sqrt{-49}$ as $4\sqrt{49}\sqrt{-1}$ and $3\sqrt{-16}$ as $3\sqrt{16}\sqrt{-1}$. Now, using the definition $\sqrt{-1} = i$, we get:

$$4\sqrt{49}\sqrt{-1} + 3\sqrt{16}\sqrt{-1} = 4(7)i + 3(4)i = 40i$$

PROBLEM 3: CIRCLE RULES

If two chords intersect in a circle, then the product of the lengths of the segments of one chord is equal to the product of the lengths of the segments of the other chord.

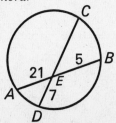

This means that $(AE)(EB) = (CE)(ED)$. If we plug in, we get:
$$(21)(5) = (CE)(7).$$
We can solve this for CE and we get:
$$(CE) = 15.$$

PROBLEM 4: TRIGONOMETRY

The sine of an angle is negative in quadrants III *and* IV; therefore, because $\csc\theta = \dfrac{1}{\sin\theta}$, the cosecant is also negative in quadrants III and IV. *The cosine of an angle is positive in quadrants* I *and* IV. Therefore, the angle must be in quadrant IV.

PROBLEM 5: RATIONAL EXPRESSIONS

Invert the bottom fraction and multiply it by the top one. We get:
$$\left(\frac{2}{x}\right)\left(\frac{2x}{1}\right)$$
Now, cancel like terms:
$$\left(\frac{2}{x}\right)\left(\frac{2x}{1}\right) = 4$$

PROBLEM 6: EXPONENTIAL EQUATIONS
$$4^x = 8^{x-1}$$
In order to solve this equation, we need to have a common base for the two sides of the equation. Because $2^3 = 8$ and $2^2 = 4$, we can rewrite the equation as $\left(2^2\right)^x = \left(2^3\right)^{x-1}$.

The rules of exponents say that *when a number raised to a power is itself raised to a power, we multiply the exponents.* Thus, we get:
$$2^{2x} = 2^{3x-3}$$
Now, we can set the powers equal to each other and solve.
$$2x = 3x - 3$$
$$x = 3$$

PROBLEM 7: CIRCLE RULES

The measure of an inscribed angle is always half of the measure of the arc that it subtends (intercepts). Here, the arc has measure $(14x - 2)$, so the angle has measure $(7x - 1)$.

PROBLEM 8: ARC LENGTH

In a circle with a central angle, θ, measured in radians; and a radius, r; the arc length, s, is found by $s = r\theta$.

Here we have $\theta = 3.5$ and $s = 24.5$. Therefore, the radius is 7 centimeters.

PROBLEM 9: RADICAL EQUATIONS

First, in order to evaluate the radical, add 2 to both sides. Now, if we square both sides of the equation, we get $2x - 3 = 49$. If we then add 3 to both sides, we get $2x = 52$ and thus $x = 26$. **BUT**, whenever we have a radical in an equation, we have to check the root to see if it satisfies the original equation. Sometimes, in the process of squaring both sides, we will get answers that are invalid. So let's check the answer.

Does $\sqrt{2(26) - 3} - 2 = 5$? $7 - 2 = 5$; so, yes, it does. Therefore the answer is $x = 26$.

PROBLEM 10: TRIGONOMETRY

Here, we substitute π for x, and we get $f(\pi) = 4\sin\frac{\pi}{3}$. You should know that $\sin\frac{\pi}{3} = \frac{\sqrt{3}}{2}$ because $\frac{\pi}{3}$ is a special angle.

Therefore, $4\sin\frac{\pi}{3} = 4\frac{\sqrt{3}}{2} = 2\sqrt{3}$.

PROBLEM 11: SUMMATIONS

To evaluate a sum, we plug each of the consecutive values for r into the expression, starting at the bottom value and ending at the top value and sum the results. We get·

$$\sum_{a=1}^{4} (a+1)^2 = \left[(1+1)^2 \right] + \left[(2+1)^2 \right] + \left[(3+1)^2 \right] + \left[(4+1)^2 \right] =$$

$$2^2 + 3^2 + 4^2 + 5^2 = 4 + 9 + 16 + 25 = 54$$

Now we multiply the result by $\dfrac{2}{3}$ and we get $\dfrac{2}{3} \cdot 54 = 36$.

Another way to do this is to use your calculator. Push **2nd STAT** (**LIST**) and under the **MATH** menu, choose **SUM**. Then, again push **2nd STAT** and under the **OPS** menu, choose **SEQ**.

We put in the following: (*expression, variable, start, finish, step*).

For *expression*, we put in the formula, using x as the variable.

For *variable*, we ALWAYS put in x.

For *start*, we put in the bottom value.

For *finish*, we put in the top value.

For *step*, we ALWAYS put in 1.

Therefore, we put in the calculator: **SUM SEQ** $\left((x+1)^2, x, 1, 4, 1 \right)$. The calculator should return the value 54. Then multiply the answer by $\dfrac{2}{3}$ and you're done.

PROBLEM 12: RATIONAL EXPRESSIONS

Whenever we have a fraction where the denominator has a radical in it, we can *rationalize* the denominator by multiplying the top and bottom by the *conjugate* of the denominator. The conjugate is merely the same as the denominator where we reverse the sign in front of the radical.

Here, we will multiply the top and bottom by the conjugate of $4 - \sqrt{13}$, which is $4 + \sqrt{13}$.

We get:

$$\left(\frac{5}{4 - \sqrt{13}} \right) \left(\frac{4 + \sqrt{13}}{4 + \sqrt{13}} \right).$$

Next, we multiply out the top and bottom using FOIL.

$$\left(\frac{5}{4-\sqrt{13}}\right)\left(\frac{4+\sqrt{13}}{4+\sqrt{13}}\right) = \frac{20+5\sqrt{13}}{16+4\sqrt{13}-4\sqrt{13}-13}$$

Now we simplify: $\dfrac{20+5\sqrt{13}}{3} = \dfrac{5\left(4+\sqrt{13}\right)}{3}$.

PROBLEM 13: PROBABILITY

*If the probability of a particular outcome is **p**, then the probability of that outcome occurring **r** times out of a possible **n** times is* $_nC_r(p)^r(1-p)^{n-r}$. *This is known as* binomial probability.

Here, we are told that the probability of Nick's winning any particular game is $\dfrac{6}{10}$ and thus the probability of Lisa's winning is $\dfrac{4}{10}$.

Therefore, the probability that Lisa wins 3 out of a possible 3 games is

$$_3C_3\left(\frac{4}{10}\right)^3\left(1-\frac{4}{10}\right)^{3-3} = {}_3C_3\left(\frac{4}{10}\right)^3\left(\frac{6}{10}\right)^0.$$

We can then evaluate $_3C_3$ according to the formula $_nC_r = \dfrac{n!}{(n-r)!\,r!}$. If we plug in, we get:

$$_3C_3 = \frac{3!}{(3-3)!\,3!} = \frac{3!}{0!\,3!} = \frac{3\cdot2\cdot1}{(3\cdot2\cdot1)(1)} = 1.$$

Therefore, the probability that Lisa wins 3 out of a possible 3 games is:

$$1\left(\frac{4}{10}\right)^3\left(\frac{6}{10}\right)^0 = \frac{64}{1000}.$$

PROBLEM 14: TRANSFORMATIONS

Here we are asked to find the results of a composite transformation $\left(r_{x-axis} \circ R_{0,90°}\right)(-3,0)$. *Whenever we have a composite transformation, always do the **right** one first.*

$R_{0,90°}$ means a rotation counterclockwise of 90° about the origin. (By the way, you will usually see $R_{0,90°}$ written as $R_{90°}$.)

You should know the following rotation rules:

$$R_{90°}(x, y) = (-y, x)$$
$$R_{180°}(x, y) = (-x, -y)$$
$$R_{270°}(x, y) = (y, -x)$$

This gives us $R_{90°}(-3, 0) \rightarrow (0, -3)$.

r_{x-axis} means a reflection of the point in the x-axis. In other words, you change the y-coordinate and leave the x-coordinate alone.

This gives us $r_{x-axis}(0, -3) \rightarrow (0, 3)$.

PROBLEM 15: BINOMIAL EXPANSIONS

The binomial theorem says that if you expand $(a + b)^n$, you get the following terms:

$$_nC_0a^n + _nC_1a^{n-1}b^1 + _nC_2a^{n-2}b^2 + ... + _nC_{n-2}a^2b^{n-2} + _nC_{n-1}a^1b^{n-1} + _nC_nb^n$$

Therefore, if we expand $(x + 1)^5$, we get:

$$_5C_0(x)^5 + _5C_1(x)^4(1)^1 + _5C_2(x)^3(1)^2 + _5C_3(x)^2(1)^3 + _5C_4(x)^1(1)^4 + _5C_5(1)^5$$

All of the powers of 1 are 1, so we can ignore them.

Next, we use the rule that $_nC_r = \dfrac{n!}{(n-r)!\,r!}$.

This gives us: $(x + 1)^5 = x^5 + 5x^4 + 10x^3 + 10x^2 + 5x + 1$

The third term is $10x^3$.

A shortcut is to know the rule that the rth term of the binomial expansion of $(a - b)^n$ is $_nC_{r-1}(a)^{n-r+1}(b)^{r-1}$.

Thus the third term is:

$$_5C_{5-3}(x)^{5-3+1}(1)^{3-1}$$

which can be simplified to $_5C_2x^3 = 10x^3$.

Multiple-Choice Problems

For problems 16–35, you will find that you can sometimes **plug-in** or **backsolve** to get the right answer. You will also find that a calculator will simplify the arithmetic. In addition, most of the problems only require knowing which formula to use.

PROBLEM 16: TRANSFORMATIONS

A *dilation* means that the x- and y-coordinates are multiplied by the same number.

In choice (4), both coordinates are multiplied by 2.

The answer is (4).

PROBLEM 17: STATISTICS

If Phyllis scored one standard deviation above the mean, and a standard deviation is 6, then the mean is 6 points lower than her score. Because Phyllis scored 84, the mean must have been 78.

The answer is (2).

PROBLEM 18: ABSOLUTE VALUE EQUATIONS

When we have an absolute value equation we break it into two equations. First, rewrite the left side without the absolute value and leave the right side alone. Second, rewrite the left side without the absolute value, and make the right side negative.

Here, we have $|3x + 2| = 5$, so we rewrite it as: $3x + 2 = 5$ or $3x + 2 = -5$.

If we solve each of these independently, we get $x = 1$ or $x = -\dfrac{7}{3}$.

The answer is (3).

PROBLEM 19: LOGARITHMS

You should know the following log rules:

(i) $\log A + \log B = \log(AB)$

(ii) $\log A - \log B = \log\left(\dfrac{A}{B}\right)$

(iii) $\log A^B = B \log A$

First, using rule (iii), we can rewrite $\log \sqrt{\dfrac{a}{b}}$ as $\dfrac{1}{2} \log \dfrac{a}{b}$.

Next, using rule (ii), we can rewrite $\dfrac{1}{2} \log \dfrac{a}{b}$ as $\dfrac{1}{2}\left(\log a - \log b\right)$.

The answer is (2).

Another way to get the right answer is to **plug-in**. First, we make up values for a and b. For example, let $a = 8$ and $b = 2$. Now, plug $a = 8$ and $b = 2$ into the problem and we get $\log \sqrt{\dfrac{8}{2}} = \log 2 \approx 0.301$.

Now we plug $a = 8$ and $b = 2$ into the answer choices and see which one matches.

Choice (1): $\dfrac{1}{2} \log 8 - \log 2 \approx 0.151$, so this is a wrong answer.

Choice (2): $\dfrac{1}{2}\left(\log 8 - \log 2\right) \approx 0.301$, so this is the right answer. Just in case . . .

Choice (3): $\dfrac{1}{2}\left(\log 8 + \log 2\right) \approx 0.602$, so this is a wrong answer.

Choice (4): $\dfrac{1}{2} \log 8 + \log 2 \approx 0.753$, so this is a wrong answer.

PROBLEM 20: ALGEBRA THEORY

The *additive inverse* of a number means the number that one adds to another number in order to get 0.

The number that you add to $a + bi$ in order to get 0 is $-a - bi$.

The answer is (4).

PROBLEM 21: TRIGONOMETRY

The expression $Arc \cos\left(\dfrac{1}{2}\right)$ means the angle θ such that $\cos \theta = \dfrac{1}{2}$.

You should know that $\cos 60° = \dfrac{1}{2}$ because 60° is a special angle.

The answer is (3).

PROBLEM 22: TRIGONOMETRY

In other words, for how many values of x, where $0 \le x \le 2\pi$, will $\sin x = \dfrac{1}{2}$.

You should know that $\sin \dfrac{\pi}{6} = \dfrac{1}{2}$ and $\sin \dfrac{5\pi}{6} = \dfrac{1}{2}$, because they are special angles.

Therefore, there are 2 points where the graphs intersect.

The answer is (2).

Another way to get the right answer is to graph the two equations.

PROBLEM 23: TRIGONOMETRIC IDENTITIES

You can find the identities for double angles on the formula sheet under *Functions of the Double Angle*. It says that $\sin 2A = 2 \sin A \cos A$. Substituting this into the expression, we get:

$$\frac{2 \sin A \cos A}{2 \cos^2 A}$$

If we cancel like terms, we get:

$$\frac{2 \sin A \cos A}{2 \cos^2 A} = \frac{\sin A}{\cos A} = \tan A$$

The answer is (2).

PROBLEM 24: QUADRATIC INEQUALITIES

When we are given a quadratic inequality, the first thing that we do is factor it. We get: $(x - 5)(x + 1) < 0$. The roots of the quadratic are $x = 5$, $x = -1$. If we plot the roots on a number line

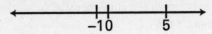

we can see that there are three regions: $x < -1$, $-1 < x < 5$, and $x > 5$. Now we try a point in each region to see whether the point satisfies the inequality.

First, we try a number less than -1. Let's try -2. Is $(-2 - 5)(-2 + 1) < 0$? No.

Now, we try a number between -1 and 5. Let's try 0. Is $(0 - 5)(0 + 1) < 0$? Yes.

Last, we try a number greater than 5. Let's try 6. Is $(6 - 5)(6 + 1) < 0$? No.

Therefore, the region that satisfies the inequality is $-1 < x < 5$.

The answer is (1).

PROBLEM 25: TRIGONOMETRY

If you look on the formula sheet, under <u>Functions of the Sum of Two Angles</u>, you will find the formula $\tan(A + B) = \dfrac{\tan A + \tan B}{1 - \tan A \tan B}$.

Now we can substitute $\tan A = 8$ and $\tan B = \dfrac{1}{2}$ into the problem and we get:

$$\tan(A + B) = \frac{8 + \dfrac{1}{2}}{1 - (8)\left(\dfrac{1}{2}\right)} = \frac{\dfrac{17}{2}}{-3} = -\frac{17}{6}.$$

The answer is (4).

PROBLEM 26: LAW OF COSINES

The Law of Cosines says that, in any triangle, with angles A, B, and C, and opposite sides of a, b, and c, respectively, $c^2 = a^2 + b^2 - 2ab\cos C$. (This is on the formula sheet.) If we have a triangle and we are given SSS (side, side, side), then the side opposite the angle we are looking for is c, and the angle is C.

In a triangle, the largest angle is opposite the largest side. Therefore, $c = 9$, and we are looking for $\cos C$.

If we substitute the values we get: $9^2 = 6^2 + 7^2 - 2(6)(7)\cos C$.

We can solve this easily.

$$81 = 36 + 49 - 84\cos C$$
$$84\cos C = 36 + 49 - 81$$
$$84\cos C = 4$$
$$\cos C = \frac{4}{84}.$$

The answer is (3).

PROBLEM 27: COMPLEX NUMBERS

In order to find a power of i, you divide the exponent by 4 and just use the remainder as the power. Then you use the following rule (which you should know):

$$i^0 = 1$$
$$i^1 = i$$
$$i^2 = -1$$
$$i^3 = -i$$

If we divide 4 into 10, we get 2 with a remainder of 2. Now we check i^2 against the rule above and we get $i^2 = -1$.

The answer is (3).

PROBLEM 28: QUADRATIC EQUATIONS

We can determine the nature of the roots of a quadratic equation of the form $ax^2 + bx + c = 0$ by using the discriminant $b^2 - 4ac$.

You should know the following rule:

If $b^2 - 4ac < 0$, the equation has two imaginary roots.

If $b^2 - 4ac = 0$, the equation has one rational root.

If $b^2 - 4ac > 0$, and $b^2 - 4ac$ is a perfect square, then the equation has two rational roots.

If $b^2 - 4ac > 0$, and $b^2 - 4ac$ is not a perfect square, then the equation has two irrational roots.

First, add 3 to both sides of the equation. We get $5x^2 - 2x + 3 = 0$.

Here, the discriminant is $b^2 - 4ac = 2^2 - 4(5)(3) = 4 - 60 = -56$.

Therefore the roots are imaginary.

The answer is (1).

PROBLEM 29: TRIGONOMETRY

If you draw a vertical line from P to the x-axis, you will get a right triangle, like this:

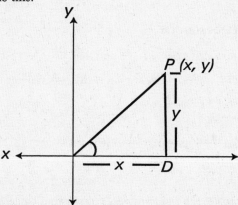

The value of the x-coordinate is the distance from the origin to the point D, which is, of course, x.

The value of the y-coordinate is the distance from the point D to the point P, which is, of course, y.

We know from the definition of sine and cosine that

$$\cos \theta = \frac{adjacent}{hypotenuse} = \frac{x}{1} = x \text{ and that } \sin \theta = \frac{opposite}{hypotenuse} = \frac{y}{1} = y.$$

Therefore, the coordinates of the point are $(\cos \theta, \sin \theta)$.

The answer is (3).

If you recognized that this was a unit circle, then you should know the rule that *the coordinates of a point in the first quadrant on the unit circle are* $(\cos \theta, \sin \theta)$.

PROBLEM 30: TRANSFORMATIONS

$R_{90°}$ means a rotation counterclockwise of 90°. This means that $R_{90°}(x, y) = (-y, x)$ (see problem 14 above).

Therefore, $R_{90°}(5, 3) = (-3, 5)$.

The answer is (1).

PROBLEM 31: TRIGONOMETRY

You should know that $\cos \frac{\pi}{3} = \frac{1}{2}$ and that $\sin \frac{3\pi}{2} = -1$ because they are both special angles. (Do you know that $\frac{\pi}{3}$ radians is 60°and that $\frac{3\pi}{2}$ radians is 270°?)

Therefore, $\cos \frac{\pi}{3} - \sin \frac{3\pi}{2} = \frac{1}{2} - (-1) = 1\frac{1}{2}$.

The answer is (1).

Another way to get the right answer is to put your calculator in **radian** mode and just put in $\cos\left(\frac{\pi}{3}\right) - \sin\left(\frac{3\pi}{2}\right)$. Make sure that you get the parentheses correct!

You should get 1.5.

PROBLEM 32: TRANSFORMATIONS

The reflection in the y-axis means to change the sign of x. In other words, to flip the left and the right sides. This would give us $y = \cos(-x)$, which is the same as $y = \cos x$.

In other words, the graph of $y = \cos(-x)$ is the same as the graph of $y = \cos x$.

The answer is (1).

PROBLEM 33: INVERSES

We can find the inverse of an equation in two simple steps.

Step 1: **Switch x and y.** This gives us $x = 3y - 2$.

Step 2: **Solve for y.** This gives us $y = \dfrac{x + 2}{3}$.

The answer is (4).

PROBLEM 34: LAW OF SINES

First, let's draw a triangle and fill in the information. Any triangle will do for now.

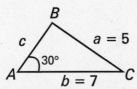

By asking for the possible number of triangles, the question is testing your knowledge of the *Ambiguous Case* of the Law of Sines. There is a simple test: Is $b \sin A < a < b$? $7 \sin 30° \approx 3.5$. This is less than side a, which is less than side b. Therefore, there are two triangles.

Another way to get this right is to use the Law of Sines and see how many triangles can be made.

The Law of Sines says that, in any triangle, with angles A, B, and C, and opposite sides of $a, b,$ and c, respectively, $\dfrac{a}{\sin A} = \dfrac{b}{\sin B} = \dfrac{c}{\sin C}$. (This is on the formula sheet.)

Here, $\dfrac{5}{\sin 30°} = \dfrac{7}{\sin B}$, so $\sin B = \dfrac{7 \sin 30°}{5} = 0.7$. (If we had gotten a number greater than 1 or less than –1 we would have NO triangles.) Therefore, $m\angle B \approx 44$

This would make $m\angle C = 180 - 30 - 44 = 106$. Now we have one triangle. But, there is a second angle for which $\sin B = 0.7$, that is $m\angle B \approx 180 - 44 = 136$. Can we make a triangle with angles of $30°$, $136°$ and C? YES, if $m\angle C = 180 - 30 - 136 = 14$. Therefore, there are two triangles.

The answer is (2).

PROBLEM 35: CONIC SECTIONS

A graph of the form $\dfrac{x^2}{a^2} + \dfrac{y^2}{b^2} = 1$, *where* ***a*** *and* ***b*** *have the same value, is the equation of a circle. If* ***a*** *and* ***b*** *have different values, the graph is that of an ellipse.*

If we rearrange the equations in the answer choices, we get the following:

Choice (1): $xy = 36$

Choice (2): $4x^2 - 9y^2 = 36$

Choice (3): $9x^2 - 4y^2 = 36$

Choice (4): $9x^2 + 4y^2 = 36$

Because the x^2 and y^2 terms must be added, **the only possible correct answer choice is (4).**

Part II

PROBLEM 36:

(a) Statistics

The method for finding a standard deviation is simple, but time-consuming.

First, find the average of the scores. We do this by multiplying each score by the number of people who obtained that score, and adding them up.

$$(40)(5) + (50)(4) + (60)(6) + (70)(3) + (80)(2) = 1130$$

Next, we divide by the total number of people $(5 + 4 + 6 + 3 + 2) = 20$, to obtain the average $\bar{x} = \dfrac{1130}{20} = 56.5$.

Second, subtract each actual score from the average.

$$56.5 - 40 = 16.5$$
$$56.5 - 50 = 6.5$$
$$56.5 - 60 = -3.5$$
$$56.5 - 70 = -13.5$$
$$56.5 - 80 = -23.5$$

Third, square each difference.

$$(16.5)^2 = 272.25$$
$$(6.5)^2 = 42.25$$
$$(-3.5)^2 = 12.25$$
$$(-13.5)^2 = 182.25$$
$$(-23.5)^2 = 552.25$$

Fourth, multiply the square of the difference by the corresponding number of people and sum.

$$(272.25)(5) = 1361.25$$
$$(42.25)(4) = 169$$
$$(12.25)(6) = 73.5$$
$$(182.25)(3) = 546.75$$
$$(552.25)(2) = 1104.5$$
$$1361.25 + 169 + 73.5 + 546.75 + 1104.5 = 3255$$

Fifth, divide this sum by the total number of people.

$$\frac{3255}{20} = 162.75.$$

Last, take the square root of this number. This is the standard deviation.

$$\sigma = \sqrt{162.75} \approx 12.757$$

Rounded to the nearest tenth, we get $\sigma \approx 12.8$.

(b) Trigonometric Graphs

A graph of the form $y = a \sin bx$ or $y = a \cos bx$ has an amplitude of $|a|$ and a period of $\dfrac{2\pi}{b}$. Therefore, the equation $y = 2 \sin\left(\dfrac{1}{2}x\right)$ has an amplitude of 2 and a period of 4π.

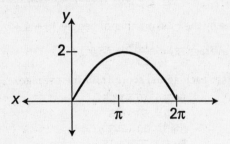

PROBLEM 37: CIRCLE RULES

(a) We know that a circle contains 360°. If we let $m\overset{\frown}{AN} = x$, then $m\overset{\frown}{NB} = 2x$, $m\overset{\frown}{CA} = 4x$, and $m\overset{\frown}{BC} = 5x$.

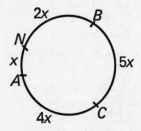

If we add up the four arcs, we have to get 360°

$$x + 2x + 4x + 5x = 360°.$$
$$12x = 360°$$
$$x = 30°.$$

Therefore, $m\overset{\frown}{BC} = 5x = 5(30) = 150$.

(b) *The measure of an inscribed angle is half of the arc it subtends (intercepts)*

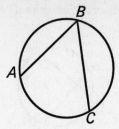

We found in part (a) that $x = 30°$, which means that:

$$m\widehat{CA} = 4(30) = 120.$$

Therefore, $m\angle ABC = \dfrac{1}{2}(120) = 60.$

(c) *The measure of an angle formed by a pair of secants, or a secant and a tangent, is equal to half of the difference between the larger and the smaller arcs that are formed by the secants, or the secant and the tangent.*

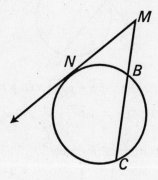

From part (a) above, we know that $m\widehat{NB} = 2(30) = 60.$

We also know that $m\widehat{NC} = m\widehat{AC} + m\widehat{AN} = 120 + 30 = 150.$

Therefore $m\angle NMC = \dfrac{m\widehat{AC} - m\widehat{NB}}{2} = \dfrac{150 - 60}{2} = 45.$

(d) From part (a) above, we know that $m\overset{\frown}{BC} = 5(30) = 150$. Now we can use the following rule: *If two chords intersect within a circle, then the angle between the two chords is the average of their intercepted arcs.*

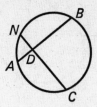

Therefore, $m\angle NDA = \dfrac{m\overset{\frown}{NA} + m\overset{\frown}{BC}}{2} = \dfrac{30 + 150}{2} = 90$.

(e) Here, we can use another rule: *An angle formed by a tangent and a chord is equal to half of its intercepted arc.*

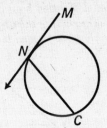

We know that $m\overset{\frown}{NC} = m\overset{\frown}{NB} + m\overset{\frown}{BC} = 60 + 150 = 210$.

Therefore, $m\angle MND = \dfrac{1}{2} m\overset{\frown}{NC} = \dfrac{1}{2} 210 = 105$.

PROBLEM 38: TRIGONOMETRIC EQUATIONS

Note that the trig equation $6\cos^2 x - 7\cos x + 2 = 0$ has the same form as a quadratic equation $6x^2 - 7x + 2 = 0$. Just as we could factor the quadratic equation, so too can we factor the trig equation. We get:

$$(3\cos x - 2)(2\cos x - 1) = 0.$$

This gives us $3\cos x - 2 = 0$ and $2\cos x - 1 = 0$. If we solve each of these equations, we get:

$$\cos x = \frac{2}{3} \text{ and } \cos x = \frac{1}{2}.$$

Therefore $x = \cos^{-1}\left(\dfrac{2}{3}\right)$ and $x = \cos^{-1}\left(\dfrac{1}{2}\right)$. We can find the solution to the first equation with the calculator. You should get $x \approx 48.19°$. This is not the only value for x. Cosine is also positive in quadrant IV, so there is another angle that satisfies the first equation. It is: $360° - 48.19° = 311.81°$. Rounding both of these to the nearest degree, we get $x = 48°, 312°$. The solution to the second equation is easy because it is a special angle. You should know that $\cos 60° = \dfrac{1}{2}$ and that

$\cos 300° = \dfrac{1}{2}$.

Therefore, the solutions are $x = 48°, 60°, 300°, 312°$.

PROBLEM 39: TRIGONOMETRY

(a) Law of Cosine

We can find the measure of the angle using the Law of Cosines. First, let's draw a picture of the situation.

The Law of Cosines says that, in any triangle, with angles $A, B,$ and C, and opposite sides of $a, b,$ and c, respectively, $c^2 = a^2 + b^2 - 2ab \cos C$. (This is on the formula sheet.) If we have a triangle and we are given SSS (side, side, side), then the side opposite the angle we are looking for is c, and that angle is C. The largest angle will be opposite the largest side.

$$(50)^2 = (22)^2 + (34)^2 - 2(22)(34) \cos C$$
$$2{,}500 = 484 + 1{,}156 - 1{,}496 \cos C$$
$$1{,}496 \cos C = 484 + 1{,}156 - 2500$$
$$1{,}496 \cos C = -860$$
$$\cos C = -\frac{860}{1496}$$
$$C = \cos^{-1}\left(-\frac{860}{1496}\right) \approx 125.09°$$

Rounded to the nearest degree, $C \approx 125°$.

(b) Trigonometric Area

The area of a triangle, if we are given the lengths of the two sides a and b, and their included angle θ, is $A = \frac{1}{2}ab\sin\theta$. In other words, if we are given SAS (side, angle, side), we find the area by multiplying $\frac{1}{2}$ by the product of the two sides by the sine of the included angle. Here, the two sides are a and b and the included angle is C. Plugging in, we get: $A = \frac{1}{2}(22)(34)\sin 125° \approx 306$.

PROBLEM 40:

(a) Complex Numbers

A complex number of the form $a + bi$ is represented graphically by a point with the coordinates (a, b). In other words, the real part of the complex number is the x-coordinate, and the imaginary part is the y-coordinate.

Thus, the point $-1 + 6i$ has the coordinates $(-1, 6)$, and the point $4 + 2i$ has the coordinates $(4, 2)$.

In order to add two complex numbers, you simply add the real parts and the imaginary parts separately. Thus,

$$(-1 + 6i) + (4 + 2i) = (-1 + 4) + (6 + 2)i = 3 + 8i$$

and has coordinates $(3, 8)$.

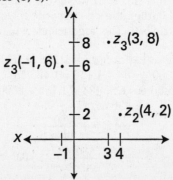

(b) Rational Expressions

First, invert the second fraction and multiply it by the first:

$$\frac{1-x^2}{6x+6} \cdot \frac{6x^2+6}{x^4-1}$$

Now, factor each of the terms:

$$\frac{(1-x)(1+x)}{6(x+1)} \cdot \frac{6(x^2+1)}{(x^2-1)(x^2+1)}$$

The denominator of the second fraction can be factored further:

$$\frac{(1-x)(1+x)}{6(x+1)} \cdot \frac{6(x^2+1)}{(x^2-1)(x^2+1)} = \frac{(1-x)(1+x)}{6(x+1)} \cdot \frac{6(x^2+1)}{(x-1)(x+1)(x^2+1)} \cdot$$

Now, cancel like terms:

$$\frac{(1-x)(1+x)}{6(x+1)} \cdot \frac{6(x^2+1)}{(x-1)(x+1)(x^2+1)} = -\frac{1}{(x+1)}$$

PROBLEM 41:

(a) Trigonometric Identities

All of the identities that you will need to know are contained on the formula sheet.

First, let's put all of the expressions on the left in terms of sine and cosine, leaving the right-hand side alone. We get:

$$\frac{\sec x + \csc x}{\tan x + \cot x} = \frac{\dfrac{1}{\cos x} + \dfrac{1}{\sin x}}{\dfrac{\sin x}{\cos x} + \dfrac{\cos x}{\sin x}}$$

Next, get a common denominator for the top and bottom separately:

$$\frac{\dfrac{1}{\cos x} + \dfrac{1}{\sin x}}{\dfrac{\sin x}{\cos x} + \dfrac{\cos x}{\sin x}} = \frac{\dfrac{\sin x}{\sin x \cos x} + \dfrac{\cos x}{\sin x \cos x}}{\dfrac{\sin^2 x}{\sin x \cos x} + \dfrac{\cos^2 x}{\sin x \cos x}} = \frac{\dfrac{\sin x + \cos x}{\sin x \cos x}}{\dfrac{\sin^2 x + \cos^2 x}{\sin x \cos x}}$$

Now, invert the bottom fraction and multiply, canceling like terms:

$$\frac{\dfrac{\sin x + \cos x}{\sin x \cos x}}{\dfrac{\sin^2 x + \cos^2 x}{\sin x \cos x}} = \frac{\sin x + \cos x}{\sin x \cos x} \cdot \frac{\sin x \cos x}{\sin^2 x + \cos^2 x} = \frac{\sin x + \cos x}{\sin^2 x + \cos^2 x} \, .$$

Finally, we can use the rule that $\sin^2 x + \cos^2 x = 1$ to substitute into the denominator of the fraction and we get: $\sin x + \cos x$ and we have proved the identity.

(b) Trigonometry

If the denominator of a fraction is zero, then it is undefined. Here, we want to find where $\sin 2x = 0$. You should know that $\sin 0 = \sin \pi = \sin 2\pi = 0$.

Therefore, $x = 0, \dfrac{\pi}{2}, \pi$ are the answers.

PROBLEM 42:

(a) Exponential Equations

The graph of an equation of the form $y = a^x$ always has the same general shape. Here, the graph goes through the points $(0, 1)$ and $(1, 2)$.

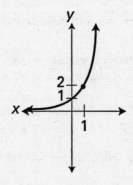

(b) Transformations

A reflection in the y-axis means that you switch the left and right sides.

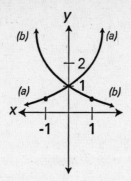

(c) Logarithms

If we take the log of both sides, we get: $\log 2^x = \log 5$. Now, we use the rule that $\log A^B = B \log A$ to rewrite the equation as: $x \log 2 = \log 5$.

Now we can solve for x:

$$x = \frac{\log 5}{\log 2} \approx 2.32.$$

Formulas

Pythagorean and Quotient Identities

$$\sin^2 A + \cos^2 A = 1 \qquad \tan A = \frac{\sin A}{\cos A}$$

$$\tan^2 A + 1 = \sec^2 A \qquad \cot A = \frac{\cos A}{\sin A}$$

$$\cot^2 A + 1 = \csc^2 A$$

Functions of the Sum of Two Angles

$$\sin (A + B) = \sin A \cos B + \cos A \sin B$$

$$\cos (A + B) = \cos A \cos B - \sin A \sin B$$

$$\tan (A + B) = \frac{\tan A + \tan B}{1 - \tan A \tan B}$$

Functions of the Difference of Two Angles

$$\sin (A - B) = \sin A \cos B - \cos A \sin B$$

$$\cos (A - B) = \cos A \cos B + \sin A \sin B$$

$$\tan (A - B) = \frac{\tan A - \tan B}{1 + \tan A \tan B}$$

Law of Sines

$$\frac{a}{\sin A} = \frac{b}{\sin B} = \frac{c}{\sin C}$$

Law of Cosines

$$a^2 = b^2 + c^2 - 2bc \cos A$$

Functions of the Double Angle

$$\sin 2A = 2 \sin A \cos A$$

$$\cos 2A = \cos^2 A - \sin^2 A$$

$$\cos 2A = 2 \cos^2 A - 1$$

$$\cos 2A = 1 - 2 \sin^2 A$$

$$\tan 2A = \frac{2 \tan A}{1 - \tan^2 A}$$

Functions of the Half Angle

$$\sin \frac{1}{2} A = \pm \sqrt{\frac{1 - \cos A}{2}}$$

$$\cos \frac{1}{2} A = \pm \sqrt{\frac{1 + \cos A}{2}}$$

$$\tan \frac{1}{2} A = \pm \sqrt{\frac{1 - \cos A}{1 + \cos A}}$$

Area of Triangle

$$K = \frac{1}{2} ab \sin C$$

Standard Deviation

$$\text{S.D.} = \sqrt{\frac{1}{n} \sum_{i=1}^{n} \left(x_i - \bar{x} \right)^2}$$

EXAMINATION
AUGUST 1995

Part I

Answer 30 questions from this part. Each correct answer will receive 2 credits. No partial credit will be allowed. Write your answers in the spaces provided on the separate answer sheet. Where applicable, answers may be left in terms of π or in radical form. [60]

1 Express the sum of $\sqrt{-81}$ and $3\sqrt{-25}$ as a monomial in terms of i.

$9i + 5 \cdot 3i$

$9i + 15i = 24i$

2 In the accompanying diagram of circle O, point O is the origin, $YO = 1$, $JO = 1$, and $\overline{TOY}$ is a diameter. If the coordinates of point J are $\left(\dfrac{\sqrt{2}}{2}, \dfrac{\sqrt{2}}{2}\right)$, how many degrees are in $m\angle JOY$?

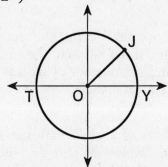

3 Solve for x: $\sqrt{x+3} - 3 = -1$.

$\sqrt{x+3} = -1 + 3$

$(\sqrt{x+3})^2 = 2^2$

$x + 3 = 4$

$x = 1$

[OVER]

4 In $\triangle ABC$, $a = 10$, $\sin A = 0.30$, and $\sin C = 0.24$.
Find c. $\dfrac{10}{0.30} = \dfrac{x}{0.24}$ $0.30x = 2.4$
$x = 8$

5 Express 315° in radian measure.
$315 \cdot \dfrac{\pi}{180} = \dfrac{7\pi}{4}$

6 In the accompanying diagram of circle O, chords
$\overline{AB}$ and $\overline{CD}$ intersect at E. If $AE = 4$, $EB = 6$,
and $CE = 3$, find ED.

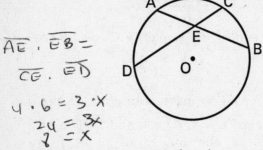

$\overline{AE} \cdot \overline{EB} =$
$\overline{CE} \cdot \overline{ED}$
$4 \cdot 6 = 3 \cdot x$
$24 = 3x$
$8 = x$

7 Find the image of $A(4, -3)$ under the transformation $r_{x=2}$.

8 In a circle, a central angle intercepts an arc of 12
centimeters. If the radius of the circle is 6 centimeters, find the number of radians in the measure of the central angle.

9 The diameter of a wheel varies inversely as the
number of revolutions that the wheel makes to
cover a certain distance. If a wheel with a 26-inch
diameter makes 10 revolutions in covering a certain distance, how many revolutions will a wheel
with a diameter of 20 inches make in covering the
same distance? $\dfrac{26}{10} = \dfrac{20}{x}$ $200 = 26x$

10 In the accompanying diagram, $\overline{PQ}$ is tangent to circle O at Q and $\overline{PRT}$ is a secant. If m$\angle P = 56$ and m$\overarc{QT} = 192$, find m$\overarc{QR}$.

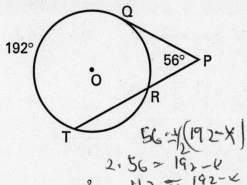

$56 = \frac{1}{2}(192-x)$

$2 \cdot 56 = 192 - x$

$112 = 192 - x$

$x = 80$

11 Find the value of $\displaystyle\sum_{x=0}^{2} 9^x$.

$9^0 + 9^1 + 9^2$

$1 + 9 + 81 = 91$

12 The area of $\triangle ABC$ is 20. If $a = 10$ and $b = 8$, find the number of degrees in the measure of acute angle C.

$20 = \frac{1}{2}(10)(8)\sin C$

$20 = 40 \sin C \quad \frac{1}{2} = \sin C$

Directions (13–35): For *each* question chosen, write on the separate answer sheet the *numeral* preceding the word or expression that best completes the statement or answers the question.

13 The amplitude of $y = 3 \sin 6x$ is

(1) $\dfrac{\pi}{3}$

(3) 3

(2) 2π

(4) 6

[OVER]

14 The expression $\log \dfrac{a^3}{b}$ is equivalent to

(1) $3 \log a - \log b$

(3) $3\left(\dfrac{\log a}{\log b}\right)$

(2) $3(\log a - \log b)$

(4) $\dfrac{1}{3}(\log a - \log b)$

15 If $\log_9 x = \dfrac{3}{2}$, what is the value of x?

(1) $\dfrac{3}{2}$

(3) $\dfrac{27}{2}$

(2) 8

(4) 27

16 For any point (x, y), which transformation is equivalent to $R_{45°} \circ R_{-135°}$?

(1) $R_{-90°}$

(3) $r_{y=x}$

(2) $R_{90°}$

(4) $r_{x\text{-axis}}$

17 If $\cos x = -\dfrac{\sqrt{3}}{2}$, in which quadrants could $\angle x$ terminate?

(1) I and IV, only

(3) II and III, only

(2) II and IV, only

(4) I and III, only

18 What is the solution set to the equation $|4x - 3| = 17$?

(1) $\{5\}$

(3) $\left\{-5, \dfrac{7}{2}\right\}$

(2) $\left\{5, -\dfrac{7}{2}\right\}$

(4) $\left\{-3\dfrac{1}{2}\right\}$

19 The expression $(1 + i)^2$ is equivalent to

(1) 1   (3) i

(2) 2   (4) $2i$

20 If a set of scores has a normal distribution and the mean is 200, which score has the greatest probability of being chosen at random?

(1) 230   (3) 176

(2) 228   (4) 168

21 The expression $1 - 2 \sin^2 45°$ has the same value as

(1) $\cos 90°$ (3) $\sin 90°$

(2) $\cos 45°$ (4) $\sin 22\frac{1}{2}°$

22 Which equation is represented by the graph in the diagram below?

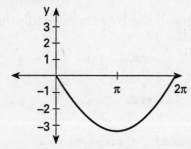

(1) $y = 3 \sin x$ (3) $y = -3 \sin x$

(2) $y = 3 \sin \frac{1}{2} x$ (4) $y = -3 \sin \frac{1}{2} x$

[OVER]

23 If the probability of hitting a target is $\frac{3}{4}$, then the probability of hitting the target *exactly* once in four tries is

(1) $\frac{3}{256}$ (3) $\frac{27}{256}$

(2) $\frac{12}{256}$ (4) $\frac{36}{256}$

24 Which relation is also a function?
(1) $x^2 + y^2 = 36$ (3) $9x^2 + 4y^2 = 36$
(2) $x^2 - y^2 = 36$ (4) $y = 4x^2$

25 If $f(x) = x^2 - 3$, then $f(a - b)$ is equivalent to
(1) $a^2 - b^2 - 3$ (3) $a^2 - 2ab + b^2 - 3$
(2) $a^2 - 2ab - b^2 - 3$ (4) $a^2 + b^2 - 3$

26 Which graph represents the solution set of $x^2 - 5x + 6 < 0$?

(1)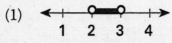

(2)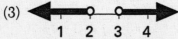

(3)

(4)

27 The expression $\dfrac{\sqrt{3}+1}{\sqrt{3}-1}$ is equal to

(1) −1   (3) $2 + \sqrt{3}$

(2) 2   (4) $5 + \sqrt{3}$

28 If $x = \text{Arc cos}\left(-\dfrac{1}{2}\right)$, then x is equal to

(1) 120   (3) 210°

(2) 150°   (4) 300°

29 If $a = 5\sqrt{2}$, $b = 8$, and $m\angle A = 45$, how many distinct triangles can be costructed?

(1) 1   (3) 3

(2) 2   (4) 0

30 The expression $\tan(180° + x)$ is equivalent to

(1) $\cot x$ (3) $-\cot x$

(2) $\tan x$ (4) $-\tan x$

31 In simplest form, the expression $\dfrac{\dfrac{1}{x}-1}{x-\dfrac{1}{x}}$ is equivalent to

(1) −1   (3) $\dfrac{1}{x-1}$

(2) $\dfrac{1}{x+1}$ (4) $-\dfrac{1}{x+1}$

[OVER]

32 Which expression is the inverse of $y = 3x$?

(1) $x = 3$

(3) $y = 3$

(2) $y = \dfrac{1}{3}x$

(4) $x = \dfrac{y}{3}$

33 If a quadratic equation with real coefficient has a discriminant of 3, then the two roots must be

(1) real and rational   (3) imaginary
(2) real and irrational   (4) equal

34 The expression $\sin A \cos A + \sin 2A$ is equivalent to

(1) $\sin A (\cos A + \sin A)$
(2) $\cos A + 2 \sin A$
(3) $3 \sin A \cos A$
(4) $\cos A + 2 \sin 2A$

35 What is the fourth term of the expansion of $\left(2x - y\right)^5$?

(1) $20x^2y^3$

(3) $40x^2y^3$

(2) $-20x^2y^3$

(4) $-40x^2y^3$

Part II

Answer four questions from this part. Clearly indicate the necessary steps, including appropriate formula substitutions, diagrams, graphs, charts, etc. Calculations that may be obtained by mental arithmetic or the calculator do not need to be shown.   [40]

36 *a* On the same set of axes, sketch and label the graphs of the equations $y = \cos 2x$ and $y = 2 \sin x$ in the interval $-\pi \leq x \leq \pi$.   [8]

   b Use the graphs sketched in part *a* to determine the number of points in the interval $-\pi \leq x \leq \pi$ that satisfies the equation $\cos 2x = 2 \sin x$.   [2]

[OVER]

37 In the accompanying diagram, chords $\overline{RT}$ and $\overline{US}$ intersect at Q, secants $\overline{PUT}$ and $\overline{PRS}$ are drawn, $m\widehat{RS} = 120$, $m\widehat{UT} = 80$, $m\angle TRS = 50$, and $\overline{VW}$ is tangent to the circle at T.

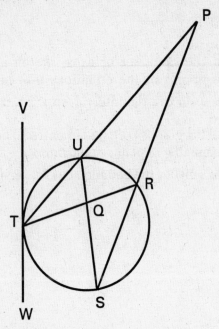

Find:

a $m\widehat{UR}$ [2]

b $m\angle SUT$ [2]

c $m\angle P$ [2]

d $m\angle RQS$ [2]

e $m\angle PTV$ [2]

38 Given: $f(x) = 2^x$

  a On graph paper, sketch the graph of $f(x)$ in the interval $-1 \le x \le 3$. Label the graph *a*.   [3]

  b On the same set of axes, sketch the image of the graph drawn in part *a* after $\left(T_{2,-1} \circ r_{x-axis}\right)$. Label the graph *b*.   [4]

  c On the same set of axes, sketch the image of the graph drawn in part *b* after D_{-1}. Label the graph *c*.   [3]

39 Find, to the *nearest degree*, all values of x between 0° and 360° that satisfy the equation $2 \sin x + 4 \cos 2x = 3$.   [10]

[OVER]

40 In the accompanying diagram, the triangular pad is divided into nine keys. The probability of pressing any key at random is the same.

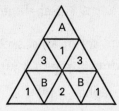

Find the probability of pressing

a a letter key   [1]

b *exactly* two number keys on three random tries   [3]

c *at least* two letter keys on three random tries   [6]

41 The sides of a triangle have lenghts 58, 92, and 124.

a Find, to the *nearest ten minutes*, the largest angle of the triangle.   [6]

b Find, to the *nearest integer*, the area of the triangle.   [4]

42 *a* Solve for x and express the roots in terms of i:
$2x^2 = 6x - 5$.   [6]

b Using logarithms, find x to the nearest tenth:
$3^{2x} = 100$.   [4]

ANSWER KEY

Part I

1. $24i$	13. (3)	25 (3)
2. 45	14. (1)	26. (1)
3. 1	15. (4)	27. (3)
4. 8	16. (1)	28. (1)
5. $\frac{7\pi}{4}$	17. (3)	29. (2)
6. 8	18. (2)	30 (2)
7. $(0, -3)$	19. (4)	31. (4)
8. 2	20. (3)	32. (2)
9. 13	21. (1)	33. (2)
10. 80	22. (4)	34. (3)
11. 91	23. (2)	35. (4)
12. 30	24. (4)	

ANSWERS AND EXPLANATIONS
AUGUST 1995

Part I

In problems 1–12, you must supply the answer. Because of this, you will not be able to **plug-in** or **backsolve**. But, you can use your calculator to simplify calculations. In addition, you will find that these problems tend to test your basic knowledge of Sequential Math III and are not tricky. Many can be solved just by knowing the right formula and by plugging in the numbers.

PROBLEM 1: COMPLEX NUMBERS

You should know that $i = \sqrt{-1}$.

This means that $\sqrt{-81} = \sqrt{81}\sqrt{-1} = 9i$ and $3\sqrt{-25} = 3\sqrt{25}\sqrt{-1} = 15i$.

Now combine the two monomials: $9i + 15i = 24i$.

PROBLEM 2: TRIGONOMETRY

This is the unit circle. You should know your special angles and that $\sin 45° = \cos 45° = \dfrac{\sqrt{2}}{2}$. Therefore, $m\angle JOY = 45$.

PROBLEM 3: RADICAL EQUATIONS

To isolate the radical, we first add 3 to both sides of the equation. $\sqrt{x+3} = 2$

Next, square both sides. $x + 3 = 4$.

Now solve for x. $x = 1$

Just in case, check the answer. Does $\sqrt{1+3} = 2$? $\sqrt{4} = 2$, so, yes, it does.

PROBLEM 4: LAW OF SINES

The Law of Sines says that, in any triangle, with angles A, B, and C, and opposite sides of a, b, and c, respectively, $\dfrac{a}{\sin A} = \dfrac{b}{\sin B} = \dfrac{c}{\sin C}$. (This is on the formula sheet.)

Here, $\dfrac{10}{0.3} = \dfrac{c}{0.24}$.

We can easily solve this. $c = \dfrac{10(0.24)}{0.3} = 8$

PROBLEM 5: CONVERTING FROM DEGREES TO RADIANS

All we have to do is multiply $315°$ by $\dfrac{\pi}{180°}$.

$$315\left(\frac{\pi}{180}\right) = \frac{7\pi}{4}.$$

PROBLEM 6: CIRCLE RULES

If two chords intersect in a circle, then the product of the lengths of the segments of one chord is equal to the product of the lengths of the segments of the other chord.

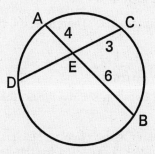

This means that $(AE)(EB) = (CE)(ED)$. If we plug in, we get: $(4)(6) = (3)(ED)$. We can solve this for ED and we get: $(ED) = 8$.

PROBLEM 7: TRANSFORMATIONS

The reflection $r_{x=2}(4,-3)$ tells you to reflect the point in the line $x = 2$. This is easy to do if we draw a picture.

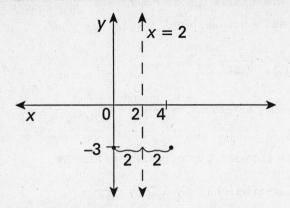

Notice that the distance from the point $(4,-3)$ to the line $x = 2$ is 2. So, the image of the point must be at a distance of 2 on the other side of the line $x = 2$. We don't change the y-coordinate. Therefore, the reflection or image of the point $(4,-3)$ is the point $(0,-3)$.

PROBLEM 8: ARC LENGTH

In a circle with a central angle, θ, measured in radians, and a radius, r; the arc length, s, is found by $s = r\theta$.

Here we have $r = 6$ and $s = 12$. Therefore, the central angle is 2 radians.

PROBLEM 9: INVERSE VARIATION

An *inverse variation* between x and y means that $xy = k$, where k is a constant.

Let x equal the diameter of the wheel and y equal the number of revolutions. If we put in the information for the first wheel, we get that $(26)(10) = k$, and thus $k = 260$. Now we have the equation $xy = 260$. If we put in the diameter of the second wheel, we can solve for the number of revolutions.

We get: $(20)y = 260$ and thus $y = 13$.

PROBLEM 10: CIRCLE RULES

The measure of an angle formed by a pair of secants, or a secant and a tangent is equal to half of the difference between the larger and the smaller arcs that are formed by the secants, or the secant and tangent.

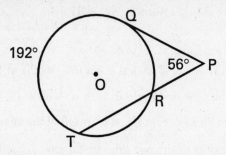

Here, the larger and smaller arcs formed by angle P are QT and QR respectively.

$$m\angle P = \frac{m\widehat{QT} - m\widehat{QR}}{2} = \frac{192 - m\widehat{QR}}{2} = 56.$$

Multiply through by 2. $192 - m\widehat{QR} = 112$

Therefore, $m\widehat{QR} = 192 - 112 = 80$

PROBLEM 11: SUMMATIONS

To evaluate a sum, we plug each of the consecutive values for k into the expression, starting at the bottom value and ending at the top value, and sum the results. We get:

$$\sum_{x=0}^{2} 9^x = 9^0 + 9^1 + 9^2 = 1 + 9 + 81 = 91$$

Another way to do this is to use your calculator. Push **2nd STAT** (**LIST**) and under the **MATH** menu, choose **SUM**. Then, again push **2nd STAT** and under the **OPS** menu, choose **SEQ**.

We put in the following: (*expression, variable, start, finish, step*).

For *expression*, we put in the formula, using x as the variable.

For *variable*, we ALWAYS put in x.

For *start*, we put in the bottom value.

For *finish*, we put in the top value.

For *step*, we ALWAYS put in 1.

Therefore, we put in the calculator: **SUM SEQ** (9^x, x, 0, 2, 1). The calculator should return the value 91.

PROBLEM 12: TRIGONOMETRIC AREA

The area of a triangle, if we are given the lengths of the two sides a and b, and their included angle θ, is $A = \frac{1}{2} ab \sin \theta$. In other words, if we are given SAS (side, angle, side), we find the area by multiplying $\frac{1}{2}$ by the product of the two sides by the sine of the included angle. Here, the two sides are a and b and the included angle is C. Plugging in, we get: $20 = \frac{1}{2}(10)(8) \sin C = 40 \sin C$.

Therefore $\sin C = \frac{1}{2}$.

You should know that C is one of the special angles and that $C = 30°$.

Multiple-Choice Problems

For problems 13–35, you will find that you can sometimes **plug-in** or **backsolve** to get the right answer. You will also find that a calculator will simplify the arithmetic. In addition, most of the problems only require knowing which formula to use.

PROBLEM 13: TRIGONOMETRY GRAPHS

A graph of the form $y = a \sin bx$ or $y = a \cos bx$ has an amplitude of $|a|$ and a period of $\frac{2\pi}{b}$.

Here, the amplitude is 3.

The answer is (3).

PROBLEM 14: LOGARITHMS

(1) You should know the following log rules:

  (i) $\log A + \log B = \log(AB)$

  (ii) $\log A - \log B = \log\left(\dfrac{A}{B}\right)$

  (iii) $\log A^B = B \log A$

First, using rule (ii), we get: $\log \dfrac{a^3}{b} = \log a^3 - \log b$.

Next, using rule (iii), we get: $\log a^3 - \log b = 3 \log a - \log b$.

The answer is (1).

Another way to get the right answer is to **plug-in**. First, make up values for a and b. Let's let $a = 2$ and $b = 3$. Then plug them into the problem. We get: $\log \dfrac{2^3}{3} = \log \dfrac{8}{3} \approx 0.426$.

Next, we plug in $a = 2$ and $b = 3$ into the answer choices and see which one matches.

Choice (1): $3 \log 2 - \log 3 \approx 0.426$. This looks like the answer, but just in case . . .

Choice (2): $3(\log 2 - \log 3) \approx -0.528$. Wrong!

Choice (3): $3\left(\dfrac{\log 2}{\log 3}\right) \approx 1.893$. Wrong!

Choice (4): $\dfrac{1}{3}(\log 2 - \log 3) \approx -0.059$. Wrong!

PROBLEM 15: LOGARITHMS

The definition of a logarithm is: $\log_b a = x$ means that $b^x = a$.

Here, we have $\log_9 x = \dfrac{3}{2}$, which means that $9^{\frac{3}{2}} = x$.

The rule for raising a number to a fractional power: $x^{\frac{b}{a}} = \left(\sqrt[a]{x}\right)^b$.

Therefore, $9^{\frac{3}{2}} = \sqrt{9^3} = 27 = x$.

The answer is (4).

Another way to get the right answer is to use the *change of base* rule for logarithms, which is: $\log_b a = \dfrac{\log a}{\log b}$. Here, we get:

$$\log_9 x = \frac{\log x}{\log 9} = \frac{3}{2}$$

If we solve for $\log x$, we get:

$$\log x = \frac{3}{2} \log 9 \approx 1.431$$

Now, take the log of each answer choice and see which one works.

Choice (1): $\log \dfrac{3}{2} \approx 0.176$. Wrong!

Choice (2): $\log 8 \approx 0.903$. Wrong!

Choice (3): $\log \dfrac{27}{2} \approx 1.130$. Wrong!

Choice (4): $\log 27 \approx 1.431$. Right!

PROBLEM 16: TRANSFORMATIONS

Here we are asked to find the results of a composite transformation $(R_{45°} \circ R_{-135°})$. *Whenever we have a composite transformation, always do the **right** one first.*

$R_{-135°}$ means a rotation clockwise of $135°$.

$R_{45°}$ means a rotation counterclockwise of $45°$.

Therefore, first we rotate clockwise $135°$, then we reverse and rotate counterclockwise $45°$, which gives us a final rotation clockwise of $90°$, which we write as $R_{-90°}$.

The answer is (1).

PROBLEM 17: TRIGONOMETRY

The cosine of an angle is negative in quadrants II *and* III.

The answer is (3).

PROBLEM 18: ABSOLUTE VALUE EQUATIONS

When we have an absolute value equation we break it into two equations. First, we rewrite the left side without the absolute value and leave the right side alone. Second, we rewrite the left side without the absolute value, and make the right side negative.

Here, we have $|4x - 3| = 17$, so we rewrite it as: $4x - 3 = 17$ or $4x - 3 = -17$. If we solve each of these independently, we get $x = 5$ or $x = -\frac{14}{4} = -\frac{7}{2}$.

The answer is (2).

PROBLEM 19: COMPLEX NUMBERS

Whenever we multiply two complex numbers, we FOIL.

Here, we have: $(1 + i)^2 = (1 + i)(1 + i) = 1 + i + i^2$.

Next, gather like terms: $1 + 2i + i^2$

Now, we use the rule that $i^2 = -1$ to get: $1 + 2i - 1 = 2i$.

The answer is (4).

PROBLEM 20: STATISTICS

You should know the following rule about normal distributions.

*The **closer** a value is to the mean, the **more likely** it is to occur.*

Just compare each score to the mean of 200 and choose the one with the smallest difference.

Choice (1): $230 - 200 = 30$

Choice (2): $228 - 200 = 28$

Choice (3): $200 - 176 = 24$. This one is the closest.

Choice (4): $200 - 168 = 32$

The answer is (3).

PROBLEM 21: TRIGONOMETRY FORMULAS

Notice that this problem has the form $1 - 2\sin^2 A$. You can find the identities for double angles on the formula sheet under *Functions of the Double Angle*. It says $\cos 2A = 1 - 2\sin^2 A$. Thus we can rewrite the problem as $1 - 2\sin^2 45° = \cos 2(45°) = \cos 90°$.

The answer is (1).

Another way to answer this question was to use your calculator to find the value of $1 - 2\sin^2 45°$. **MAKE SURE THAT YOUR CALCULATOR IS IN DEGREE MODE.** You should get 0. Now find the value of each of the answer choices with your calculator.

Choice (1): $\cos 90° = 0$, but just in case, check the other answers.

Choice (2): $\cos 45° \approx 0.707$, so (2) is not the answer.

Choice (3): $\sin 90° = 1$, so (3) is not the answer.

Choice (4): $\sin 22\frac{1}{2}° \approx 0.383$, so (4) is not the answer.

PROBLEM 22: TRIGONOMETRY GRAPHS

You should know that the graph of $y = \sin x$ looks like this:

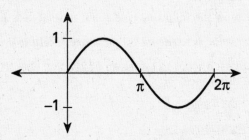

The given curve looks like the first half of the graph, turned upside down.

The amplitude of the given graph is 3, but all of the choices contain 3 sin x, so this doesn't help us much.

We turn a graph upside down by making it negative, so we can eliminate choices (1) and (2).

Next, you should note that only *half* of the graph is contained in the interval $0 \le x \le 2\pi$, so the period of the graph must be double this interval, or 4π.

A graph of the form $y = a \sin bx$ has a period of $\dfrac{2\pi}{b}$.

If we set $\dfrac{2\pi}{b} = 4\pi$, we get $b = \dfrac{1}{2}$. Therefore, the graph is

$$y = -3 \sin \frac{1}{2} x.$$

The answer is (4).

PROBLEM 23: PROBABILITY

If the probability of a particular outcome is p, then the probability of that outcome occurring r times out of a possible n times is $_nC_r(p)^r(1 - p)^{n-r}$. This is known as *binomial probability*.

Here, we are told that the probability of hitting the target is $p = \dfrac{3}{4}$. The probability of hitting the target once out of a possible 4 tries is

$_4C_1\left(\dfrac{3}{4}\right)^1\left(1 - \dfrac{3}{4}\right)^{4-1}$. This can be simplified to $_4C_1\left(\dfrac{3}{4}\right)\left(\dfrac{1}{4}\right)^3$.

We can then evaluate $_4C_1$ according to the formula $_nC_r = \dfrac{n!}{(n-r)!\,r!}$.

If we plug in, we get $_4C_1 = \dfrac{4!}{(4-1)!\,1!} = \dfrac{4!}{3!\,1!} = \dfrac{4 \cdot 3 \cdot 2 \cdot 1}{(3 \cdot 2 \cdot 1)(1)} = 4$.

Therefore, the probability of hitting the target once out of a possible

4 tries is $4\left(\dfrac{3}{4}\right)\left(\dfrac{1}{4}\right)^3 = \dfrac{12}{256}$.

The answer is (2).

PROBLEM 24: FUNCTIONS

A relation is a function if, for every x, there is only one y. One clue that a relation is <u>not</u> a function is if it contains a y^2 term.

All of the answers here contain y^2 except the last one.

The answer is (4).

PROBLEM 25: FUNCTIONS

Here, we are being asked to substitute $a - b$ for x in $f(x)$.

We get $f(a - b) = (a - b)^2 - 3$.

If we expand the binomial term, we get:

$$(a - b)^2 - 3 = (a - b)(a - b) - 3 =$$
$$a^2 - ab - ab + b^2 =$$
$$a^2 - 2ab + b^2 - 3$$

The answer is (3).

Another way to get this right is to **plug-in**. First, make up values of a and b.

Let's let $a = 5$ and $b = 3$.

Next, plug these values into the problem. $f(5 - 3) = f(2) = 2^2 - 3 = 1$.

Now plug $a = 5$ and $b = 3$ into the answer choices and see which one gives us 1.

Choice (1): $5^2 - 3^2 - 3 = 25 - 9 - 3 = 13$. This is not the right answer.

Choice (2): $5^2 - 2(5)(3) - 3^2 - 3 = 25 - 30 - 9 - 3 = -17$. This is not the right answer.

Choice (3): $5^2 - 2(5)(3) + 3^2 - 3 = 25 - 30 + 9 - 3 = 1$. Correct!

Choice (4): $5^2 + 3^2 - 3 = 25 + 9 - 3 = 31$. This is not the right answer.

PROBLEM 26: QUADRATIC INEQUALITIES

When we are given a quadratic inequality, the first thing that we do is factor it. We get: $(x - 2)(x - 3) < 0$. The roots of the quadratic are $x = 2$, $x = 3$. If we plot the roots on a number line

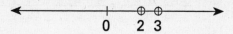

we can see that there are three regions: $x < 2$, $2 < x < 3$, and $x > 3$. Now we try a point in each region to see whether the point satisfies the inequality.

First, we try a number less than 2. Let's try 0. Is $(0 - 2)(0 - 3) < 0$? No.

Now, we try a number between 2 and 3. Let's try 2.5. Is $(2.5 - 2)(2.5 - 3) < 0$? Yes

Last, we try a number greater than 3. Let's try 4. Is $(4 - 2)(4 - 3) < 0$? No.

Therefore, the region that satisfies the inequality is $2 < x < 3$.

When graphing this on a number line, remember that you put an open circle for the inequalities < and >; and a closed circle for the inequalities ≤ and ≥.

The answer is (1).

PROBLEM 27: RATIONAL EXPRESSIONS

Whenever we have a fraction where the denominator has a radical in it, we can *rationalize* the denominator by multiplying the top and bottom by the *conjugate* of the denominator. The conjugate is merely the same as the denominator where we reverse the sign in front of the radical.

Here, we will multiply the top and bottom by the conjugate of $\sqrt{3} - 1$, which is $\sqrt{3} + 1$.

We get: $\left(\dfrac{\sqrt{3} + 1}{\sqrt{3} - 1} \right)\left(\dfrac{\sqrt{3} + 1}{\sqrt{3} + 1} \right)$.

Next, we multiply out the top and bottom using FOIL.

$$\left(\frac{\sqrt{3}+1}{\sqrt{3}-1}\right)\left(\frac{\sqrt{3}+1}{\sqrt{3}+1}\right) = \frac{3+\sqrt{3}+\sqrt{3}+1}{3+\sqrt{3}-\sqrt{3}-1}$$

Now we simplify. $\dfrac{3+2\sqrt{3}+1}{3-1} = \dfrac{4+2\sqrt{3}}{2} = \dfrac{4}{2} + \dfrac{2\sqrt{3}}{2} = 2+\sqrt{3}$

The answer is (3).

Another way to get the right answer is to find the value of $\dfrac{\sqrt{3}+1}{\sqrt{3}-1}$ on

your calculator. You should get: $\dfrac{\sqrt{3}+1}{\sqrt{3}-1} \approx 3.732$.

Now, evaluate each of the answers to see which one matches.

Choice (1): -1. Wrong.

Choice (2): 2. Wrong.

Choice (3): $2+\sqrt{3} \approx 3.732$. **Right.**

Choice (4): $5+\sqrt{3} \approx 6.732$. Wrong.

PROBLEM 28: TRIGONOMETRY

$Arc\cos\left(-\dfrac{1}{2}\right)$ means "what angle θ has $\cos\theta = -\dfrac{1}{2}$?" Although there are an infinite number of answers, when the A in Arccosine is capitalized, you only use the principal angle. This means that, if the Arc-cosine is of a positive number, use the **first** quadrant answer; and, if the Arccosine is of a negative number, use the **second** quadrant answer. Here, $Arc\cos\dfrac{1}{2} = 60°$, so $Arc\cos\left(-\dfrac{1}{2}\right)$ is the reference angle of $60°$ in the second quadrant, which is $180° - 60° = 120°$.

The answer is (1).

PROBLEM 29: LAW OF SINES

First, let's draw a triangle and fill in the information. Any triangle will do for now.

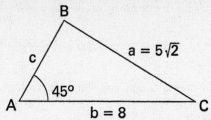

By asking for the possible number of triangles, the question is testing your knowledge of the *Ambiguous Case* of the Law of Sines. There is a simple test: Is $b \sin A < a < b$? $8 \sin 45° \approx 5.66$. Side a is $5\sqrt{2} \approx 7.07$ which is less than 8. Therefore, $8 \sin 45° < 5\sqrt{2} < 8$, and there are two triangles.

The answer is (2).

Another way to get this right is to use the Law of Sines and see how many triangles can be made.

The Law of Sines says that, in any triangle, with angles A, B, and C, and opposite sides of a, b, and c, respectively, $\dfrac{a}{\sin A} = \dfrac{b}{\sin B} = \dfrac{c}{\sin C}$. (This is on the formula sheet.)

Here, $\dfrac{5\sqrt{2}}{\sin 45°} = \dfrac{8}{\sin B}$, so $\sin B = \dfrac{8 \sin 45°}{5\sqrt{2}} = 0.8$. (If we had gotten a number greater than 1 or less than –1 we would have NO triangles.) Therefore, $m\angle B \approx 53$. This would make $m\angle C = 82$. Now we have one triangle.

But, there is a second angle for which $\sin B = 0.8$, that is $m\angle B \approx 180 - 53 = 127$.

Can we make a triangle with angles of $45°, 127°$ and C? Yes, if $m\angle C = 8$. Therefore, there are two triangles.

PROBLEM 30: TRIGONOMETRY

If you look on the formula sheet, under <u>Functions of the Sum of Two Angles</u>, you will find the formula $\tan(A + B) = \dfrac{\tan A + \tan B}{1 - \tan A \tan B}$. Thus

we can rewrite the problem as $\tan(180° + x) = \dfrac{\tan 180° + \tan x}{1 - \tan 180° \tan x}$.

Because $180°$ is a special angle, you should know that $\tan 180° = 0$.

Therefore, $\dfrac{\tan 180° + \tan x}{1 - \tan 180° \tan x} = \dfrac{0 + \tan x}{1 - (0)(\tan x)} = \tan x$.

The answer is (2).

Another way to get the right answer is to **plug-in**. First, make up a value for x. Let's use $x = 30°$.

Plug $x = 30°$ into the problem for x and we get:

$$\tan(180° + 30°) = \tan 210° \approx 0.577.$$

Now plug $x = 30°$ into the answer choices and see which one matches.

Choice (1): $\cot 30° \approx 1.73$. Wrong.

Choice (2): $\tan 30° \approx 0.577$. **Right**.

The last two choices can't be correct because they are negative.

PROBLEM 31: RATIONAL EXPRESSIONS

First, we find common denominators for the top and bottom fractions separately.

$$\frac{1}{x} - 1 = \frac{1}{x} - \frac{x}{x} = \frac{1 - x}{x} \quad \text{and} \quad x - \frac{1}{x} = \frac{x^2}{x} - \frac{1}{x} = \frac{x^2 - 1}{x}.$$

Now, because we have one fraction over another fraction, we invert the bottom one and multiply: $\dfrac{1 - x}{x} \cdot \dfrac{x}{x^2 - 1}$.

Next, factor the denominator: $\dfrac{1-x}{x} \cdot \dfrac{x}{x^2-1} = \dfrac{1-x}{x} \cdot \dfrac{x}{(x-1)(x+1)}$

Next, cancel like terms: $\dfrac{1-x}{x} \cdot \dfrac{x}{(x-1)(x+1)} = -\dfrac{1}{x+1}$

The answer is (4).

Another way to get this right is by plugging in. First make up a value for x. For example, let $x = 2$.

If we plug $x = 2$ into the problem, we get

$$\dfrac{\dfrac{1}{2}-1}{2-\dfrac{1}{2}} = \dfrac{-\dfrac{1}{2}}{\dfrac{3}{2}} = -\dfrac{1}{2} \cdot \dfrac{2}{3} = -\dfrac{1}{3}$$

Now, plug $x = 2$ into each of the answer choices and pick the one that matches.

Choice (1): -1; so answer (1) is wrong.

Choice (2): $\dfrac{1}{2+1} = \dfrac{1}{3}$; so answer (2) is wrong.

Choice (3): $\dfrac{1}{2-1} = 1$; so answer (3) is wrong.

Choice (4): $-\dfrac{1}{2+1} = -\dfrac{1}{3}$; so answer (4) is correct.

PROBLEM 32: INVERSES

We can find the inverse of an equation in two simple steps.

Step 1: Switch x and y. This gives us $x = 3y$.

Step 2: Solve for y. This gives us $y = \dfrac{x}{3}$

The answer is (2).

PROBLEM 33: QUADRATIC EQUATIONS

We can determine the nature of the roots of a quadratic equation of the form $ax^2 + bx + c = 0$ by using the discriminant $b^2 - 4ac$.

You should know the following rule:

If $b^2 - 4ac < 0$, the equation has two imaginary roots.

If $b^2 - 4ac = 0$, the equation has one rational root.

If $b^2 - 4ac > 0$, and $b^2 - 4ac$ is a perfect square, then the equation has two rational roots.

If $b^2 - 4ac > 0$, and $b^2 - 4ac$ is not a perfect square, then the equation has two irrational roots.

Here, the discriminant is positive and not a perfect square, so there are two irrational roots.

The answer is (2).

PROBLEM 34: TRIGONOMETRY IDENTITIES

You can find the identities for double angles on the formula sheet under *Functions of the Double Angle*. It says $\sin 2A = 2\sin A \cos A$. Substituting this into the expression, we get:

$$\sin A \cos A + 2\sin A \cos A = 3\sin A \cos A.$$

The answer is (3).

Another way to get this right is to **plug-in**. First, we make up a value for A. How about $A = 30°$? Now we plug $A = 30°$ into the problem and we get:

$$\sin 30° \cos 30° + \sin(2 \cdot 30°) = \sin 30° \cos 30° + \sin 60° \approx 1.299$$

Next, plug $A = 30°$ into each answer choice and see which one matches.

Choice (1): $\sin 30°(\cos 30° + \sin 30°) \approx 0.656$; so this is a wrong answer.

Choice (2): $\cos 30° + 2\sin 30° \approx 1.866$; so this is a wrong answer.

Choice (3): $3 \sin 30° \cos 30° \approx 1.299$; so this is the **right** answer.

Choice (4): $\cos 30° + 2 \sin(2 \cdot 30°) \approx 2.598$; so this is a wrong answer.

PROBLEM 35: BINOMIAL EXPANSIONS

The binomial theorem says that if you expand $(a + b)^n$, you get the following terms:

$$_nC_0a^n +_n C_1a^{n-1}b^1 +_n C_2a^{n-2}b^2 + ... +_n C_{n-2}a^2b^{n-2} +_n C_{n-1}a^1b^{n-1} +_n C_nb^n .$$

Therefore, if we expand $(2x - y)^5$, we get:

$$_5C_0(2x)^5 +_5 C_1(2x)^4(-y)^1 +_5 C_2(2x)^3(-y)^2 +_5 C_3(2x)^2(-y)^3 +_5 C_4(2x)^1(-y)^4 +_5 C_5(-y)^5$$

Next, we use the rule that $_nC_r = \dfrac{n!}{(n - r)!\, r!}$. This gives us:

$$(2x - y)^5 = (2x)^5 + 5(2x)^4(-y) + 10(2x)^3(-y)^2 + 10(2x)^2(-y)^3 + 5(2x)(-y)^4 + (-y)^5$$

which simplifies to

$$(2x - y)^5 = 32x^5 - 80x^4y + 80x^3y^2 - 40x^2y^3 + 10xy^4 - y^5.$$

The fourth term is $-40x^2y^3$.

The answer is (4).

A shortcut is to know the rule that the *rth* term of the binomial expansion of $(a - b)^n$ is $_nC_{r-1}(a)^{n-r+1}(b)^{r-1}$. Thus the fourth term is·

$$_5C_{4-1}(2x)^{5-4+1}(-y)^{4-1},$$

which can be simplified to

$$_5C_3(2x)^2(-y)^3 = 10(4x^2)(-y^3) = -40x^2y^3.$$

Part II

PROBLEM 36: TRIGONOMETRIC GRAPHS

(a) A graph of the form $y = a \sin bx$ or $y = a \cos bx$ has an amplitude of $|a|$ and a period of $\frac{2\pi}{b}$. Therefore, the graph of the equation $y = \cos 2x$ has an amplitude of 1 and a period of π, and the graph of the equation $y = 2 \sin x$ has an amplitude of 2 and a period of 2π.

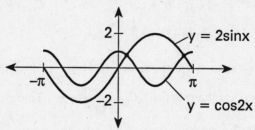

Remember, in graphing sines and cosines, the shape of the graph doesn't change, but it can be stretched or shrunk depending on the amplitude or period.

(b) The graphs intersect in two places.

PROBLEM 37: CIRCLE RULES

(a) *The measure of an inscribed angle is half of the arc it subtends (intercepts)*

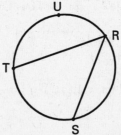

Thus, because $m\angle TRS = 50$, we know that the measure of arc $\overset{\frown}{ST}$ is 100.

Now, because the number of degrees in a circle is 360°, we know that the measure of arc $\overset{\frown}{UR}$ is $360 - 120 - 100 - 80 = 60$.

(b) We know from the rule in (a) that $m\angle SUT$ is half the measure of arc $\overset{\frown}{ST}$. Therefore, $m\angle SUT = \dfrac{100}{2} = 50$.

(c) *The measure of an angle formed by a pair of secants, or a secant and a tangent, is equal to half of the difference between the larger and the smaller arcs that are formed by the secants, or the secant and tangent.*

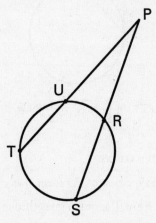

Here, the larger and smaller arcs formed by angle P are $\overset{\frown}{ST}$ and $\overset{\frown}{UR}$, respectively.

$$m\angle P = \frac{m\overset{\frown}{ST} - m\overset{\frown}{UR}}{2} = \frac{100 - 60}{2} = 20.$$

(d) *If two chords intersect within a circle, then the angle between the two chords is the average of their intercepted arcs.*

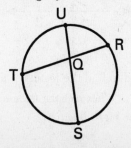

Therefore, $m\angle RQS = \dfrac{m\overset{\frown}{RS} + m\overset{\frown}{UT}}{2} = \dfrac{120 + 80}{2} = 100$.

(e) *An angle formed by a tangent and a chord is equal to half of its intercepted arc.*

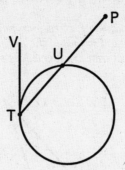

Therefore, $m\angle PTV = \dfrac{1}{2} m\overset{\frown}{UT} = \dfrac{1}{2} 80 = 40$.

PROBLEM 38: TRANSFORMATIONS

(a) The graph of an equation of the form $y = a^x$ always has the same general shape. Here, the graph goes through the points $(0, 1)$ and $(1, 2)$.

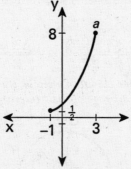

(b) Here we are asked to find the results of a composite transformation $\left(T_{2,-1} \circ r_{x-axis}\right)$. *Whenever we have a composite transformation, always do the **right** one first.*

r_{x-axis} means that we reflect the graph in the x-axis; that is, we turn it upside down. It looks like this:

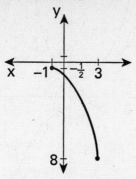

Now we do the other transformation to *the graph that we just reflected*.

T_{2-1} means that we add 2 to each x-coordinate, shifting the graph 2 to the right, and that we add -1 to each y-coordinate, shifting the graph down 1. It looks like this

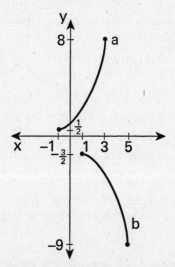

(c) Now we are asked to do the transformation D_{-1}. This is a dilation of -1 and means that we multiply both the x- and y-coordinates by -1. It looks like this:

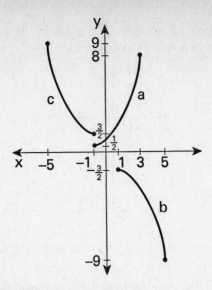

PROBLEM 39: TRIGONOMETRIC EQUATIONS

First, put all of the terms on the left side of the equals sign. We get $2 \sin x + 4 \cos 2x - 3 = 0$. Now we replace $\cos 2x$ using the Double Angle Formula (which can be found on the formula sheet) $\cos 2x = 1 - 2 \sin^2 x$.

This gives us: $2 \sin x + 4(1 - 2 \sin^2 x) - 3 = 0$. This can be simplified to:

$$2 \sin x + 4 - 8 \sin^2 x - 3 = 0$$
$$2 \sin x + 1 - 8 \sin^2 x = 0$$
$$8 \sin^2 x - 2 \sin x - 1 = 0$$

Note that the trig equation $8 \sin^2 x - 2 \sin x - 1 = 0$ has the same form as a quadratic equation $8x^2 - 2x - 1 = 0$. Just as we could factor the quadratic equation, so too can we factor the trig equation. We get: $(4 \sin x + 1)(2 \sin x - 1) = 0$.

This gives us $4 \sin x + 1 = 0$ and $2 \sin x - 1 = 0$. If we solve each of these equations, we get: $\sin x = -\dfrac{1}{4}$ and $\sin x = \dfrac{1}{2}$. Therefore

$x = \sin^{-1}\left(-\dfrac{1}{4}\right)$ and $x = \sin^{-1}\left(\dfrac{1}{2}\right)$. The solutions to the latter equation are easy because they are special angles.

You should know that $\sin 30° = \dfrac{1}{2}$ and that $\sin 150° = \dfrac{1}{2}$.

There is not a special angle whose sine is $-\dfrac{1}{4}$, so we need to use the calculator to find the value. Using the calculator, push **2nd SIN (–1/4) ENTER**. (**MAKE SURE THAT THE CALCULATOR IS IN DEGREE MODE!**) This gives us –14.47°, which rounds to –14°, so we use a reference angle of 14°. But, we are asked to find values of θ between 0° and 360°.

We can find the required values of θ by making a small graph. Note the positions of the reference angles in quadrants III and IV, where sine is negative.

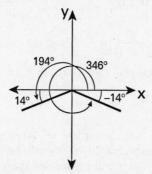

Notice that –14° is the same as 346°. The other place where sine is negative and has a reference angle of 14° is in quadrant III. The other angle then is 180° + 14° = 194°.

PROBLEM 40: PROBABILITY

(a) Given that there are 9 keys and 3 of them have letters on them, the probability that a letter key will be pressed on any given flip is

$\dfrac{3}{9} = \dfrac{1}{3}$.

(b) This problem requires that you know something called *binomial probability*. The rule is: *If the probability of a particular outcome is **p**, then the probability of that outcome occurring **r** times out of a possible **n** times is* $_nC_r(p)^r(1-p)^{n-r}$.

Given that there are 9 keys and 6 of them have numbers on them, the probability that a number key will be pressed on any given flip is $p = \dfrac{6}{9} = \dfrac{2}{3}$.

Therefore, the probability that a number key will be pressed two times out of a possible three is: $_3C_2\left(\dfrac{2}{3}\right)^2\left(1-\dfrac{2}{3}\right)^{3-2}$, which can be simplified to $_3C_2\left(\dfrac{2}{3}\right)^2\left(\dfrac{1}{3}\right)$

The rule for finding $_nC_r$ is $_nC_r = \dfrac{n!}{(n-r)!\,r!}$. Therefore,

$_3C_2 = \dfrac{3!}{1!\,2!} = \dfrac{3\cdot 2\cdot 1}{(1)(2\cdot 1)} = 3$ Thus, the probability that a number key will be pressed two times out of a possible three is:

$3\left(\dfrac{2}{3}\right)^2\left(\dfrac{1}{3}\right) = \dfrac{4}{9}$.

(c) The probability that a letter key will be pressed at least two times is the sum of the probability that it will be pressed two times and the probability that it will be pressed three times.

The probability that a letter key will be pressed two out of three times is: $_3C_2\left(\dfrac{1}{3}\right)^2\left(1-\dfrac{1}{3}\right)^{3-2}$, which can be simplified to $_3C_2\left(\dfrac{1}{3}\right)^2\left(\dfrac{2}{3}\right)$, and the probability that a letter key will be pressed all three times is: $_3C_3\left(\dfrac{1}{3}\right)^3\left(1-\dfrac{1}{3}\right)^{3-3}$, which can be simplified to $_3C_3\left(\dfrac{1}{3}\right)^3\left(\dfrac{2}{3}\right)^0$.

We already found $_3C_2 = 3$ in (b) and $_3C_3 = \dfrac{3!}{0!\,3!} = \dfrac{3 \cdot 2 \cdot 1}{(1)(3 \cdot 2 \cdot 1)} = 1$
(By the way, $0! = 1$).

Thus, the probability that a letter key will be pressed at least two times is:

$$3\left(\frac{1}{3}\right)^2\left(\frac{2}{3}\right) + \left(\frac{1}{3}\right)^3\left(\frac{2}{3}\right)^0 = \frac{2}{9} + \frac{1}{27} = \frac{7}{27}.$$

PROBLEM 41: TRIGONOMETRY

(a) Law of Cosines

First, let's draw a picture of the situation.

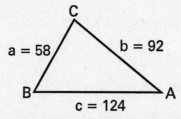

We can use the Law of Cosines. The Law of Cosines says that, in any triangle, with angles $A, B,$ and C, and opposite sides of a, b, and c, respectively, $c^2 = a^2 + b^2 - 2ab \cos C$. (This is on the formula sheet.) If we have a triangle and we are given SSS (side, side, side), then the side opposite the angle we seek is c, and the angle is C. *In a triangle, the largest angle is opposite the largest side.*

In this case, the largest side is 124, so we have:

$$124^2 = 58^2 + 92^2 - 2(58)(92) \cos C$$
$$15376 = 3364 + 8464 - 10672 \cos C$$
$$10672 \cos C = 3364 + 8464 - 15376 = -3548$$

$$\cos C = -\frac{3548}{10672}$$

$$C = \cos^{-1}\left(-\frac{3548}{10672}\right) \approx 109.418°$$

Notice that we are asked to give the answer to the *nearest ten minutes*. In order to do this, simply multiply the decimal part of the answer by 60, and round.

$(0.418)(60) = 25.08$ minutes, which rounds up to 30 minutes.

Therefore, the answer is $C = 109°30'$.

(b) Trigonometric Area

The area of a triangle, if we are given the lengths of the two sides a and b, and their included angle θ, is $A = \frac{1}{2}ab\sin\theta$. In other words, if we are given SAS (side, angle, side), we find the area by multiplying $\frac{1}{2}$ by the product of the two sides by the sine of the included angle. Here, the two sides are a and b, and the included angle is C. Plugging in, we get:

$$A = \frac{1}{2}(58)(92)\sin 109.5° \approx 2514.97,$$ which rounds to 2515.

PROBLEM 42:

(a) Quadratic Equations

If we move all of the terms to the left side, we get $2x^2 - 6x + 5 = 0$. We can now use the quadratic formula to solve for x.

The formula says that, given a quadratic equation of the form $ax^2 + bx + c = 0$, the roots of the equation are: $x = \frac{-b \pm \sqrt{b^2 - 4ac}}{2a}$. Plugging in to the formula we get:

$$x = \frac{-(-6) \pm \sqrt{(-6)^2 - 4(2)(5)}}{2(2)} = \frac{6 \pm \sqrt{-4}}{4} = \frac{6 \pm 2i}{4} = \frac{3 \pm i}{2}.$$

(b) Logarithms

First, take the logarithm of both sides. We get: $\log 3^{2x} = \log 100$.

You should know that $\log 100 = 2$, so we now have: $\log 3^{2x} = 2$.

Next, use the log rule $\log a^b = b \log a$ to rewrite the equation: $2x \log 3 = 2$

Divide both sides by $2\log 3$ and we get: $x = \frac{2}{2\log 3} = \frac{1}{\log 3} \approx 2.1$.

Formulas

Pythagorean and Quotient Identities

$$\sin^2 A + \cos^2 A = 1 \qquad \tan A = \frac{\sin A}{\cos A}$$

$$\tan^2 A + 1 = \sec^2 A \qquad \cot A = \frac{\cos A}{\sin A}$$

$$\cot^2 A + 1 = \csc^2 A$$

Functions of the Sum of Two Angles

$$\sin (A + B) = \sin A \cos B + \cos A \sin B$$

$$\cos (A + B) = \cos A \cos B - \sin A \sin B$$

$$\tan (A + B) = \frac{\tan A + \tan B}{1 - \tan A \tan B}$$

Functions of the Difference of Two Angles

$$\sin (A - B) = \sin A \cos B - \cos A \sin B$$

$$\cos (A - B) = \cos A \cos B + \sin A \sin B$$

$$\tan (A - B) = \frac{\tan A - \tan B}{1 + \tan A \tan B}$$

Law of Sines

$$\frac{a}{\sin A} = \frac{b}{\sin B} = \frac{c}{\sin C}$$

Law of Cosines

$$a^2 = b^2 + c^2 - 2bc \cos A$$

Functions of the Double Angle

$$\sin 2A = 2 \sin A \cos A$$

$$\cos 2A = \cos^2 A - \sin^2 A$$

$$\cos 2A = 2 \cos^2 A - 1$$

$$\cos 2A = 1 - 2 \sin^2 A$$

$$\tan 2A = \frac{2 \tan A}{1 - \tan^2 A}$$

Functions of the Half Angle

$$\sin \frac{1}{2} A = \pm \sqrt{\frac{1 - \cos A}{2}}$$

$$\cos \frac{1}{2} A = \pm \sqrt{\frac{1 + \cos A}{2}}$$

$$\tan \frac{1}{2} A = \pm \sqrt{\frac{1 - \cos A}{1 + \cos A}}$$

Area of Triangle

$$K = \frac{1}{2} ab \sin C$$

Standard Deviation

$$\text{S.D.} = \sqrt{\frac{1}{n} \sum_{i=1}^{n} \left(x_i - \bar{x} \right)^2}$$

EXAMINATION
JANUARY 1996

Part I

Answer 30 questions from this part. Each correct answer will receive 2 credits. No partial credit will be allowed. Write your answers in the spaces provided on the separate answer sheet. Where applicable, answers may be left in terms of π or in radical form.

1 Solve for x: $\sqrt{x + 3} + 2 = 6$.

$x + 3 = 16 \mid x = 13$

2 Express 240° in radian measure.

$240 \cdot \dfrac{\pi}{180} = \dfrac{4\pi}{3}$

3 Find the coordinates of the image of (–3,4) under the transformation $T_{-2,3}$.

$(-5, 7)$

4 Solve for x: $3^x = 27^{\frac{2}{3}}$

5 Find the value of $\sum\limits_{n=1}^{4} (3n - 2)$.

22

$\dfrac{3}{2}(4)$ $(3(1) - 2) + (3(2) - 2) + (3(3) - 2) + 3(4) - 2$

$1 + 4 + 7 + 10$

6 If $f(x) = \sin^2 x + \cos^2 x$, find $f\left(\dfrac{\pi}{4}\right)$.

$F(\pi/4) = \sin^2 45 + \cos^2 45$ $\dfrac{\sqrt{2}}{2} + \dfrac{\sqrt{2}}{2}$ $F(\pi/4) = \dfrac{2\sqrt{2}}{2}$

7 Simplify and express in $a + bi$ form:

$(12 + 3i) - (3 - i)$

$12 + 3i + (-3 + i)$

$9 + 4i = 13i$

8 If $a = 14$, $e = 16$, and m $\angle C = 30$, find the area of $\triangle ACE$.

$$K = \frac{1}{2}(14)(16) \cdot \sin 30$$
$$K = 112 \cdot \sin 30$$
$$K = 112 \cdot \frac{1}{2}$$
$$K = 56$$

9 In a circle, two tangents from an external point intercept a major arc of 240°. Find the number of degrees in the angle formed by the tangents.

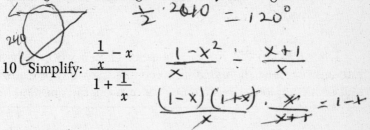

$$\frac{1}{2} \cdot 240 = 120°$$

240

10 Simplify: $\dfrac{\dfrac{1}{x} - x}{1 + \dfrac{1}{x}}$

$$\frac{1-x^2}{x} \div \frac{x+1}{x}$$

$$\frac{(1-x)(1+x)}{x} \cdot \frac{x}{x+1} = 1-x$$

11 If θ terminates in Quadrant II and $\sin \theta = \dfrac{12}{13}$, find $\cos \theta$.

12 In the accompanying diagram, $\overline{AB}$ is tangent to circle O and secant $\overline{ACD}$ is drawn. If $AB = 4$ and $AD = 8$, find AC.

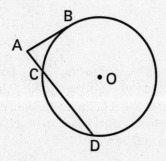

13 Find the coordinates of point $N(-1,3)$ under the composite $r_{y\text{-axis}} \circ R_{90°}$.

[OVER]

14 Find the value of x in the domain $0° \leq x° < 90°$ that satisfies the equation $2 \sin x - \sqrt{2} = 0$.

15 An arc of a circle measures 30 centimeters and the radius measures 10 centimeters. In radians, what is the measure of the central angle that subtends the arc?

Directions (16–35): For *each* question chosen, write on the separate answer sheet the *numeral* preceding the word or expression that best completes the statement or answers the question.

16 For which value of x is $f(x) = \dfrac{1}{3^x - 1}$ undefined?

(1) 1   (3) 3
(2) –1   (4) 0

17 Which number is *not* an element of the range of $y = \sin x$?

(1) 1   (3) –1
(2) 2   (4) 0

18 The scores on a examination have a normal distribution. The mean of the scores is 50, and the standard deviation is 4. What is the best approximation of the percentage of students who can be expected to score between 46 and 54?

(1) 95%   (3) 50%
(2) 68%   (4) 34%

19 The expression $\log \sqrt[4]{\dfrac{a^2}{b}}$ is equivalent to

(1) $\dfrac{1}{4}\left(\dfrac{\log a^2}{\log b}\right)$

(2) $4(\log a^2 - \log b)$

(3) $\dfrac{1}{2}(4 \log a - \log b)$

(4) $\dfrac{1}{4}(2 \log a - \log b)$

20 The expression $\dfrac{2 + \sqrt{3}}{2 - \sqrt{3}}$ is equivalent to

(1) $11\sqrt{3}$ (3) $7 + 4\sqrt{3}$

(2) $7 - 4\sqrt{3}$ (4) $\dfrac{7 + 4\sqrt{3}}{7}$

21 In $\triangle ABC$, $a = 8$, $b = 5$, and $c = 9$. What is the value of $\cos A$?

(1) $-\dfrac{1}{4}$ (3) $-\dfrac{7}{15}$

(2) $\dfrac{1}{4}$ (4) $\dfrac{7}{15}$

22 The expression $(3 - i)^2$ is equivalent to

(1) 8   (3) 10

(2) $8 - 6i$ (4) $8 + 6i$

[OVER]

23 In a family of six children, what is the probability
    that *exactly* one child is female?

    (1) $\dfrac{6}{64}$ (3) $\dfrac{32}{64}$

    (2) $\dfrac{7}{64}$ (4) $\dfrac{58}{64}$

24 What is the solution set for the equation
    $|3x - 1| = x + 5$?

    (1) $\{-1\}$ (3) $\{3\}$
    (2) $\{-1,3\}$ (4) $\{1,-3\}$

25 What is the value of y if $y = \cos\left(\text{Arc}\sin\dfrac{1}{2}\right)$?

    (1) $30°$ (3) $\dfrac{1}{2}$

    (2) $60°$ (4) $\dfrac{\sqrt{3}}{2}$

26 What is the inverse of the equation $y = 3x + 2$?

    (1) $3y = x + 2$ (3) $y = \dfrac{1}{3}x - 2$

    (2) $x = 3y + 2$ (4) $x = \dfrac{1}{3}y + \dfrac{2}{3}$

27 What is the period of the graph of the equation
    $y = 3\cos 4x$?

    (1) $\dfrac{\pi}{4}$ (3) 3

    (2) $\dfrac{\pi}{2}$ (4) 4

28 If $b^2 - 4ac < 0$, then the roots of the equation $ax^2 + bx + c = 0$ must be

(1) real, irrational, and unequal
(2) real, rational, and unequal
(3) real, rational, and equal
(4) imaginary

29 Which expression is equivalent to $\sin 22° \cos 18° + \cos 22° \sin 18°$?

(1) $\sin 4°$ (3) $\sin 40°$
(2) $\cos 4°$ (4) $\cos 40°$

30 The expression $\cos \theta (\sec \theta - \cos \theta)$ is equivalent to

(1) 1   (3) $\sin^2 \theta$
(2) $\sin \theta$ (4) $-\cos^2 \theta$

31 Which graph represents the reflection in the x-axis of the curve $y = \cos x$?

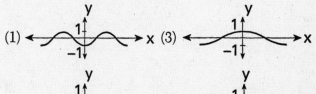

32 The fifth term of the expansion of $(1 - \pi)^6$ is

(1) $15\pi^4$ (3) $-6\pi^5$
(2) $6\pi^4$ (4) $-30\pi^5$

[OVER]

33 What is the solution set of $x^2 - 3x - 28 \geq 0$?

    (1) $x \geq 7$ or $x \leq -4$ (3) $-4 \leq x \leq 7$

    (2) $x \leq 7$ or $x \geq -4$ (4) $-4 < x < 7$

34 The graph of the equation $x^2 + 2y^2 = 5$ is

    (1) a parabola   (3) an ellipse

    (2) a hyperbola   (4) a circle

35 The expression $\sin(180° - x)$ is equivalent to

    (1) $\sin x$ (3) $-\sin x$

    (2) $\cos x$ (4) $-\cos x$

Part II

Answer four questions from this part. Clearly indicate the necessary steps, including appropriate formula substitutions, diagrams, graphs, charts, etc. Calculations that may be obtained by mental arithmetic or the calculator do not need to be shown. [40]

36 *a* The table below shows the scores of 40 students on an advanced placement mathematics examination. Find the standard deviation to the *nearest tenth*. [6]

Score	Number of Students
5	8
4	12
3	14
2	4
1	2

b Express in simplest form:

$$\frac{x^2 + 2x}{x^2 + 2x - 15} \cdot \frac{2x - 6}{4} \div \frac{x^2 + x - 2}{x^2 + 4x - 5}$$ [4]

37 *a* For all values of θ for which the expressions are defined, prove the following is an identity:

$$\frac{\sin 2\theta + \sin \theta}{\cos 2\theta + \cos \theta + 1} = \tan \theta$$ [6]

b If $\log 7 = x$ and $\log 3 = y$, express in terms of x and y:

(1) $\log \sqrt{\dfrac{3}{7}}$ [2]

(2) $\log 63$ [2]

[OVER]

38 *a* Express the roots of the equation $x^2 + 1 = 8(x - 3)$ in $a + bi$ form.   [5]

b Find, to the nearest degree, all values of θ in the interval $0° \le \theta < 360°$ that satisfy the equation $4 \cos^2\theta - 3 \cos \theta = 1$.   [5]

39 In the accompanying diagram of circle O, diameter $\overline{CA}$ intersects chord $\overline{BD}$ at F; $\overline{AE}$ is a tangent; $\overline{EDC}$ is a secant; $\overline{CB}$, $\overline{BA}$, and $\overline{AD}$ are chords; m$\widehat{BC}$ = 100; and m$\widehat{AD}$ = 70.

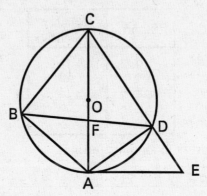

Find:

a m$\widehat{AB}$ [2]
b m$\angle$AEC   [2]
c m$\angle$BCA   [2]
d m$\angle$DFA   [2]
e m$\angle$DAE   [2]

40 *a* On the same set of axes, sketch and label the graphs of the equations $y = -\cos x$ and $y = \sin 2x$ in the interval $0 \leq x \leq 2\pi$. [4,4]

b Using the graphs from part a, determine which value in the interval $0 \leq x \leq \pi$ satisfies the equation $-\cos x = \sin 2x$. [2]

41 Given: $f(x) = 11x + 3$ and $g(x) = \sqrt{x}$.

Find:

a f(2) [2]
b g(f(2)) [2]
c g(100) [2]
d f⁻¹(x) [2]
e g⁻¹(3) [2]

42 An airplane traveling at a level altitude of 2,050 feet sights the top of a 50-foot tower at an angle of depression of 28° from point A. After continuing in level flight to point B, the angle of depression to the same tower is 34°. Find, to the *nearest foot*, the distance that the plane traveled from point A to point B. [10]

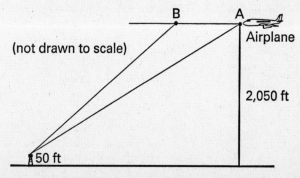

B   A

Airplane

(not drawn to scale)

2,050 ft

50 ft

[OVER]

ANSWER KEY

Part I

1. 13
2. $\dfrac{4\pi}{3}$
3. (−5, 7)
4. 2
5. 22
6. 1
7. 9 + 4i
8. 56
9. 60
10. 1 − x
11. $-\dfrac{5}{13}$
12. 2

13. (3, −1)
14. 45
15. 3
16. (4)
17. (2)
18. (2)
19. (4)
20. (3)
21. (4)
22. (2)
23. (1)
24. (2)

25. (4)
26. (2)
27. (2)
28. (4)
29. (3)
30. (3)
31. (1)
32. (1)
33. (1)
34. (3)
35. (1)

ANSWERS AND EXPLANATIONS JANUARY 1996

Part I

In problems 1–15, you must supply the answer. Because of this, you will not be able to **plug-in** or **backsolve**. But, you can use your calculator to simplify calculations. In addition, you will find that these problems tend to test your basic knowledge of Sequential Math III and are not tricky. Many can be solved just by knowing the right formula and by plugging in the numbers.

PROBLEM 1: RADICAL EQUATIONS

First, subtract 2 from both sides. Now, if we square both sides of the equation, we get $x + 3 = 16$. If we then subtract 3 from both sides, we get $x = 13$. **BUT**, whenever we have a radical in an equation, we have to check the root to see if it satisfies the original equation. Sometimes, in the process of squaring both sides, we will get answers that are invalid. So let's check the answer.

Does $\sqrt{13 + 3} + 2 = 6$? $4 + 2 = 6$; so, yes, it does. Therefore the answer is $x = 13$.

PROBLEM 2: CONVERTING FROM DEGREES TO RADIANS

All we have to do is multiply $240°$ by $\dfrac{\pi}{180°}$.

$$240\left(\frac{\pi}{180}\right) = \frac{4\pi}{3}.$$

PROBLEM 3: TRANSFORMATIONS

The translation $T_{(x,y)}$ tells you to add x to the x-coordinate of the point being translated, and to add y to the y-coordinate of the point being translated. So, here, $T_{(-2,3)}(-3,4)$ means $-3 + -2 = -5$ and $4 + 3 = 7$. Therefore, the translation is $T_{(-2,3)}(-3,4) \rightarrow (-5,7)$.

PROBLEM 4: EXPONENTIAL EQUATIONS

In order to solve this equation, we need to have a common base for the two sides of the equation. Because $3^3 = 27$, we can rewrite the right side as $3^x = \left(3^3\right)^{\frac{2}{3}}$.

The rules of exponents say that *when a number raised to a power is itself raised to a power, we multiply the exponents.* Thus, we get:

$$3^x = 3^{\left(3 \cdot \frac{2}{3}\right)} = 3^2.$$

Now, we can set the powers equal to each other and we get $x = 2$.

PROBLEM 5: SUMMATIONS

To evaluate a sum, we plug each of the consecutive values for n into the expression, starting at the bottom value and ending at the top value, and sum the results. We get:

$$\sum_{n=1}^{4} (3n - 2) = [3(1) - 2] + [3(2) - 2] + [3(3) - 2] + [3(4) - 2]$$
$$= 1 + 4 + 7 + 10 = 22$$

Another way to do this is to use your calculator. Push **2nd STAT** (**LIST**) and under the **MATH** menu, choose **SUM**. Then, again push **2nd STAT** and under the **OPS** menu, choose **SEQ**.

We put in the following: *(expression, variable, start, finish, step)*.

For *expression*, we put in the formula, using x as the variable.

For *variable*, we ALWAYS put in x.

For *start*, we put in the bottom value.

For *finish*, we put in the top value.

For *step*, we ALWAYS put in 1.

Therefore, we put in the calculator: **SUM SEQ** $(3x - 2, x, 1, 4, 1)$. The calculator should return the value 22.

PROBLEM 6: TRIGONOMETRY

$\sin^2 x + \cos^2 x = 1$ no matter what angle is used for x, so the answer is 1.

If you didn't know this rule, you would just plug in $x = \dfrac{\pi}{4}$ and you would get:

$$\sin^2 \frac{\pi}{4} + \cos^2 \frac{\pi}{4} = \left(\frac{\sqrt{2}}{2}\right)^2 + \left(\frac{\sqrt{2}}{2}\right)^2 = \frac{1}{2} + \frac{1}{2} = 1.$$

PROBLEM 7: COMPLEX NUMBERS

To find the difference of $12 + 3i$ and $3 - i$ we subtract the real components and the imaginary components separately. This gives us:

$$(12 - 3) + (3 - (-1))i = 9 + 4i.$$

PROBLEM 8: TRIGONOMETRIC AREA

The area of a triangle, if we are given the lengths of the two sides a and b, and their included angle θ, is $A = \dfrac{1}{2}\,ab\sin\theta$. In other words, if we are given SAS (side, angle, side), we find the area by multiplying $\dfrac{1}{2}$ by the product of the two sides by the sine of the included angle. Here, the two sides are a and e, and the included angle is C. Plugging in, we get: $A = \dfrac{1}{2}(14)(16)\sin 30° = 56$.

PROBLEM 9: CIRCLE RULES

You should know this rule: *The measure of an angle formed by a pair of tangents is equal to half of the difference between the major and the minor arcs that are formed by the tangents.*

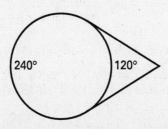

Here, we know that the major arc is 240°, so the minor arc is 360° − 240° = 120°. Therefore, the angle formed by the tangents is

$$\frac{240° - 120°}{2} = 60°.$$

PROBLEM 10: RATIONAL EXPRESSIONS

First, we obtain common denominators for the top and bottom separately. The common denominator for the top is x. The top then becomes

$$\frac{1}{x} - x\left(\frac{x}{x}\right) = \frac{1 - x^2}{x}$$

The common denominator for the bottom is also x. The bottom then becomes

$$1\left(\frac{x}{x}\right) + \frac{1}{x} = \frac{x + 1}{x}$$

Next, we invert the bottom fraction and multiply

$$\frac{\dfrac{1 - x^2}{x}}{\dfrac{x + 1}{x}} = \left(\frac{1 - x^2}{x}\right)\left(\frac{x}{x + 1}\right) = \frac{1 - x^2}{x + 1}$$

Now factor the numerator and cancel like terms.

$$\frac{1 - x^2}{x + 1} = \frac{(1 - x)(1 + x)}{x + 1} = 1 - x$$

PROBLEM 11: TRIGONOMETRY

You should know that the sine of an angle in a right triangle is

$$\sin = \frac{\text{opposite}}{\text{hypotenuse}}$$

If $\sin \theta = \dfrac{12}{13}$, then we can draw a triangle in the second quadrant whose opposite side is 12 and whose hypotenuse is 13. Now we use the Pythagorean Theorem to find the third (adjacent) side. $12^2 + a^2 = 13^2$.

$$a^2 = 13^2 - 12^2 = 25$$
$$a = \pm 5$$

Next, you should know that the cosine of an angle is

$$\cos = \frac{\text{adjacent}}{\text{hypotenuse}}$$

Therefore, $\cos\theta$ could equal $\pm\frac{5}{13}$. Because *the cosine of an angle is negative in quadrants* II *and* III the answer must be $\cos\theta = -\frac{5}{13}$.

PROBLEM 12: CIRCLE RULES

If a tangent and a secant are drawn from an external point, then the square of the measure of the tangent segment is equal to the product of the measures of the secant and its external segment. In other words, if you look at the diagram:

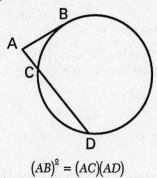

$$(AB)^2 = (AC)(AD)$$

If we plug in the values, we get:

$$(4)^2 = (AC)(8).$$

If we solve this, we get $AC = 2$.

PROBLEM 13: TRANSFORMATIONS

Here we are asked to find the results of a composite transformation $(r_{y\text{-axis}} \circ R_{90°})(-1,3)$. *Whenever we have a composite transformation, always do the **right** one first.* $R_{90°}$ means a rotation counterclockwise of 90°. You should know the following rotation rules:

$$R_{90°}(x,y) = (-y,x)$$
$$R_{180°}(x,y) = (-x,-y)$$
$$R_{270°}(x,y) = (y,-x)$$

This gives us $R_{90°}(-1,3) \rightarrow (-3,-1)$. r_{y-axis} means a reflection of the point in the y-axis. In other words, you change the sign of the x-coordinate and leave the y-coordinate alone. This gives us $r_{y-axis}(-3,-1) \rightarrow (3,-1)$.

PROBLEM 14: TRIGONOMETRIC EQUATIONS

First, add $\sqrt{2}$ to both sides. We get: $2\sin x = \sqrt{2}$. Now divide both sides by 2 and we get:

$$\sin x = \frac{\sqrt{2}}{2}.$$

You should know from your special angles that this is true for $x = 45°$, $135°$. Therefore, because of the domain restriction, the answer is $x = 45°$.

PROBLEM 15: ARC LENGTH

In a circle with a central angle, θ, measured in radians; and a radius, r; the arc length, s, is found by $s = r\theta$.

Here we have $r = 10$ and $s = 30$. Plug these into the formula and we get:

$$30 = 10\theta.$$

Therefore, the central angle is $\theta = 3$ radians.

Multiple-Choice Problems

For problems 16–35, you will find that you can sometimes **plug-in** or **backsolve** to get the right answer. You will also find that a calculator will simplify the arithmetic. In addition, most of the problems only require knowing which formula to use.

PROBLEM 16: EXPONENTS

A function is undefined when the denominator equals zero. Here, the denominator will equal zero when $3^x = 1$. *Any number raised to the zero power is equal to* 1. Therefore, $x = 0$.

The answer is (4).

PROBLEM 17: TRIGONOMETRY GRAPHS

A graph of the form $y = a \sin bx$ or $y = a \cos bx$ has a range of $-a \leq y \leq a$.
Thus, the range of $y = \sin x$ is $-1 \leq y \leq 1$.

The answer is (2).

PROBLEM 18: STATISTICS

You should know the following rule about normal distributions.

> In a normal distribution, with a mean of $\bar{x}$ and a standard deviation of σ:
> - approximately 68% of the outcomes will fall between $\bar{x} - \sigma$ and $\bar{x} + \sigma$.
> - approximately 95% of the outcomes will fall between $\bar{x} - 2\sigma$ and $\bar{x} + 2\sigma$.
> - approximately 99.5% of the outcomes will fall between $\bar{x} - 3\sigma$ and $\bar{x} + 3\sigma$.

Here, we are told that $\bar{x} = 50$ and $\sigma = 4$. Thus $\bar{x} - \sigma = 50 - 4 = 46$ and $\bar{x} + \sigma = 50 + 4 = 54$. Therefore, 68% of the scores lie between 46 and 54.

The answer is (2).

PROBLEM 19: LOGARITHMS

You should know the following log rules:

(i) $\log A + \log B = \log(AB)$

(ii) $\log A - \log B = \log\left(\dfrac{A}{B}\right)$

(iii) $\log A^B = B \log A$

First, using rule (iii), we can rewrite $\log \sqrt[4]{\dfrac{a^2}{b}}$ as $\dfrac{1}{4} \log \dfrac{a^2}{b}$.

Next, using rule (ii), we can rewrite $\dfrac{1}{4} \log \dfrac{a^2}{b}$ as $\dfrac{1}{4}\left(\log a^2 - \log b\right)$.

Finally, again using rule (iii), we can rewrite $\frac{1}{4}\left(\log a^2 - \log b\right)$ as

$$\frac{1}{4}\left(2\log a - \log b\right).$$

The answer is (4).

Another way to get the right answer is to **plug-in**. First, let's make up values for a and b. Let $a = 2$ and $b = 3$.

Then we use our calculator to find $\log \sqrt[4]{\dfrac{2^2}{3}} = \log \sqrt[4]{\dfrac{4}{3}} \approx 0.0312$.

Next, we plug $a = 2$ and $b = 3$ into each of the answer choices to see which one matches.

Choice (1): Does $\dfrac{1}{4}\left(\dfrac{\log 4}{\log 3}\right) \approx 0.0312$? If you plug it into the calculator, you get ≈ 0.3155, so choice (1) is not the answer.

Choice (2): Does $4(\log 4 - \log 3) \approx 0.0312$? If you plug it into the calculator, you get ≈ 0.4998, so choice (2) is not the answer.

Choice (3): Does $\dfrac{1}{2}\left(4\log 2 - \log 3\right) \approx 0.0312$? If you plug it into the calculator, you get ≈ 0.3635, so choice (3) is not the answer.

Choice (4): Does $\dfrac{1}{4}\left(2\log 2 - \log 3\right) = 0.0312$? If you plug it into the calculator, you get ≈ 0.0312, **so choice (4) is the answer.**

PROBLEM 20: RATIONAL EXPRESSIONS

Whenever we have a fraction where the denominator has a radical in it, we can *rationalize* the denominator by multiplying the top and bottom by the *conjugate* of the denominator. The conjugate is merely the same as the denominator where we reverse the sign in front of the radical.

Here, we will multiply the top and bottom by the conjugate of $2 - \sqrt{3}$, which is $2 + \sqrt{3}$.

We get:

$$\left(\frac{2+\sqrt{3}}{2-\sqrt{3}}\right)\left(\frac{2+\sqrt{3}}{2+\sqrt{3}}\right).$$

Next, we multiply out the top and bottom using FOIL.

$$\left(\frac{2+\sqrt{3}}{2-\sqrt{3}}\right)\left(\frac{2+\sqrt{3}}{2+\sqrt{3}}\right) = \frac{4+2\sqrt{3}+2\sqrt{3}+3}{4+2\sqrt{3}-2\sqrt{3}-3}$$

Now we simplify.

$$\frac{4+2\sqrt{3}+2\sqrt{3}+3}{4+2\sqrt{3}-2\sqrt{3}-3} = \frac{7+4\sqrt{3}}{1} = 7+4\sqrt{3}$$

The answer is (3).

Another way to get the right answer is to find the value of $\dfrac{2+\sqrt{3}}{2-\sqrt{3}}$ on your calculator. You should get:

$$\frac{2+\sqrt{3}}{2-\sqrt{3}} \approx 13.928$$

Now, evaluate each of the answers to see which one matches.

Choice (1): $11\sqrt{3} \approx 19.053$. Wrong.

Choice (2): $7-4\sqrt{3} \approx 0.072$. Wrong.

Choice (3): $7+4\sqrt{3} \approx 13.928$. **Right**

Choice (4): $\dfrac{7+4\sqrt{3}}{7} \approx 1.990$. Wrong.

PROBLEM 21: LAW OF COSINES

The Law of Cosines says that, in any triangle, with angles A, B, and C, and opposite sides of a, b, and c, respectively, $c^2 = a^2 + b^2 - 2ab \cos C$. Here, if we are looking for $\cos A$, we rewrite the formula as

$$a^2 = b^2 + c^2 - 2bc \cos A.$$

When we substitute the values we get:

$$8^2 = 5^2 + 9^2 - 2(5)(9) \cos A.$$

If we solve this for cos A, we get:

$$64 = 25 + 81 - 90 \cos A$$
$$90 \cos A = 25 + 81 - 64 = 42$$
$$\cos A = \frac{42}{90} = \frac{7}{15}.$$

The answer is (4).

PROBLEM 22: COMPLEX NUMBERS

When we multiply two complex numbers, we FOIL.

$$(3 - i)^2 = (3 - i)(3 - i) = 9 - 3i - 3i + i^2.$$

Next, we combine like terms. $9 - 3i - 3i + i^2 = 9 - 6i + i^2$.

You should know that $i^2 = -1$, we get:

$$9 - 3i - 3i + i^2 = 9 - 6i - 1 = 8 - 6i.$$

The answer is (2).

PROBLEM 23: PROBABILITY

*If the probability of a particular outcome is **p**, then the probability of that outcome occurring **r** times out of a possible **n** times is* $_nC_r(p)^r(1-p)^{n-r}$. This is known as *binomial probability*.

Here, the probability of having a female child is $\frac{1}{2}$. The probability of having one female child out of a possible 6 is $_6C_1\left(\frac{1}{2}\right)^1\left(1-\frac{1}{2}\right)^{6-1}$.

This can be simplified to $_6C_1\left(\frac{1}{2}\right)^1\left(\frac{1}{2}\right)^5$

We can then evaluate $_6C_1$ according to the formula $_nC_r = \dfrac{n!}{(n-r)!\,r!}$.

If we plug in, we get:

$$_6C_1 = \frac{6!}{(6-1)!\,1!} = \frac{6!}{5!\,1!} = \frac{6 \cdot 5 \cdot 4 \cdot 3 \cdot 2 \cdot 1}{(5 \cdot 4 \cdot 3 \cdot 2 \cdot 1)(1)} = 6.$$

Therefore, one female child out of a possible 6 is $6\left(\dfrac{1}{2}\right)^1\left(\dfrac{1}{2}\right)^5 = \dfrac{6}{64}$

The answer is (1).

PROBLEM 24: ABSOLUTE VALUE EQUATIONS

When we have an absolute value equation, we break it into two equations. First, rewrite the left side without the absolute value and leave the right side alone. Second, rewrite the left side without the absolute value, and change the sign of the right side.

Here, we have $|3x - 1| = x + 5$, so we rewrite it as:

$$3x - 1 = x + 5 \text{ or } 3x - 1 = -(x + 5).$$

If we solve each of these independently, we get $x = 3$ or $x = -1$.

We should double-check each of the solutions.

Does $|3(3) - 1| = 3 + 5$? $8 = 8$, so, yes, it does.

Does $|3(-1) - 1| = -1 + 5$? $4 = 4$, so, yes, it does.

The answer is (2).

PROBLEM 25: TRIGONOMETRY

$Arc\sin\dfrac{1}{2}$ means "what angle θ has $\sin\theta = \dfrac{1}{2}$?" Although there are an infinite number of answers, when the A in Arcsine is capitalized, you only use the principal angle. This means that, if the Arcsine is of a positive number, use the **first** quadrant answer; and, if the Arc sine is of a negative number, use the **fourth** quadrant answer. Here, $Arc\sin\dfrac{1}{2} = 30°$.

Now, we just have to find $\cos 30°$. This is a special angle, so you should know that $\cos 30° = \dfrac{\sqrt{3}}{2}$

The answer is (4).

Another way to get the right answer is to find the value of $\cos\left(Arc\sin\dfrac{1}{2}\right)$ on your calculator. You should get:

$$\cos\left(Arc\sin\dfrac{1}{2}\right) \approx 0.866.$$

Now, find the value of each answer to see which one matches. It is obviously not (1), (2), or (3), **so the answer must be (4).**

PROBLEM 26: INVERSES

We can find the inverse of an equation in two simple steps.

Step 1: Switch x and y. This gives us $x = 3y + 2$.

Step 2: Solve for y. This gives us $y = \dfrac{x-2}{3}$.

Here, the Regents exam doesn't even ask us to carry out step 2.

The answer is (2).

PROBLEM 27: TRIGONOMETRY GRAPHS

A graph of the form $y = a\sin bx$ or $y = a\cos bx$ has an amplitude of $|a|$ and a period of $\dfrac{2\pi}{b}$.

Therefore, the period of $y = 3\cos 4x = \dfrac{2\pi}{4} = \dfrac{\pi}{2}$

The answer is (2).

PROBLEM 28: QUADRATIC EQUATIONS

We can determine the nature of the roots of a quadratic equation of the form $ax^2 + bx + c = 0$ by using the discriminant $b^2 - 4ac$.

You should know the following rule:

If $b^2 - 4ac < 0$, the equation has two imaginary roots.

If $b^2 - 4ac = 0$, the equation has one rational root.

If $b^2 - 4ac > 0$, and $b^2 - 4ac$ is a perfect square, then the equation has two rational roots.

If $b^2 - 4ac > 0$, and $b^2 - 4ac$ is not a perfect square, then the equation has two irrational roots.

Here, the discriminant is $b^2 - 4ac < 0$, **so the answer is (4)**.

PROBLEM 29: TRIGONOMETRY

Notice that this problem has the form $\sin A \cos B + \cos A \sin B$. Each Regents exam comes with a formula sheet that contains all of the trig formulas that you will need to know. If you look under <u>Functions of the Sum of Two Angles</u>, you will find the formula $\sin (A+B) = \sin A \cos B + \cos A \sin B$. Thus we can rewrite the problem as

$$\sin 22° \cos 18° + \cos 22° \sin 18° = \sin\left(22° + 18°\right) = \sin 40°$$

The answer is (3).

Another way to answer this question was to use your calculator to find the value of $\sin 22° \cos 18° + \cos 2° \sin 18°$. **MAKE SURE THAT YOUR CALCULATOR IS IN DEGREE MODE.** You should get $\sin 22° \cos 18° + \cos 22° \sin 18° \approx 0.6428$. Now find the value of each of the answer choices with your calculator.

$\sin 4° \approx 0.0698$, so (1) is not the answer.

$\cos 4° \approx 0.9976$, so (2) is not the answer

$\sin 40° \approx 0.6428$, so (3) **is the answer**.

$\cos 40° \approx 0.7660$, so (4) is not the answer.

PROBLEM 30: TRIGONOMETRIC IDENTITIES

If we replace $\sec \theta$ with $\dfrac{1}{\cos \theta}$ and distribute, we get:

$$\cos \theta \left(\frac{1}{\cos \theta} - \cos \theta \right) = 1 - \cos^2 \theta.$$

Next, we replace 1 with $\sin^2\theta + \cos^2\theta$ (which is on the formula sheet) and we get:

$$1 - \cos^2\theta = \sin^2\theta + \cos^2\theta - \cos^2\theta = \sin^2\theta.$$

The answer is (3).

Another way to get this right is to **plug-in**. Make up a number for θ. Let's let $\theta = 30°$. Now, we plug $\theta = 30°$ into the problem and we get:

$$\cos 30°\left(\sec 30° - \cos 30°\right) = \frac{1}{4}.$$ (You find $\sec 30°$ by finding $\dfrac{1}{\cos 30°}$.)

Now, we plug $\theta = 30°$ into the answer choices and pick the one that matches.

Choice (1): is obviously not it.

Choice (2): $\sin 30° = \dfrac{1}{2}$. Wrong answer.

Choice (3): $\sin^2 30° = \dfrac{1}{4}$. **Right!**

Choice (4): $-\cos^2 30° = -\dfrac{3}{4}$. Wrong answer.

PROBLEM 31: TRANSFORMATIONS

The reflection in the x-axis means to change the sign of y. In other words, to turn the graph upside down. This would give you $-y = \cos x$ or $y = -\cos x$. This looks like:

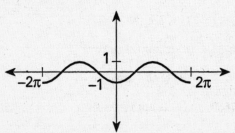

The answer is (1).

PROBLEM 32: BINOMIAL EXPANSIONS

The binomial theorem says that if you expand $(a + b)^n$, you get the following terms:

$$_nC_0a^n +_n C_1a^{n-1}b^1 +_n C_2a^{n-2}b^2 + ... +_n C_{n-2}a^2b^{n-2} +_n C_{n-1}a^1b^{n-1} +_n C_nb^n.$$

Therefore, if we expand $(1 - \pi)^6$, we get:

$$_6C_0(1)^6 +_6 C_1(1)^5(-\pi)^1 +_6 C_2(1)^4(-\pi)^2 +_6 C_3(1)^3(-\pi)^3 +$$
$$_6C_4(1)^2(-\pi)^4 +_6 C_5(1)^1(-\pi)^5 +_6 C_6(-\pi)^6$$

Next, we use the rule that $_nC_r = \dfrac{n!}{(n-r)!\, r!}$. This gives us:

$$(1 - \pi)^6 = (1)^6 + 6(1)^5(-\pi) + 15(1)^4(-\pi)^2 + 20(1)^3(-\pi)^3 +$$
$$15(1)^2(-\pi)^4 + 6(1)(-\pi)^5 + (-\pi)^6$$

which simplifies to

$$(1 - \pi)^6 = 1 - 6\pi + 15\pi^2 - 20\pi^3 + 15\pi^4 - 6\pi^5 + \pi^6.$$

The fifth term is $15\pi^4$.

The answer is (1).

A shortcut is to know the rule that the *fifth* term of the binomial expansion of $(a - b)^n$ is $_nC_{r-1}(a)^{n-r+1}(b)^{r-1}$. Thus the fifth term is:

$$_6C_{5-1}(1)^{6-5+1}(-\pi)^{5-1},$$

which can be simplified to $_6C_4(1)^2(-\pi)^4 = 15\pi^4$.

PROBLEM 33: QUADRATIC INEQUALITIES

When we are given a quadratic inequality, the first thing that we do is factor it. We get:

$$(x - 7)(x + 4) \geq 0$$

The roots of the quadratic are $x = 7$, $x = -4$. If we plot the roots on a number line

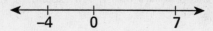

we can see that there are three regions: $x < -4$rf, $-4 < x < 7$, and $x > 7$ Now we try a point in each region to see whether the point satisfies the inequality.

First, we try a number less than –4. Let's try –5. Is $(-5 - 5)(-5 + 4) \geq 0$? Yes.

Now, we try a number between –4 and 7. Let's try 0. Is $(0 - 7)(0 + 4) \geq 0$? No.

Last, we try a number greater than 7. Let's try 8. Is $(8 - 7)(8 + 4) \geq 0$? Yes.

Therefore, the regions that satisfy the inequality are $x \leq -4$ or $x \geq 7$

The answer is (1).

PROBLEM 34: CONIC SECTIONS

First, divide the equation through by 5. We get·

$$\frac{x^2}{5} + \frac{2y^2}{5} = 1$$

A graph of the form $\frac{x^2}{a^2} + \frac{y^2}{b^2} = 1$, *where* **a** *and* **b** *have the same value, is the equation of a circle. If* **a** *and* **b** *have different values, the graph is that of an ellipse.*

The answer is (3).

PROBLEM 35: TRIGONOMETRY

We can evaluate $\sin(180° - x)$ using the <u>Functions of the Difference of Two Angles</u>, which is on the formula sheet. $\sin(A - B) = \sin A \cos B - \cos A \sin B$. Thus, we can rewrite the problem as $\sin(180° - x) = \sin 180° \cos x - \cos 180° \sin x$.

You should know from that $\sin 180° = 0$ and $\cos 180° = -1$ because they are special angles. Therefore, $\sin 180° \cos x - \cos 180° \sin x = \sin x$.

The answer is (1).

Another way to get this right is to **plug-in**. Make up a value for x. Let $x = 30°$. If we plug it into x in the problem, we get $\sin(180° - 30°) = \sin 150°$. Using the calculator, we find that $\sin 150° = 0.5$. Now, plug $x = 30°$ into each answer choice to see which one matches.

Choice (1): Does $\sin 30° = 0.5$? Yes.

Choice (2): Does $\cos 30° = 0.5$? No.

Choice (3): Does $-\sin 30° = 0.5$? No.

Choice (4): Does $-\cos 30° = 0.5$? No.

Part II

PROBLEM 36:

(a) Statistics

The method for finding a standard deviation is simple, but time-consuming.

First, find the average of the scores. We do this by multiplying each score by the number of people who obtained that score, and adding them up.

$(8)(5) + (12)(4) + (14)(3) + (4)(2) + (2)(1) = 140$. Next, we divide by the total number of people $(8 + 12 + 14 + 4 + 2) = 40$, to obtain the average $\overline{x} = 3.5$.

Second, subtract the average from each actual score.

$$5 - 3.5 = 1.5$$
$$4 - 3.5 = 0.5$$
$$3 - 3.5 = -0.5$$
$$2 - 3.5 = -1.5$$
$$1 - 3.5 = -2.5$$

Third, square each difference.

$$(1.5)^2 = 2.25$$

$$(0.5)^2 = 0.25$$

$$(-0.5)^2 = 0.25$$

$$(-1.5)^2 = 2.25$$

$$(-2.5)^2 = 6.25$$

Fourth, multiply the square of the difference by the corresponding number of people and sum.

$$(2.25)(8) = 18$$
$$(0.25)(12) = 3$$
$$(0.25)(14) = 3.5$$
$$(2.25)(4) = 9$$
$$(6.25)(2) = 12.5$$
$$18 + 3 + 3.5 + 9 + 12.5 = 46$$

Fifth, divide this sum by the total number of people.

$$\frac{46}{40} = 1.15.$$

Last, take the square root of this number. This is the standard deviation.

$$\sigma = \sqrt{1.15} \approx 1.072$$

Rounded to the nearest tenth, we get $\sigma \approx 1.1$

(b) Rational Expressions

First, invert the last expression and multiply instead of divide.

$$\frac{x^2 + 2x}{x^2 + 2x - 15} \cdot \frac{2x - 6}{4} \div \frac{x^2 + x - 2}{x^2 + 4x - 5} =$$

$$\frac{x^2 + 2x}{x^2 + 2x - 15} \cdot \frac{2x - 6}{4} \cdot \frac{x^2 + 4x - 5}{x^2 + x - 2}$$

Next, factor the numerators and denominators of each expression. We get:

$$\frac{x^2+2x}{x^2+2x-15} \cdot \frac{2x-6}{4} \cdot \frac{x^2+4x-5}{x^2+x-2} =$$

$$\frac{(x)(x+2)}{(x+5)(x-3)} \cdot \frac{(2)(x-3)}{4} \cdot \frac{(x+5)(x-1)}{(x+2)(x-1)}$$

Finally, cancel like terms.

$$\frac{(x)(x+2)}{(x+5)(x-3)} \cdot \frac{(2)(x-3)}{4} \cdot \frac{(x+5)(x-1)}{(x+2)(x-1)} = \frac{2x}{4} = \frac{x}{2}$$

PROBLEM 37:

(a) Trigonometric Identities

All of the identities that you will need to know are contained on the formula sheet.

First, let's get rid of the double-angle terms.

Using $\cos 2\theta = 2\cos^2\theta - 1$ and $\sin 2\theta = 2\sin\theta\cos\theta$, we can rewrite the left side of the equation as:

$$\frac{2\sin\theta\cos\theta + \sin\theta}{2\cos^2\theta - 1 + \cos\theta + 1} = \tan\theta.$$

Next, simplify the denominator:

$$\frac{2\sin\theta\cos\theta + \sin\theta}{2\cos^2\theta + \cos\theta} = \tan\theta.$$

Next, factor the numerator and the denominator:

$$\frac{\sin\theta(2\cos\theta + 1)}{\cos\theta((2\cos\theta + 1))} = \tan\theta.$$

Now, we cancel like terms: $\dfrac{\sin\theta}{\cos\theta} = \tan\theta.$

Finally, we can rewrite the left side as $\tan\theta$ and set the two sides equal, and we have proved the identity.

(b) Logarithms

You should know the following log rules:

(i) $\log A + \log B = \log(AB)$

(ii) $\log A - \log B = \log\left(\dfrac{A}{B}\right)$

(iii) $\log A^B = B \log A$

(1) First, we can use rule (iii) to rewrite $\log\sqrt{\dfrac{3}{7}}$ as $\dfrac{1}{2}\log\dfrac{3}{7}$.

Next, we can use rule (ii) to rewrite $\dfrac{1}{2}\log\dfrac{3}{7}$ as $\dfrac{1}{2}(\log 3 - \log 7)$.

Finally, we substitute $\log 7 = x$ and $\log 3 = y$, and we get:

$$\frac{1}{2}(\log 3 - \log 7) = \frac{1}{2}(y - x)$$

(2) If we can write 63 in terms of 3's and 7's, we can use the laws of logarithms to find the answer.

We know that $63 = 7 \cdot 9 = 7 \cdot 3^2$, so we can rewrite $\log 63$ as $\log(7 \cdot 3^2)$.

Now, we can use rule (i) to rewrite $\log(7 \cdot 3^2)$ as $\log 7 + \log 3^2$.

Next, we can use rule (iii) to rewrite $\log 7 + \log 3^2$ as $\log 7 + 2\log 3$.

Finally, we substitute $\log 7 = x$ and $\log 3 = y$, and we get:

$$\log 7 + 2\log 3 = x + 2y$$

PROBLEM 38:

(a) Quadratic Equations

If we move all of the terms to the left side, we get $x^2 + 1 - 8(x - 3) = 0$.

Next, we distribute the 8:

$$x^2 + 1 - 8x + 24 = 0$$

Simplify:

$$x^2 - 8x + 25 = 0$$

We can now use the quadratic formula to solve for x.

The formula says that, given a quadratic equation of the form $ax^2 + bx + c = 0$, the roots of the equation are:

$$x = \frac{-b \pm \sqrt{b^2 - 4ac}}{2a}.$$

Plugging in to the formula we get:

$$x = \frac{-(-8) \pm \sqrt{(-8)^2 - 4(1)(25)}}{2(1)} = \frac{8 \pm \sqrt{64 - 100}}{2} =$$

$$\frac{8 \pm \sqrt{-36}}{2} = \frac{8 \pm 6i}{2} = \frac{8}{2} \pm \frac{6}{2} i = 4 \pm 3i$$

(b) Trigonometric Equations

First, subtract 1 from both sides. $4\cos^2\theta - 3\cos\theta - 1 = 0$.

Note that the trig equation $4\cos^2\theta - 3\cos\theta - 1 = 0$ has the same form as a quadratic equation $4x^2 - 3x - 1 = 0$. Just as we could factor the quadratic equation, so too can we factor the trig equation. We get:

$$(4\cos\theta + 1)(\cos\theta - 1) = 0.$$

This gives us $4\cos\theta + 1 = 0$ and $\cos\theta - 1 = 0$. If we solve each of these equations, we get:

$$\cos\theta = -\frac{1}{4} \quad \text{and} \quad \cos\theta = 1.$$

Therefore $\theta = \cos^{-1}\left(-\frac{1}{4}\right)$ and $\theta = \cos^{-1}(1)$. The answer to the second equation is easy. This is a special angle and you should know that $\cos 0°$ is 1, so one answer is $\theta = 0°$.

There is not a special angle whose cosine is $-\frac{1}{4}$, so we need to use the calculator to find the value. Using the calculator, push **2nd COS (–1/4) ENTER. (MAKE SURE THAT THE CALCULATOR IS IN DEGREE MODE!)** This gives us $104.48° \approx 104°$ as an answer, but, we are asked to find all of the values of θ between $0°$ and $360°$.

We can find the required values of θ by making a small graph and plotting the reference angle of 180° − 104° = 76°.

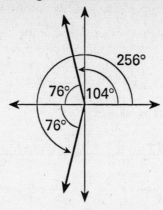

The other place where cosine is negative and has a reference angle of 76° is in quadrant III. The other angle then is 180 + 76° = 256°.

Therefore, the answers are θ = 0°, 104°, and 256°.

PROBLEM 39: CIRCLE RULES

(a) We are given that AC is a diameter, therefore $m\overset{\frown}{AC} = 180$. Because $m\overset{\frown}{BC} = 100$, we know that $m\overset{\frown}{AB} = 180 - 100 = 80$.

(b) *The measure of an angle formed by a pair of secants, or a secant and a tangent, is equal to half of the difference between the larger and the smaller arcs that are formed by the secants, or the secant and the tangent.*

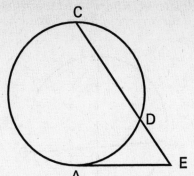

Here, the larger and smaller arcs formed by angle AEC are $\overset{\frown}{AC}$ and $\overset{\frown}{AD}$, respectively.

$$m\angle AEC = \frac{m\overset{\frown}{AC} - m\overset{\frown}{AD}}{2} = \frac{180 - 70}{2} = 55.$$

(c) *The measure of an inscribed angle is half of the arc it subtends (intercepts).*

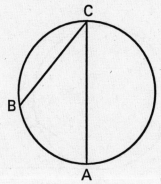

Therefore, $m\angle BCA = \frac{1}{2} m\overset{\frown}{AB} = \frac{1}{2} 80 = 40.$

(d) *If two chords intersect within a circle, then the angle between the two chords is the average of their intercepted arcs.*

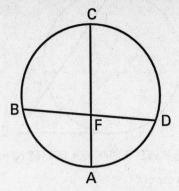

Therefore, $m\angle DFA = \dfrac{m\widehat{CB} + m\widehat{AD}}{2} = \dfrac{100 + 70}{2} = 85$.

(e) *An angle formed by a tangent and a chord is equal to half of its intercepted arc.*

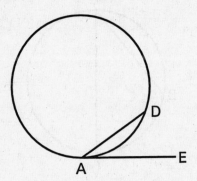

Therefore, $m\angle DAE = \dfrac{1}{2}\, m\widehat{AD} = \dfrac{1}{2}\, 70 = 35$.

Formulas

Pythagorean and Quotient Identities

$$\sin^2 A + \cos^2 A = 1 \qquad \tan A = \frac{\sin A}{\cos A}$$

$$\tan^2 A + 1 = \sec^2 A \qquad \cot A = \frac{\cos A}{\sin A}$$

$$\cot^2 A + 1 = \csc^2 A$$

Functions of the Sum of Two Angles

$$\sin (A + B) = \sin A \cos B + \cos A \sin B$$

$$\cos (A + B) = \cos A \cos B - \sin A \sin B$$

$$\tan (A + B) = \frac{\tan A + \tan B}{1 - \tan A \tan B}$$

Functions of the Difference of Two Angles

$$\sin (A - B) = \sin A \cos B - \cos A \sin B$$

$$\cos (A - B) = \cos A \cos B + \sin A \sin B$$

$$\tan (A - B) = \frac{\tan A - \tan B}{1 + \tan A \tan B}$$

Law of Sines

$$\frac{a}{\sin A} = \frac{b}{\sin B} = \frac{c}{\sin C}$$

Law of Cosines

$$a^2 = b^2 + c^2 - 2bc \cos A$$

Functions of the Double Angle

$$\sin 2A = 2 \sin A \cos A$$

$$\cos 2A = \cos^2 A - \sin^2 A$$

$$\cos 2A = 2 \cos^2 A - 1$$

$$\cos 2A = 1 - 2 \sin^2 A$$

$$\tan 2A = \frac{2 \tan A}{1 - \tan^2 A}$$

Functions of the Half Angle

$$\sin \frac{1}{2} A = \pm\sqrt{\frac{1 - \cos A}{2}}$$

$$\cos \frac{1}{2} A = \pm\sqrt{\frac{1 + \cos A}{2}}$$

$$\tan \frac{1}{2} A = \pm\sqrt{\frac{1 - \cos A}{1 + \cos A}}$$

Area of Triangle

$$K = \frac{1}{2} ab \sin C$$

Standard Deviation

$$S.D. = \sqrt{\frac{1}{n} \sum_{i=1}^{n} \left(x_i - \bar{x}\right)^2}$$

[OVER]

EXAMINATION
JUNE 1996

Part I

Answer 30 questions from this part. Each correct answer will receive 2 credits. No partial credit will be allowed. Write your answers in the spaces provided. Where applicable, answers may be left in terms of π or in radical form. [60]

1 If $f(x) = \sqrt{25 - x^2}$, find the value of $f(3)$.

$F(3) = \sqrt{25 - 3^2}$ $F(3) = \sqrt{16}$
$F(3) = \sqrt{25 - 9}$ $F(3) = 4$

2 An angle that measures $\dfrac{5\pi}{6}$ radians is drawn in standard position. In which quadrant does the terminal side of the angle lie?

$\dfrac{5\pi}{6} \cdot \dfrac{180^{30}}{\pi} = 150$ II

3 In the accompanying diagram, isosceles triangle ABC is inscribed in circle O and $m\angle BAC = 40$.

Find $m\widehat{AC}$.

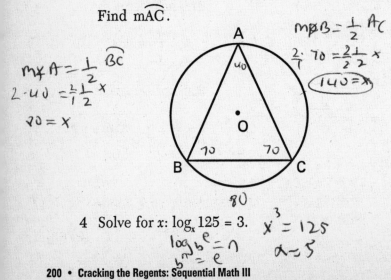

$m\angle A = \frac{1}{2} \widehat{BC}$
$2 \cdot 40 = \frac{2}{1} \frac{1}{2} x$
$80 = x$

$m\angle B = \frac{1}{2} \widehat{AC}$
$2 \cdot 70 = \frac{2}{2} \frac{1}{2} x$
$\boxed{140 = x}$

4 Solve for x: $\log_x 125 = 3$. $x^3 = 125$
$\log_b e = n$ $x = 5$
$b^n = e$

5 Point (–3,4) is rotated 180° about the origin in a counterclockwise direction. What are the coordinates of its image?

(handwritten) r^{180} $(-x, -y)$
$(3, -4)$

6 For which positive value of x is the function

$$f(x) = \frac{5x}{x^2 - 4x - 45} \text{ undefined?}$$

7 Solve for x: $8^x = 2^{(x+6)}$.

(handwritten)
$2^{3x} = 2^{x+6}$ $3x = x+6$
$-x + 3x = 6$
$2x = 6$ $x = 3$

8 If $h(x) = 2x - 1$ and $g(x) = 3x + 1$, what is $(h \circ g)(2)$?

(handwritten)
$g(2) = 3(2) + 1$ $g(2) = 7$ $h(7) = 2(7) - 1$
$g(2) = 6 + 1$ $h(7) = 13$

9 Subtract $(3 - 2i)$ from $(-2 + 3i)$, and express in $a + bi$ form.

(handwritten)
$-2 + 3i$
$3 - 2i$ $-2 + 3i + (-3 + 2i)$
$-5 + 5i$

10 In $\triangle ABC$, $a = 8$, $b = 7$, and $m\angle C = 30$. What is the area of $\triangle ABC$?

(handwritten)
$K = \frac{1}{2}(8)(7) \cdot \sin 30$
$K = \frac{56}{2} \cdot \sin 30$ $K = 28 \cdot \sin 30$
$K = 28 \cdot \frac{1}{2}$
$K = 14$

11 Evaluate: $\displaystyle\sum_{r=1}^{3} r^{(r-1)}$.

(handwritten)
$1^0 + 2^1 + 3^2$
$1 + 2 + 27 = 30$

12 Chords $\overline{XY}$ and $\overline{ZW}$ intersect in a circle at P. If $XP = 7$, $PY = 12$, and $WP = 14$, find PZ.

(handwritten circle diagram)

13 Find the number of degrees in the measure of the *smallest* positive angle that satisfies the equation $2 \cos x + 1 = 0$.

[OVER]

14 Find the complete solution set of $|2x - 4| = 8$.

15 In the accompanying diagram, $\overline{AFB}$, $\overline{AEC}$, and $\overline{BGC}$ are tangent to circle O at F, E, and G, respectively. If $AB = 32$, $AE = 20$, and $EC = 24$, find BC.

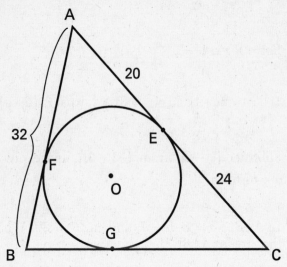

Directions (16–35): For *each* question chosen, write the *numeral* preceding the word or expression that best completes the statement or answers the question.

16 The expression $\dfrac{\sqrt{-36}}{-\sqrt{36}}$ is equivalent to

(1) $6i$ (3) $-i$

(2) i (4) $-6i$

17 If $\sin \theta < 0$ and $\tan \theta = -\dfrac{4}{5}$, in which quadrant does θ terminate?

(1) I   (3) III
(2) II   (4) IV

18 The expression $\dfrac{\dfrac{a-1}{a}}{\dfrac{a^2-1}{a^2}}$ is equivalent to

(1) $\dfrac{a}{a+1}$ (3) $\dfrac{a}{a-1}$

(2) $\dfrac{a+1}{a}$ (4) $\dfrac{a-1}{a}$

19 The roots of the equation $x^2 + 7x - 8 = 0$ are

(1) real, rational, and equal
(2) real, rational, and unequal
(3) real, irrational, and unequal
(4) imaginary

20 The product of $(-2 + 6i)$ and $(3 + 4i)$ is

(1) $-6 + 24i$ (3) $18 + 10i$
(2) $-6 - 24i$ (4) $-30 + 10i$

21 If $\sin 2A = \cos 3A$, then $m\angle A$ *is*

(1) $1\dfrac{1}{2}$ (3) 18

(2) 5   (4) 36

[OVER]

22 The graph of the equation $x = \dfrac{2}{y}$ is best described as

(1) a circle   (3) a hyperbola
(2) an ellipse   (4) a parabola

23 The accompanying diagram represents the graph of f(x)

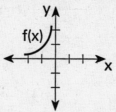

Which graph below represents $f^{-1}(x)$?

(1)
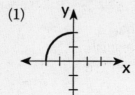

(3)
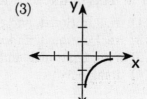

(2)
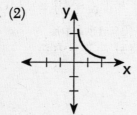

(4)
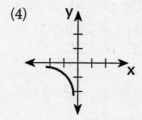

24 What is the domain of $f(x) = \sqrt{x - 4}$ over the set of real numbers?

(1) $\{x \mid x \leq 4\}$ (3) $\{x \mid x > 4\}$

(2) $\{x \mid x \geq 4\}$ (4) $\{x \mid x = 4\}$

25 What is the solution set of the equation $\sqrt{5 - x} + 3 = x$?

(1) $\{1\}$ (3) $\{\ \}$

(2) $\{4, 1\}$ (4) $\{4\}$

26 If $\log 28 = \log 4 + \log x$, what is the value of x?

(1) 7   (3) 24

(2) 14   (4) 32

27 If $a = 4$, $b = 6$, and $\sin A = \dfrac{3}{5}$ in $\triangle ABC$, then $\sin B$ equals

(1) $\dfrac{3}{20}$ (3) $\dfrac{8}{10}$

(2) $\dfrac{6}{10}$ (4) $\dfrac{9}{10}$

28 What is the image of $(5, -2)$ under the transformation $r_{y=x}$?

(1) $(-5, 2)$ (3) $(2, 5)$

(2) $(5, 2)$ (4) $(-2, 5)$

[OVER]

29 Each day the probability of rain on a tropical island is $\frac{7}{8}$. Which expression represents the probability that it will rain on the island exactly n days in the next 3 days?

(1) $_3C_n\left(\frac{7}{8}\right)^n\left(\frac{1}{8}\right)^{3-n}$

(2) $_3C_3\left(\frac{7}{8}\right)^3\left(\frac{1}{8}\right)^n$

(3) $_nC_3\left(\frac{7}{8}\right)^3\left(\frac{1}{8}\right)^n$

(4) $_8C_7(3)^n(3)^{8-n}$

30 Which graph represents the solution of the inequality $x^2 + 4x - 21 < 0$?

(1)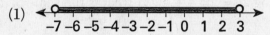

(2)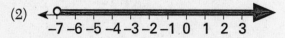

(3)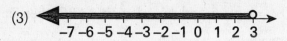

(4)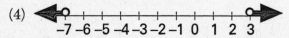

31 On a standardized test, the mean is 68 and the standard deviation is 4.5. What is the best approximation of the percentage of scores that will fall in the range 59–77?

(1) 34% (3) 95%

(2) 68% (4) 99%

32 In $\triangle ABC$, $a = 6$, $b = 4$, and $c = 9$. The value of $\cos C$ is

(1) $\dfrac{61}{72}$ (3) $\dfrac{2}{3}$

(2) $\dfrac{-29}{48}$ (4) $\dfrac{4}{9}$

33 If $m\angle A = 125$, AB = 10, and BC = 12, what is the number of distinct triangles that can be constructed?

(1) 1   (3) 3
(2) 2   (4) 0

34 The graph of which function has an amplitude of 2 and a period of 4π?

(1) $y = 2 \sin \dfrac{1}{2} x$ (3) $y = 4 \sin \dfrac{1}{2} x$

(2) $y = 2 \sin 4x$ (4) $y = 4 \sin 2x$

35 What is the sum of the roots of the equation $2x^2 + 6x - 7 = 0$?

(1) $-\dfrac{7}{2}$ (3) 3

(2) -3 (4) $\dfrac{7}{2}$

[OVER]

Part II

Answer *four* questions from this part. Clearly indicate the necessary steps, including appropriate formula substitutions, diagrams, graphs, etc. Calculations that may be obtained by mental arithmetic or the calculator do not need to be shown.

36 In the accompanying diagram of circle O with inscribed isosceles triangle ABC, $\overline{AB} \cong \overline{AC}$, $m\widehat{CB} = 60$, $\overline{FC}$ is a tangent, and secant $\overline{FBA}$ intersects diameter $\overline{CD}$ at E.

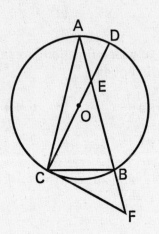

Find:

a	$m\angle ABC$	[2]
b	$m\widehat{AD}$	[2]
c	$m\angle DEB$	[2]
d	$m\angle AFC$	[2]
e	$m\angle BCF$	[2]

37 *a* On graph paper, sketch the graph of the equation $y = 2 \cos x$ in the interval $-\pi \le x \le \pi$. [4]

    b On the same set of axes, reflect the graph drawn in part *a* in the *x*-axis and label it *b*. [2]

    c Write an equation of the graph drawn in part *b*. [2]

    d Using the equation from part *c*, find the value of *y* when $x = \dfrac{\pi}{6}$. [2]

38 Find, *to the nearest degree*, all values of *x* in the interval $0° \le x \le 360°$ that satisfy the equation $3 \cos 2x + \cos x + 2 = 0$. [10]

39 In a contest, the probability of the Alphas beating the Betas is $\dfrac{3}{5}$. The teams compete four times a season and each contest has a winner. Find the probability that

    a the Betas win all four contests [2]

    b each team wins two contests during the season [2]

    c the Alphas win *at least* two contests during the season [3]

    d the Betas win *at most* one contest during the season [3]

[OVER]

40 Answer both a and b.

a For all values of x for which the expressions are defined, prove that the following is an identity:

$$\tan x + \cot x = 2 \csc 2x \quad [6]$$

b Given: $\log 2 = x$ and $\log 3 = y$.

(1) Express $\log \dfrac{\sqrt{2}}{9}$ in terms of x and y. [2]

(2) Express $\log \sqrt[3]{6}$ in terms of x and y. [2]

41 Answer both a and b.

a Expand and express in simplest form:

$$\left(x - \frac{1}{x} \right)^4 \quad [7]$$

b Solve for x to the *nearest tenth*:

$$5^{3x} = 1000 \quad [3]$$

42 The lengths of the sides of $\triangle ABC$ are 9.5, 12.8, and 13.7.

a Find, to the *nearest hundredth of a degree* or the *nearest ten minutes*, the measure of the *smallest* angle in the triangle. [6]

b Find, to the *nearest tenth*, the area of $\triangle ABC$. [4]

ANSWER KEY

Part I

1. 4	13. 120	25. (4)
2. II	14. −2, 6	26. (1)
3. 140	15. 36	27. (4)
4. 5	16. (3)	28. (4)
5. (3, −4)	17. (4)	29. (1)
6. 9	18. (1)	30. (1)
7. 3	19. (2)	31. (3)
8. 13	20. (4)	32. (2)
9. −5 + 5i	21. (3)	33. (1)
10. 14	22. (3)	34. (1)
11. 12	23. (3)	35. (2)
12. 6	24. (2)	

[OVER]

ANSWERS AND EXPLANATIONS
JUNE 1996

Part I

In problems 1–15, you must supply the answer. Because of this, you will not be able to **plug-in** or **backsolve**. But, you can use your calculator to simplify calculations. In addition, you will find that these problems tend to test your basic knowledge of Sequential Math III and are not tricky. Many can be solved just by knowing the right formula and by plugging in the numbers.

PROBLEM 1: RADIACAL EQUATIONS

All that you have to do here is to plug in 3 for x. You get:

$$\sqrt{25 - x^2} = \sqrt{25 - 3^2} = \sqrt{16} = 4$$

PROBLEM 2: CONVERTING FROM RADIANS TO DEGREES

First, let's convert the angle from radians to degrees. We simply multiply the angle by $\dfrac{180°}{\pi}$. We get:

$$\frac{5\pi}{6} \frac{180°}{\pi} = 150°$$

Next, to figure out which quadrant the angle is in, begin at the positive x-axis and move counterclockwise. The angle will terminate in quadrant II.

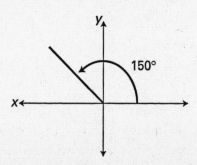

PROBLEM 3: CIRCLE RULES

The measure of an inscribed angle is always half of the measure of the arc that it subtends (intercepts). So, to find the measure of arc AC, we need to find $m\angle ABC$.

Here, we are given that $m\angle BAC = 40$, so $m\angle ABC + m\angle ACB = 180 - 40 = 140$. Next, because triangle ABC is isosceles $m\angle ACB = m\angle ABC$. This means that $m\angle ABC = 70$. Therefore, $m\overarc{AC} = 140$.

PROBLEM 4: LOGARITHMS

The definition of a logarithm is that $\log_b x = a$ means $b^a = x$.

Here, $\log_x 125 = 3$ means that $x^3 = 125$ and that $x = 5$.

PROBLEM 5: TRANSFORMATIONS

$R_{180°}$ means a rotation counterclockwise of 180°. You should know the following rotation rules:

$$R_{90°}(x, y) = (-y, x)$$
$$R_{180°}(x, y) = (-x, -y)$$
$$R_{270°}(x, y) = (y, -x)$$

Therefore, $R_{180°}(-3, 4) = (3, -4)$.

PROBLEM 6: RATIONAL EXPRESSIONS

A function is undefined where its denominator equals zero. First, let's factor the denominator. We get:

$$\frac{5x}{x^2 - 4x - 45} = \frac{5x}{(x-9)(x+5)}.$$

If we set the denominator equal to zero we get:

$$(x-9)(x+5) = 0 \text{ and } x = 9 \text{ or } x = -5.$$

Therefore, the positive value for which the function is undefined is $x = 9$.

Problem 7: Exponential Equations

$$8^x = 2^{x+6}.$$

In order to solve this equation, we need to have a common base for the two sides of the equation. Because $2^3 = 8$, we can rewrite the left side as $(2^3)^x = 2^{x+6}$.

The rules of exponents say that *when a number raised to a power is itself raised to a power, we multiply the exponents*. Thus, we get:

$$2^{3x} = 2^{x+6}$$

Now, we can set the powers equal to each other and solve.

$$3x = x + 6$$
$$x = 3$$

Problem 8: Composite Functions

With composite functions, first we work with the function on the right and then the function on the left.

Here, we first plug 2 into $g(x)$. We obtain $g(2) = 3(2) + 1 = 7$.

Next, we plug 7 into $h(x)$. We obtain $h(7) = 2(7) - 1 = 13$.

Problem 9: Complex Numbers

To find $(-2 + 3i) - (3 - 2i)$ we subtract the real components and the imaginary components separately. This gives us:

$$((-2) - 3) + (3 - (-2))i = -5 + 5i.$$

Problem 10: Trigonometric Area

The area of a triangle, if we are given the lengths of the two sides a and b, and their included angle θ, is $A = \frac{1}{2}ab\sin\theta$. In other words, if we are given SAS (side, angle, side), we find the area by multiplying $\frac{1}{2}$ by the product of the two sides by the sine of the included angle. Here, the two sides are a and b and the included angle is C. Plugging in, we get:

$$A = \frac{1}{2}(8)(7)\sin 30° = 28\left(\frac{1}{2}\right) = 14.$$

Problem 11: Summations

To evaluate a sum, we plug each of the consecutive values for r into the expression, starting at the bottom value and ending at the top value and sum the results. We get:

$$\sum_{r=1}^{3} r^{(r-1)} = \left[1^{(1-1)}\right] + \left[2^{(2-1)}\right] + \left[3^{(3-1)}\right] = 1^0 + 2^1 + 3^2 = 1 + 2 + 9 = 12$$

Another way to do this is to use your calculator. Push **2nd STAT (LIST)** and under the **MATH** menu, choose **SUM**. Then, again push **2nd STAT** and under the **OPS** menu, choose **SEQ**.

We put in the following: (*expression, variable, start, finish, step*).

For *expression*, we put in the formula, using x as the variable.

For *variable*, we ALWAYS put in x.

For *start*, we put in the bottom value.

For *finish*, we put in the top value.

For *step*, we ALWAYS put in 1.

Therefore, we put in the calculator: **SUM SEQ** (x^$(x-1), x, 1, 3, 1$). The calculator should return the value 12.

Problem 12: Circle Rules

If two chords intersect in a circle, then the product of the lengths of the segments of one chord is equal to the product of the lengths of the segments of the other chord.

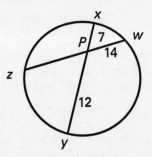

This means that $(XP)(PY) = (WP)(PZ)$. If we plug in, we get:

$$(7)(12) = (14)(PZ)$$

We can solve this for PZ and we get:

$$(PZ) = 6.$$

PROBLEM 13: TRIGONOMETRIC EQUATIONS

Note that the trig equation $2 \cos x + 1 = 0$ has the same form as the equation $2x + 1 = 0$. Just as we can solve the latter equation, so too can we solve the former.

$$\cos x = -\frac{1}{2}$$

You should know the values of x between $0°$ and $360°$ for which this is true because they are special angles. They are $x = 120°$ and $x = 240°$. We are asked for the smallest positive angle, so the answer is $x = 120°$.

PROBLEM 14: ABSOLUTE VALUE EQUATIONS

When we have an absolute value equation we break it into two equations. First, rewrite the left side without the absolute value and leave the right side alone. Second, rewrite the left side without the absolute value, and change the sign of the right side.

Here, we have $|2x - 4| = 8$, so we rewrite it as:

$$2x - 4 = 8 \text{ or } 2x - 4 = -8.$$

If we solve each of these independently, we get·

$$x = 6 \text{ or } x = -2.$$

PROBLEM 15: CIRCLE RULES

The two tangents drawn to a circle from the same point are congruent. This means that $AF = AE$, $BF = BG$, and $CE = CG$.

Thus, if $AE = 20$ then $AF = 20$; and if $CE = 24$ then $CG = 24$.

We also know that, because $AB = 32$ and $AF = 20$, then $BF = BG = 12$.

Therefore, $BC = BG + CG = 24 + 12 = 36$

Multiple Choice Problems

For problems 16–35, you will find that you can sometimes **plug-in** or **backsolve** to get the right answer. You will also find that a calculator will simplify the arithmetic. In addition, most of the problems only require knowing which formula to use.

PROBLEM 16: COMPLEX NUMBERS

In complex numbers, also referred to as *imaginary numbers*, $\sqrt{-1} = i$.

This means that $\sqrt{-36} = \sqrt{36}\sqrt{-1} = 6i$

Now we can simplify the fraction to $\dfrac{6i}{-\sqrt{36}} = \dfrac{6i}{-6} = -i$.

The answer is (3).

PROBLEM 17: TRIGONOMETRY

The sine of an angle is negative in quadrants III and IV.

The tangent of an angle is negative in quadrants II and IV.

Therefore, the angle must be in quadrant IV.

The answer is (4).

PROBLEM 18: RATIONAL EXPRESSIONS

First, invert the bottom fraction and multiply it by the top one. We get:

$$\left(\frac{a-1}{a}\right)\left(\frac{a^2}{a^2-1}\right)$$

Next, factor the denominator:

$$\left(\frac{a-1}{a}\right)\left(\frac{a^2}{(a-1)(a+1)}\right)$$

Now, cancel like terms:

$$\left(\frac{a-1}{a}\right)\left(\frac{a^2}{(a-1)(a+1)}\right) = \frac{a}{a+1}$$

The answer is (1).

Another way to get this right is to **plug-in**. First, make up a number for a. Let's use $a = 3$. If we plug it into the problem, we get:

$$\frac{\left(\dfrac{(3)-1}{(3)}\right)}{\left(\dfrac{(3)^2-1}{(3)^2}\right)} = \frac{\dfrac{2}{3}}{\dfrac{8}{9}} = \left(\frac{2}{3}\right)\left(\frac{9}{8}\right) = \frac{18}{24} = \frac{3}{4}$$

Now, plug $a = 3$ into the answer choices to see which one matches. Always plug into all four answer choices because it's possible that more than one will give you the same answer. If so, repeat the plug in with a different value for a.

Choice (1): $\dfrac{a}{a+1} = \dfrac{3}{4}$

Choice (2): $\dfrac{a+1}{a} = \dfrac{4}{3}$

Choice (3): $\dfrac{a}{a-1} = \dfrac{3}{2}$

Choice (4): $\dfrac{a-1}{a} = \dfrac{2}{3}$

Choice (1) matches so the answer is (1). Wasn't that easy? Notice how we were able to get the question right without doing the algebra!

PROBLEM 19: QUADRATIC EQUATIONS

We can determine the nature of the roots of a quadratic equation of the form $ax^2 + bx + c = 0$ by using the discriminant $b^2 - 4ac$.

You should know the following rule:

If $b^2 - 4ac < 0$, the equation has two imaginary roots.

If $b^2 - 4ac = 0$, the equation has one rational root.

If $b^2 - 4ac > 0$, and $b^2 - 4ac$ is a perfect square, then the equation has two rational roots.

If $b^2 - 4ac > 0$, and $b^2 - 4ac$ is not a perfect square, then the equation has two irrational roots.

Here, the discriminant is $b^2 - 4ac = 7^2 - 4(1)(-8) = 49 + 32 = 81$.

81 is a perfect square, **so the answer is (2).**

PROBLEM 20: COMPLEX NUMBERS

When we want to find the product or two complex numbers, the first thing we do is FOIL:

$$(-2 + 6i)(3 + 4i) = -6 - 8i + 18i + 24i^2.$$

Next we combine the two middle terms:

$$-6 - 8i + 18i + 24i^2 = -6 + 10i + 24i^2.$$

Next, because $i^2 = -1$, the last term becomes $-6 + 10i - 24$.

This can be simplified to $-30 + 10i$.

The answer is (4).

PROBLEM 21: TRIGONOMETRY

You should know that $\sin x = \cos(90 - x)$ and that $\cos x = \sin(90 - x)$. We could use either substitution, so if we use the latter substitution in the equation we get:

$$\sin(2A) = \sin(90 - 3A).$$

Therefore $(2A) = (90 - 3A)$. If we solve for A, we get $A = 18°$.

The answer is (3).

Another way to answer this question is **backsolve.**

Try each of the answer choices in the equation to see which one works.

Does $\sin(2(1.5)) = \cos(3(1.5))$? $0.0523 \neq 0.9969$

Does $\sin(2(5)) = \cos(3(5))$? $0.1736 \neq 0.9659$

Does $\sin(2(18)) = \cos(3(18))$? $0.5878 = 0.5878$

Does $\sin(2(36)) = \cos(3(36))$? $0.9511 \neq -0.3090$

Choice (3) works, so that is the answer.

PROBLEM 22: CONIC SECTIONS

A graph of the form $y = \dfrac{k}{x}$ or $x = \dfrac{k}{y}$, where k is a constant, is a hyperbola.

The answer is (3).

PROBLEM 23: INVERSES

The graph of the inverse of a function $f(x)$ is the reflection of the graph in the line $y = x$.

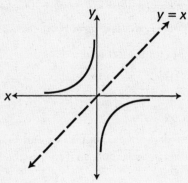

The answer is (3).

PROBLEM 24: RADICAL EQUATIONS

The domain of a square root is found by setting the inside of the radical (called the *radicand*) greater than or equal to zero, so that the square root will be real.

Here we have: $x - 4 \geq 0$, so the domain is $x \geq 4$.

The answer is (2).

PROBLEM 25: RADICAL EQUATIONS

First, subtract 3 from both sides of the equation. $\sqrt{5 - x} = x - 3$

Next, square both sides. $5 - x = (x - 3)^2 = x^2 - 6x + 9$.

Put all of the terms on one side of the equals sign: $x^2 - 5x + 4 = 0$

Factor: $(x-4)(x-1) = 0$

This means that $x - 4 = 0$ or $x - 1 = 0$.

Therefore, $x = 4$ or $x = 1$.

HOWEVER, whenever we have a radical in an equation, we have to check the roots to see if they satisfy the original equation. Sometimes, in the process of squaring both sides, we will get answers that are invalid. So let's check the answer.

Does $\sqrt{5-4} + 3 = 4$? $1 + 3 = 4$; so, yes, it does.

Does $\sqrt{5-1} + 3 = 1$? $2 + 3 \neq 1$; so, no, it does not.

Therefore, the only solution is $x = 4$.

The answer is (4).

Another way to get the right answer is to **backsolve**. Try each of the values for x in the answer choices to see which one works.

Does $\sqrt{5-1} + 3 = 1$? $2 + 3 \neq 1$. No, it does not.

We don't need to try choice (2) because we already know that 1 is not a solution.

Next, we try choice (4). If it doesn't work, the answer has to be (3).

Does $\sqrt{5-4} + 3 = 4$? $1 + 3 = 4$; yes, it does.

PROBLEM 26: LOGARITHMS

One of the log rules says that $\log A + \log B = \log(AB)$. So, here, we can rewrite the right side of the equation as: $\log 4 + \log x = \log(4x)$.

This gives us:

$$\log 28 = \log(4x). \text{ Thus, } 28 = 4x \text{ and } x = 7.$$

The answer is (1).

Another way to get the right answer is to **backsolve**. Try each of the values for x in the answer choices to see which one works.

Does $\log 28 = \log 4 + \log 7$? $1.447 = 0.602 + 0.845 = 1.447$. Yes, it does.

Does $\log 28 = \log 4 + \log 14$? $1.447 = 0.602 + 1.146$. No, it does not.

Does $\log 28 = \log 4 + \log 24$? $1.447 = 0.602 + 1.380$. No, it does not.

Does $\log 28 = \log 4 + \log 32$? $1.447 = 0.602 + 1.505$. No, it does not.

We didn't really need to test all four choices here. Since each logarithm has a unique value, once the first answer choice worked, the others couldn't have. But you should develop the habit of trying all of the choices anyway.

PROBLEM 27: LAW OF SINES

The Law of Sines says that, in any triangle, with angles A, B, and C, and opposite sides of a, b, and c, respectively, $\dfrac{a}{\sin A} = \dfrac{b}{\sin B}$. (This is on the formula sheet.)

Here, $\dfrac{4}{\dfrac{3}{5}} = \dfrac{6}{\sin B}$.

We can easily solve this.

$$\frac{20}{3} = \frac{6}{\sin B}$$
$$20 \sin B = 18$$
$$\sin B = \frac{18}{20} = \frac{9}{10}$$

The answer is (4).

PROBLEM 28: TRANSFORMATIONS

Here we are asked to find the reflection of a point about the line $y = x$. You should know this rule: $r_{y=x}(x, y) \rightarrow (y, x)$. In other words, you just switch x and y.

Therefore, $r_{y=x}(5, -2) \rightarrow (-2, 5)$.

The answer is (4).

PROBLEM 29: PROBABILITY

*If the probability of a particular outcome is **p**, then the probability of that outcome occurring **r** times out of a possible **n** times is* $_nC_r(p)^r(1-p)^{n-r}$. This is known as *binomial probability.*

Here, we are told that the probability of rain is $\frac{7}{8}$. Therefore, the probability of rain n out of the next 3 days is

$$_3C_n\left(\frac{7}{8}\right)^n\left(1-\frac{7}{8}\right)^{3-n} =\, _3C_n\left(\frac{7}{8}\right)^n\left(\frac{1}{8}\right)^{3-n}.$$

The answer is (1).

PROBLEM 30: QUADRATIC INEQUALITIES

When we are given a quadratic inequality, the first thing that we do is factor it. We get:

$$(x+7)(x-3) < 0.$$

The roots of the quadratic are $x = 3$, $x = -7$. If we plot the roots on a number line

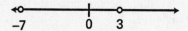

$$-7 \qquad 0 \quad 3$$

we can see that there are three regions:

$$x < -7, \ -7 < x < 3, \text{ and } x > 3$$

Now we try a point in each region to see whether the point satisfies the inequality.

First, we try a number less than -7. Let's try -8. Is $(-8+7)(-8-3) < 0$? No.

Now, we try a number between -7 and 3. Let's try 0. Is $(0+7)(0-3) < 0$? Yes.

Last, we try a number greater than 3. Let's try 4. Is $(4+7)(4-3) < 0$? No.

Therefore, the region that satisfies the inequality is $-7 < x < 3$.

The answer is (1).

PROBLEM 31: STATISTICS

You should know the following rule about normal distributions.

> In a normal distribution, with a mean of $\bar{x}$ and a standard deviation of σ:
> - approximately 68% of the outcomes will fall between $\bar{x} - \sigma$ and $\bar{x} + \sigma$.
> - approximately 95% of the outcomes will fall between $\bar{x} - 2\sigma$ and $\bar{x} + 2\sigma$.
> - approximately 99.5% of the outcomes will fall between $\bar{x} - 3\sigma$ and $\bar{x} + 3\sigma$.

Here, $\bar{x} = 68$ and $\sigma = 4.5$, so $59 = \bar{x} - 2\sigma$ and $77 = \bar{x} + 2\sigma$. Therefore, 95% of the scores should lie between 59 and 77.

The answer is (3).

PROBLEM 32: LAW OF COSINES

The Law of Cosines says that, in any triangle, with angles A, B, and C, and opposite sides of a, b, and c, respectively, $c^2 = a^2 + b^2 - 2ab \cos C$. (This is on the formula sheet.) If we have a triangle and we are given SSS (side, side, side), then the side opposite the angle we are looking for is c, and the angle is C.

If we substitute the values we get:

$$9^2 = 6^2 + 4^2 - 2(6)(4) \cos C.$$

We can solve this easily.

$$81 = 36 + 16 - 48 \cos C$$
$$48 \cos C = 36 + 16 - 81 = -29$$
$$\cos C = -\frac{29}{48}.$$

The answer is (2).

PROBLEM 33: LAW OF SINES

First, let's draw a triangle and fill in the information. Any triangle will do for now.

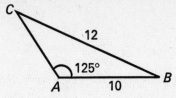

By asking for the possible number of triangles, the question is testing your knowledge of the *Ambiguous Case* of the Law of Sines. There is a simple test: Is $c \sin A < a < c$? $10 \sin 125 \approx 8.19$. This is less than side a, but side a is greater than side c; therefore, there is only one triangle.

The answer is (1).

Another way to get this right is to use the Law of Sines and see how many triangles can be made.

The Law of Sines says that, in any triangle, with angles A, B, and C, and opposite sides of a, b, and c, respectively, $\dfrac{a}{\sin A} = \dfrac{b}{\sin B} = \dfrac{c}{\sin C}$ (This is on the formula sheet.)

Here, $\dfrac{12}{\sin 125°} = \dfrac{10}{\sin C}$, so $\sin C = \dfrac{10 \sin 125°}{12} \approx 0.6826$. (If we had gotten a number greater than 1 or less than –1 we would have NO triangles.) Therefore, $m\angle C \approx 43$ This would make $m\angle B = 12°$. Now we have one triangle. But, there is a second angle for which $\sin C \approx 0.6826$, that is $m\angle C \approx 137$. Can we make a triangle with angles of 125°, 137° and B ? NO, because the sum of the angles is greater than 180°. Therefore, there is only one triangle.

PROBLEM 34: TRIGONOMETRY GRAPHS

A graph of the form $y = a \sin bx$ *or* $y = a \cos bx$ *has an amplitude of* $|a|$ *and a period of* $\dfrac{2\pi}{b}$.

Thus we have: $\dfrac{2\pi}{b} = 4\pi$

If we solve for b, we get: $b = \dfrac{1}{2}$.

The answer is (1).

PROBLEM 35: QUADRATIC EQUATIONS

Given an equation of the form $ax^2 + bx + c = 0$, *the product of the roots is* $\dfrac{c}{a}$, *and the sum of the roots is* $-\dfrac{b}{a}$.

Here, we have the equation $2x^2 + 6x - 7 = 0$, so the sum of the roots is $-\dfrac{b}{a} = -\dfrac{6}{2} = -3$.

The answer is (2).

Another way to get the answer is to actually find the roots and add them. The equation doesn't factor, so we use the quadratic formula

$$x = \frac{-b \pm \sqrt{b^2 - 4ac}}{2a}.$$

Plugging in, we get:

$$x = \frac{-6 \pm \sqrt{(6)^2 - 4(2)(-7)}}{2(2)} = \frac{-6 \pm \sqrt{92}}{4} =$$

The roots are thus $\dfrac{-6 + \sqrt{92}}{4} \approx 0.8979$ and $\dfrac{-6 - \sqrt{92}}{4} \approx -3.8979$. If we add them, we get -3, which is answer (2).

Part II

PROBLEM 36: CIRCLE RULES

(a) *The measure of an inscribed angle is half of the arc it subtends (intercepts).* Thus, $m\angle BAC = 30$. Next, because chord AB is congruent to chord AC, we know that $m\angle ABC = m\angle ACB$, and that $m\angle ABC + m\angle ACB = 180 - 30 = 150$. Therefore, $m\angle ABC = 75$ (and $m\overset{\frown}{AC} = 150$).

(b) We are given that CD is a diameter of the circle; therefore, the measure of arc $\overset{\frown}{DAC}$ is 180.

We found in part (a) that $m\overset{\frown}{AC} = 150$. Therefore, $m\overset{\frown}{AD} = 30$.

(c) Because $m\angle ACB = 75$, $m\overset{\frown}{AB} = 150$. Furthermore, because $m\overset{\frown}{AD} = 30$, $m\overset{\frown}{DB} = 150 - 30 = 120$. Now we can use the following rule: *If two chords intersect within a circle, then the angle between the two chords is the average of their intercepted arcs.*

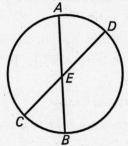

Therefore, $m\angle DEB = \dfrac{m\overset{\frown}{DB} + m\overset{\frown}{AC}}{2} = \dfrac{120 + 150}{2} = 135$.

(d) Here, we can use another rule: *The measure of an angle formed by a pair of secants, or a secant and a tangent, is equal to half of the difference between the larger and the smaller arcs that are formed by the secants, or the secant and the tangent.*

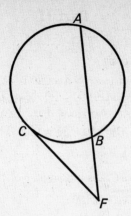

Here, the larger and smaller arcs formed by angle F are $\overset{\frown}{AC}$ and $\overset{\frown}{CB}$, respectively.

$$m\angle F = \frac{m\overset{\frown}{AC} - m\overset{\frown}{CB}}{2} = \frac{150 - 60}{2} = 45.$$

(e) $\angle ABC$ and $\angle FBC$ form a line, so their sum is 180°. $m\angle ABC = 75$ so $m\angle FBC = 180 - 75 = 105$.

Next, because the sum of the angles in a triangle is 180°, $m\angle BCF = 180 - (m\angle F + m\angle FBC) = 180 - (45 + 105) = 30$.

PROBLEM 37: TRIGONOMETRIC GRAPHS

(a) A graph of the form $y = a\sin bx$ or $y = a\cos bx$ has an amplitude of $|a|$ and a period of $\dfrac{2\pi}{b}$. Therefore, the equation $y = 2\cos x$ has an amplitude of 2 and a period of 2π.

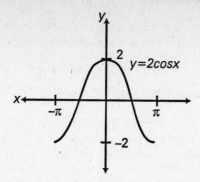

(b) When you reflect a graph in the x-axis, you change the sign of all of the y values. In other words, you turn the graph upside down.

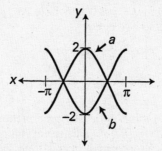

(c) Because the equation for the reflection in the x-axis means changing the sign of all of the y values, the equation for the graph becomes $-y = 2\cos x$. If we then multiply through by -1, we get:

$$y = -2\cos x.$$

(d) Here, you need to know your special angles: $\cos\dfrac{\pi}{6} = \dfrac{\sqrt{3}}{2}$, so using the equation from part (c) above, the value of y at $x = \dfrac{\pi}{6}$

is: $y = -2\cos\dfrac{\pi}{6} = -2\left(\dfrac{\sqrt{3}}{2}\right) = -\sqrt{3}$.

PROBLEM 38: TRIGONOMETRIC EQUATIONS

(a) First, we use the trigonometric identity $\cos 2x = 2\cos^2 x - 1$ (it's on the formula sheet) and substitute it into the equation. We get:

$$3(2\cos^2 x - 1) + \cos x + 2 = 0.$$

We then simplify and get:

$$6\cos^2 x + \cos x - 1 = 0.$$

Note that the trig equation $6\cos^2 x + \cos x - 1 = 0$ has the same form as a quadratic equation $6x^2 + x - 1 = 0$. Just as we could factor the quadratic equation, so too can we factor the trig equation. We get:

$$(3\cos x - 1)(2\cos x + 1) = 0.$$

This gives us $3\cos x - 1 = 0$ and $2\cos x + 1 = 0$. If we solve each of these equations, we get:

$$\cos x = \frac{1}{3} \text{ and } \cos x = -\frac{1}{2}.$$

Therefore $x = \cos^{-1}\left(\frac{1}{3}\right)$ and $x = \cos^{-1}\left(-\frac{1}{2}\right)$.

We can find the solution to the first equation with the calculator. You should get $x \approx 70.53°$. This is not the only value for x, however. Cosine is also positive in quadrant IV, so there is another angle that satisfies the first equation. It is: $360° - 70.53° = 289.47°$. Rounding both of these to the nearest degree, we get $x = 71°$, $289°$. The solution to the second equation is easy because it is a special angle. You should know that $\cos 120° = -\frac{1}{2}$ and that $\cos 240° = -\frac{1}{2}$.

Therefore, the solutions are $x = 71°, 120°, 240°, 289°$.

PROBLEM 39: PROBABILITY

(a) This problem requires that you know something called *binomial probability*. The rule is: *If the probability of a particular outcome is **p**, then the probability of that outcome occurring **r** times out of a possible **n** times is* $_nC_r(p)^r(1-p)^{n-r}$.

The probability that the Alphas will win is $p = \dfrac{3}{5}$; therefore, the probability that the Betas will win is $p = 1 - \dfrac{3}{5} = \dfrac{2}{5}$.

The probability that the Betas will win all four contests is:

$${}_4C_4\left(\dfrac{2}{5}\right)^4\left(1 - \dfrac{2}{5}\right)^{4-4} = {}_4C_4\left(\dfrac{2}{5}\right)^4\left(\dfrac{3}{5}\right)^0.$$

The rule for finding ${}_nC_r$ is ${}_nC_r = \dfrac{n!}{(n-r)!\, r!}$, so,

$${}_4C_4 = \dfrac{4!}{0!\,4!} = \dfrac{4 \cdot 3 \cdot 2 \cdot 1}{(1)(4 \cdot 3 \cdot 2 \cdot 1)} = 1.$$

(By the way, $0! = 1$.) Thus, the probability that the Betas will win all four contests is:

$$1\left(\dfrac{2}{5}\right)^4\left(\dfrac{3}{5}\right)^0 = \dfrac{16}{625}.$$

(b) We can find the probability that each team will win two games by finding the probability that the Betas win exactly two games (because then the Alphas must win the other two).

The probability is:

$${}_4C_2\left(\dfrac{2}{5}\right)^2\left(\dfrac{3}{5}\right)^2.$$

$${}_4C_2 = \dfrac{4!}{2!\,2!} = \dfrac{4 \cdot 3 \cdot 2 \cdot 1}{(2 \cdot 1)(2 \cdot 1)} = 6.$$

Thus, the probability that the Betas will win two games is:

$$6\left(\dfrac{2}{5}\right)^2\left(\dfrac{3}{5}\right)^2 = \dfrac{216}{625}.$$

(c) The probability that the Alphas will win at least two games is the probability that they will win two games *plus* the probability that they will win three games *plus* the probability that they will win all four games. We already know the probability that they will win two games from part (b).

The probability that they will win three games is:

$$_4C_3\left(\frac{3}{5}\right)^3\left(\frac{2}{5}\right)^{4-3} = {_4C_3}\left(\frac{3}{5}\right)^3\left(\frac{2}{5}\right)^1.$$

The probability that they will win four games is:

$$_4C_4\left(\frac{3}{5}\right)^4\left(\frac{2}{5}\right)^{4-4} = {_4C_4}\left(\frac{3}{5}\right)^4\left(\frac{2}{5}\right)^0.$$

$_4C_3 = \dfrac{4!}{1!\,3!} = \dfrac{4\cdot3\cdot2\cdot1}{(1)(3\cdot2\cdot1)} = 4$, and we already know that $_4C_4 = 1$; therefore, the probability that they will win three games is:

$$4\left(\frac{3}{5}\right)^3\left(\frac{2}{5}\right)^1 = \frac{216}{625}.$$

and the probability that they will win four games is:

$$1\left(\frac{3}{5}\right)^4\left(\frac{2}{5}\right)^0 = \frac{81}{625}.$$

The sum of the three probabilities is:

$$\frac{216}{625} + \frac{216}{625} + \frac{81}{625} = \frac{513}{625}.$$

(d) The probability that the Betas win at most one game is the probability that they win one game *plus* the probability that they win zero games. The probability that the Betas win one game is the same as the probability that the Alphas win three games, which is: $\frac{216}{625}$. The probability that they win zero games is the same as the probability that the Alphas win four games, which is: $\frac{81}{625}$.

Therefore, the probability that the Betas win at most one game is:

$$\frac{216}{625} + \frac{81}{625} = \frac{297}{625}.$$

PROBLEM 40:

(a) Trigonometric Identities

Almost all of the identities that you will need to know are contained on the formula sheet.

First, let's get rid of the double angle terms.

Using $\csc 2x = \dfrac{1}{\sin 2x}$ and $\sin 2x = 2\sin x \cos x$, we can rewrite the right side of the equation as: $\dfrac{2}{2\sin x \cos x} = \dfrac{1}{\sin x \cos x}$.

Next, we use the identities $\tan x = \dfrac{\sin x}{\cos x}$ and $\cot x = \dfrac{\cos x}{\sin x}$, to rewrite the equation as: $\dfrac{\sin x}{\cos x} + \dfrac{\cos x}{\sin x} = \dfrac{1}{\sin x \cos x}$ Now, we find a common denominator for the left side and combine the two fractions:

$$\frac{\sin x}{\cos x}\left(\frac{\sin x}{\sin x}\right) + \frac{\cos x}{\sin x}\left(\frac{\cos x}{\cos x}\right) = \frac{1}{\sin x \cos x}$$

$$\frac{\sin^2 x}{\sin x \cos x} + \frac{\cos^2 x}{\sin x \cos x} = \frac{1}{\sin x \cos x}$$

$$\frac{\sin^2 x + \cos^2 x}{\sin x \cos x} = \frac{1}{\sin x \cos x}$$

Finally, we use the identity $\sin^2 x + \cos^2 x = 1$ and we get:

$$\frac{1}{\sin x \cos x} = \frac{1}{\sin x \cos x}$$

and we have proved the identity.

(b) Logarithms

(1) You should know the following log rules.

(i) $\log A + \log B = \log(AB)$

(ii) $\log A - \log B = \log\left(\dfrac{A}{B}\right)$

(iii) $\log A^B = B\log A$

Using rule (ii), we can rewrite $\log\dfrac{\sqrt{2}}{9} = \log\sqrt{2} - \log 9$

Next, because $\sqrt{2} = 2^{\frac{1}{2}}$ and $9 = 3^2$, we can rewrites:

$$\log \sqrt{2} - \log 9 = \frac{1}{2}\log 2 - 2\log 3.$$

Finally, we can substitute $\log 2 = x$ and $\log 3 = y$ and we get:

$$\frac{x}{2} - 2y.$$

(2) First, because $\sqrt[3]{6} = 6^{\frac{1}{3}}$, we can rewrite $\log \sqrt[3]{6} = \frac{1}{3}\log 6$.

Next, we know that $6 = 2 \cdot 3$, so we can rewrite $\frac{1}{3}\log(2 \cdot 3)$.

Now, using rule (i), we get: $\frac{1}{3}\log(2 \cdot 3) = \frac{1}{3}\left(\log 2 + \log 3\right)$.

Finally, we can substitute $\log 2 = x$ and $\log 3 = y$ and we get:

$$\frac{1}{3}(x + y).$$

PROBLEM 41:

(a) Binomial Expansions

The binomial theorem says that if you expand $(a + b)^n$, you get the following terms·

$$_nC_0 a^n + {_nC_1} a^{n-1}b^1 + {_nC_2} a^{n-2}b^2 + ... + {_nC_{n-2}} a^2 b^{n-2} + {_nC_{n-1}} a^1 b^{n-1} + {_nC_n} b^n$$

Therefore, if we expand $\left(x - \dfrac{1}{x}\right)^4$, we get·

$$_4C_0(x)^4 + {_4C_1}(x)^3\left(-\frac{1}{x}\right)^1 + {_4C_2}(x)^2\left(-\frac{1}{x}\right)^2 + {_4C_3}(x)^1\left(-\frac{1}{x}\right)^3 + {_4C_4}\left(-\frac{1}{x}\right)^4$$

Next, we use the rule that $_nC_r = \dfrac{n!}{(n-r)!\,r!}$. This gives us·

$$1(x)^4 + 4(x)^3\left(-\frac{1}{x}\right)^1 + 6(x)^2\left(-\frac{1}{x}\right)^2 + 4(x)^1\left(-\frac{1}{x}\right)^3 + 1\left(-\frac{1}{x}\right)^4,$$

which simplifies to

$$\left(x - \frac{1}{x}\right)^4 = x^4 - 4x^2 + 6 - \frac{4}{x^2} + \frac{1}{x^4}$$

(b) Exponential Equations

We can use logarithms to solve this equation. If we take the log of both sides, we get: $\log 5^{3x} = \log 1000$. Next, because $10^3 = 1000$, we can rewrite the equation as: $\log 5^{3x} = \log 10^3$

Now, we use the rule that $\log A^B = B \log A$ to rewrite the equation as. $3x \log 5 = 3 \log 10$.

Now we can solve for x:

$$x \log 5 = \log 10$$

$$x = \frac{\log 10}{\log 5} = \frac{1}{\log 5} \approx 1.4$$

PROBLEM 42: TRIGONOMETRY

(a) We can find the measure of the angle using the Law of Cosines.

First, let's draw a picture of the situation.

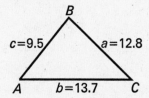

The Law of Cosines says that, in any triangle, with angles A, B, and C, and opposite sides of a, b, and c, respectively, $c^2 = a^2 + b^2 - 2ab \cos C$. (This is on the formula sheet.) If we have a triangle and we are given SSS (side, side, side), then the side opposite the angle we are looking for is c, and that angle is C. The smallest angle will be opposite the smallest side.

$$(9.5)^2 = (13.7)^2 + (12.8)^2 - 2(13.7)(12.8)\cos C$$

$$90.25 = 187.69 + 163.84 - 350.72 \cos C$$

$$350.72 \cos C = 187.69 + 163.84 - 90.25$$

$$350.72 \cos C = 261.28$$

$$\cos C = \frac{261.28}{350.72}$$

$$C = \cos^{-1}\left(\frac{261.28}{350.72}\right) \approx 41.84° \text{ or } 41°50'.$$

(b) Trigonometric Area

The area of a triangle, if we are given the lengths of the two sides a and b, and their included angle θ, is $A = \frac{1}{2}ab\sin\theta$. In other words, if we are given SAS (side, angle, side), we find the area by multiplying $\frac{1}{2}$ by the product of the two sides by the sine of the included angle. Here, the two sides are a and b and the included angle is C. Plugging in, we get:

$$A = \frac{1}{2}(13.7)(12.8)\sin 41.84° \approx 58.5.$$

Formulas

Pythagorean and Quotient Identities

$$\sin^2 A + \cos^2 A = 1 \qquad \tan A = \frac{\sin A}{\cos A}$$

$$\tan^2 A + 1 = \sec^2 A \qquad \cot A = \frac{\cos A}{\sin A}$$

$$\cot^2 A + 1 = \csc^2 A$$

Functions of the Sum of Two Angles

$$\sin (A + B) = \sin A \cos B + \cos A \sin B$$

$$\cos (A + B) = \cos A \cos B - \sin A \sin B$$

$$\tan (A + B) = \frac{\tan A + \tan B}{1 - \tan A \tan B}$$

Functions of the Difference of Two Angles

$$\sin (A - B) = \sin A \cos B - \cos A \sin B$$

$$\cos (A - B) = \cos A \cos B + \sin A \sin B$$

$$\tan (A - B) = \frac{\tan A - \tan B}{1 + \tan A \tan B}$$

Law of Sines

$$\frac{a}{\sin A} = \frac{b}{\sin B} = \frac{c}{\sin C}$$

Law of Cosines

$$a^2 = b^2 + c^2 - 2bc \cos A$$

Functions of the Double Angle

$$\sin 2A = 2 \sin A \cos A$$

$$\cos 2A = \cos^2 A - \sin^2 A$$

$$\cos 2A = 2 \cos^2 A - 1$$

$$\cos 2A = 1 - 2 \sin^2 A$$

$$\tan 2A = \frac{2 \tan A}{1 - \tan^2 A}$$

Functions of the Half Angle

$$\sin \frac{1}{2} A = \pm\sqrt{\frac{1 - \cos A}{2}}$$

$$\cos \frac{1}{2} A = \pm\sqrt{\frac{1 + \cos A}{2}}$$

$$\tan \frac{1}{2} A = \pm\sqrt{\frac{1 - \cos A}{1 + \cos A}}$$

Area of Triangle

$$K = \frac{1}{2} ab \sin C$$

Standard Deviation

$$\text{S.D.} = \sqrt{\frac{1}{n} \sum_{i=1}^{n} \left(x_i - \bar{x}\right)^2}$$

EXAMINATION
AUGUST 1996

Part I

Answer 30 questions from this part. Each correct answer will receive 2 credits. No partial credit will be allowed. Write your answers in the spaces provided on the separate answer sheet. Where applicable, answers may be left in terms of π or in radical form.   [60]

1 In the accompanying diagram, $\overline{AB}$ and $\overline{AC}$ are tangents to circle O, and chord $\overline{BC}$ is drawn. If $m\angle ABC = 72$, what is $m\angle A$?

$72 + 72 + x = 180$
$144 + x = 180$
$x = 36$

$m\angle A = 36°$

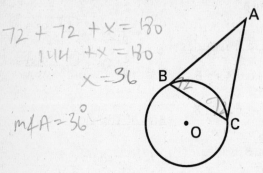

2 Express the product in simplest form:

$\dfrac{(a-3)(a+3)}{a(a-3)} \cdot \dfrac{a(a+1)}{a+3}$ $\dfrac{a^2-9}{a^2-3a} \cdot \dfrac{a^2+a}{a+3}.$ $= a+1$

3 Express 225° in radian measure.

$225 \cdot \dfrac{\pi}{180} = \dfrac{5\pi}{4}$

4 Solve for x: $\sqrt{2x-2} - 2 = 0$.

$2x - 2 = 2^2$ $2x = 6$
$2x = 4 + 2$ $x = 3$

238 • **Cracking the Regents: Sequential Math III**

5 If the transformation $T_{(x, y)}$ maps point $A(1,-3)$ onto point $A'(-4,8)$, what is the value of x?

$x = -5$

6 A set of boys' heights is distributed normally with a mean of 58 inches and a standard deviation of 2 inches. Expressed in inches, between which two heights should 95% of the heights fall?

$62, 54$

7 If $\sin A > 0$ and $\sec A < 0$, in which quadrant does the terminal side of $\angle A$ lie?

8 In which quadrant does the sum of $3 + 2i$ and $-4 - 5i$ lie?

$(3 + 2i) + (-4 - 5i) = -1 - 3i$

9 Solve for x: $2^{x+2} = 4^{x-1}$.

$x + 2 = 2x - 2$
$2^{x+2} = 2^{2x-2}$
$2 + 2 = 2x - x$
$4 = x$

10 In $\triangle CAT$, $a = 4$, $c = 5$, and $\cos T = \dfrac{1}{8}$. What is the length of side t?

$t^2 = a^2 + c^2 - 2ac \cdot \cos T$
$t^2 = 4^2 + 5^2 - 2(4)(5) \cdot \frac{1}{8}$
$16 + 25 - 40 \cdot \frac{1}{8}$
$41 - 5 = 36$
$\sqrt{t^2} = \sqrt{36}$
$t = \sqrt{36}$

11 Evaluate: $\displaystyle\sum_{i=3}^{7} (3k + 2)$.

$(3(3)+2) + (3(4)+2) + (3(5)+2) + (3(6)+2) + (3(7)+2)$
$11 + 14 + 15 + 20 + 23$

12 Express $\dfrac{5}{2-i}$ in simplest $a + bi$ form.

$\frac{5}{2-i} \cdot \frac{(2+i)}{(2+i)} = \frac{10 + 5i}{4 + 2i - 2i - i^2} = \frac{10 + 5i}{5}$
$2 + i$

13 Solve for y: $y^{-\frac{1}{2}} = \dfrac{1}{3}$.

14 In $\triangle ABC$, $a = 6$, $b = 7$, and m $\angle B = 30$. Find $\sin A$.

15 Solve for all values of x: $|3x - 2| = 16$. [OVER]

Directions (16–35): For *each* question chosen, write on the separate answer sheet the *numeral* preceding the word or expression that best completes the statement or answers the question.

16 In the accompanying figure of circle O, m $\angle ABC = 38$.

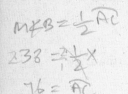

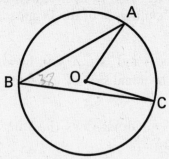

What is m $\angle AOC$?

(1) 19   (3) 76
(2) 38   (4) 152

17 In a circle, diameter $\overline{AB}$ is perpendicular to chord $\overline{CD}$ at L. Which statement will always be true about this circle?

(1) $CL = LD$ (3) $(CL) \times (LD) = AB$
(2) $AL > LB$ (4) $BL > LA$

18 After which tranformation of $\triangle ABC$ could the image $\triangle A'B'C'$ *not* have the same area?

(1) translation   (3) point reflection
(2) rotation   (4) dilation

19 The expression sin 50° cos 40° + cos 50° sin 40° is equivalent to

(1) sin 10° (3) sin 90°
(2) cos 10° (4) cos 90°

20 If $\log_b x = y$, then $\log_b x^2$ is

(1) $y + 2$ (3) $y - 2$
(2) $2y$ (4) y

21 Which graph represents an inverse variation between all values of x and y?

(1)
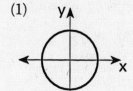

(3)
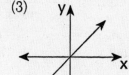

(2)
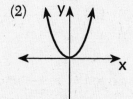

(4)
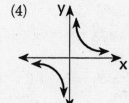

22 The graph of any function and the graph of its inverse are symmetric with respect to the

(1) x-axis
(2) y-axis
(3) graph of the equation $y = -x$
(4) graph of the equation $y = x$

[OVER]

23 If m $\angle B = 60$, $a = 6$, and $c = 10$, what is the area of $\triangle ABC$?

(1) 15   (3) $15\sqrt{3}$

(2) 30   (4) $30\sqrt{3}$

24 What is the period of the equation $y = -6 \sin 2x$?

(1) $-\dfrac{2}{6}$ (3) 2π

(2) -6π (4) π

25 The expression $\dfrac{1 + \cos 2x}{\sin 2x}$ is equivalent to

(1) $\tan x$ (3) $-\sin x$

(2) $\cot x$ (4) $-\cos x$

26 If $f(x) = 4\cos 3x$, what is the value of $f\left(\dfrac{\pi}{4}\right)$?

(1) $-\sqrt{2}$ (3) 135

(2) $-2\sqrt{2}$ (4) 4

27 For which value of θ is the fraction $\dfrac{6}{\cos\theta}$ undefined?

(1) $0°$ (3) $60°$

(2) $30°$ (4) $90°$

28 Which equation has roots of $3 + \sqrt{2}$ and $3 - \sqrt{2}$?

(1) $x^2 + 6x + 7 = 0$ (3) $x^2 - 7x - 4 = 0$

(2) $x^2 - 6x + 7 = 0$ (4) $x^2 - 7x + 6 = 0$

29 In basketball, Nicole makes 4 baskets for every 10 shots. If she takes 3 shots, what is the probability that *exactly* 2 of them will be baskets?

(1) 0.288   (3) 0.600
(2) 0.432   (4) 0.960

30 When two resistors are connected in a parallel circuit, the total resistence is $\dfrac{1}{\dfrac{1}{R_1} + \dfrac{1}{R_2}}$. This complex fraction is equivalent to

(1) $R_1 + R_2$ (3) $R_1 R_2$

(2) $\dfrac{R_1 + R_2}{R_1 R_2}$ (4) $\dfrac{R_1 R_2}{R_1 + R_2}$

31 Which graph represents the inequality $x^2 - 4 > 0$?

(1)

(2)

(3)

(4)

32 For which value of c will the roots of the equation $4x^2 - 4x + c = 0$ be real numbers?

(1) 1   (3) 3
(2) 2   (4) 4

[OVER]

33 If $\cos (2x - 1)° = \sin (3x + 6)°$, then the value of x is

(1) –7   (3) 35
(2) 17   (4) 71

34 Square $ABCD$ is inscribed in a circle with the center at O.

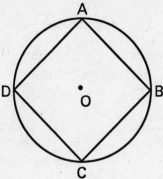

What is $(R_{-180°} \circ R_{90°})(B)$?

(1) A (3) C
(2) B (4) D

35 The graph of the equation $\dfrac{x^2 + y^2}{2} = 5$ is

(1) a circle   (3) a hyperbola
(2) an ellipse   (4) a parabola

Part II

Answer four questions from this part. Clearly indicate the necessary steps, including appropriate formula substitutions, diagrams, graphs, charts, etc. Calculations that may be obtained by mental arithmetic or the calculator do not need to be shown.

36 In the accompanying diagram, $\overrightarrow{PA}$ is tangent to circle O at point A, secant $\overline{PBD}$ intersects diameter $\overline{AC}$ at point E, chord $\overline{AB}$ is drawn, m $\angle P = 40$, and m$\overparen{CD}$: m$\overparen{DA}$ = 1: 8 .

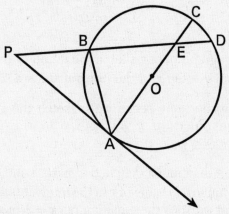

Find:

a   m$\overparen{DA}$ [2]
b   m$\overparen{AB}$ [2]
c   m$\angle BEA$ [2]
d   m$\angle BAC$ [2]
e   m$\angle PBA$ [2]

[OVER]

37 *a* Solve for x: $x + \sqrt{2x - 1} = 8$. [5]

  b Solve for y: $\dfrac{y}{y-1} = \dfrac{8}{y} + \dfrac{1}{y-1}$. [5]

38 *a* Find, to the *nearest degree*, all values of x in the interval $0° \leq x < 360°$ that satisfy the equation $2 \sin^2 x = 1 + \sin x$. [5]

  b For all the values of x for which the expression is defined, prove that the following is an identity:

$$\cot x = \frac{\sin 2x}{1 - \cos 2x} \quad [5]$$

39 *a* On graph paper, sketch the graph of the equation $y = \tan x$ in the interval $0 \leq x \leq 2\pi$. [4]

  b On the same set of axes, sketch the graph of the equation $y = 2 \sin x$ in the interval $0 \leq x \leq 2\pi$. [2]

  c Use the graphs sketched in parts *a* and *b* to determine *one* value of x in the interval $0 \leq x \leq 2\pi$ that satisfies the equation $\tan x = 2\sin x$. [2]

40 Two forces act on a body at an angle of 100°. The forces are 30 pounds and 40 pounds.

a Find the magnitude of the resultant force to the *nearest tenth* of a pound.   [6]

b Find the angle formed by the greater of the two forces and the resultant force to the *nearest degree.*   [4]

41 *a* On the same set of axes, sketch and label the graphs of the equations $xy = 8$ and $y = \log_2 x$ in the interval $-6 \le x \le 6$.   [8]

b Using the graphs sketched in part *a*, find an integer value of x for which $\log_2 x > \dfrac{8}{x}$.   [2]

42 *a* The probability of a biased coin coming up heads is $\dfrac{3}{4}$.

    (1) When the coin is flipped three times, what is the probability of *at least* two heads?   [3]

    (2) When the coin is flipped four times, what is the probability of *at most* one head?   [3]

b Find the standard deviation, to the *nearest hundredth*, for the following test scores:

100, 99, 99, 97, 96, 96, 95, 94, 93, 91.   [4]

[OVER]

ANSWER KEY

Part I

1. 36

2. $a + 1$

3. $\dfrac{5\pi}{4}$

4. 3

5. –5

6. 54 and 62

7. II

8. III

9. 4

10. 6

11. 85

12. $2 + i$

13. 9

14. $\dfrac{3}{7}$

15. $6, -\dfrac{14}{3}$

16. (3)

17. (1)

18. (4)

19. (3)

20. (2)

21. (4)

22. (4)

23. (3)

24. (4)

25. (2)

26. (2)

27. (4)

28. (2)

29. (1)

30. (4)

31. (3)

32. (1)

33. (2)

34. (3)

35. (1)

ANSWERS AND EXPLANATIONS
AUGUST 1996

Part I

In problems 1–15, you must supply the answer. Because of this, you will not be able to **plug-in** or **backsolve**. But, you can use your calculator to simplify calculations. In addition, you will find that these problems tend to test your basic knowledge of Sequential Math III and are not tricky. Many can be solved just by knowing the right formula and by plugging in the numbers.

PROBLEM 1: CIRCLE RULES

The two tangents drawn to a circle from the same point are congruent. This means that triangle ABC is isosceles. *In an isosceles triangle, equal opposite sides have equal opposite angles.* This means that $m\angle ABC = m\angle ACB = 72$. Therefore, because the sum of the angles in a triangle is 180°,

$$m\angle A = 180° - (m\angle ABC + m\angle ACB) = 180 - (72 + 72) = 36.$$

PROBLEM 2: RATIONAL EXPRESSIONS

$$\frac{a^2 - 9}{a^2 - 3a} \cdot \frac{a^2 + a}{a + 3}$$

Factor the top and bottom of the left side to obtain

$$\frac{(a + 3)(a - 3)}{a(a - 3)} \cdot \frac{a^2 + a}{a + 3}.$$

Factor the top of the right side to obtain

$$\frac{(a + 3)(a - 3)}{a(a - 3)} \cdot \frac{a(a + 1)}{a + 3}.$$

Now, we cancel like terms and we get $a + 1$.

PROBLEM 3: CONVERTING FROM DEGREES TO RADIANS

All we have to do is multiply 225° by $\dfrac{\pi}{180°}$.

$$225\left(\frac{\pi}{180}\right) = \frac{5\pi}{4}.$$

PROBLEM 4: RADICAL EQUATIONS

First, add 2 to both sides of the equation. $\sqrt{2x-2} = 2$

Next, square both sides. $2x - 2 = 4$.

Now solve for x.
$$2x = 6$$
$$x = 3$$

HOWEVER, whenever we have a radical in an equation, we have to check the roots to see if they satisfy the original equation. Sometimes, in the process of squaring both sides, we will get answers that are invalid. So let's check the answer.

Does $\sqrt{2(3) - 2} = 2$? $\sqrt{4} = 2$; so, yes, it does.

PROBLEM 5: TRANSFORMATIONS

The translation $T_{(x,y)}$ tells you to add x to the x-coordinate of the point being translated, and to add y to the y-coordinate of the point being translated. So, here, $T_{(x,y)}(1,-3) \to (-4,8)$ means that $1 + x = -4$ and $-3 + y = 8$. Therefore, the translation is $T_{(-5,11)}$ and the value of x is -5.

PROBLEM 6: STATISTICS

You should know the following rule about normal distributions.

> In a normal distribution, with a mean of $\bar{x}$ and a standard deviation of σ:
> - approximately 68% of the outcomes will fall between $\bar{x} - \sigma$ and $\bar{x} + \sigma$.
> - approximately 95% of the outcomes will fall between $\bar{x} - 2\sigma$ and $\bar{x} + 2\sigma$.
> - approximately 99.5% of the outcomes will fall between $\bar{x} - 3\sigma$ and $\bar{x} + 3\sigma$.

Here, 95% of the scores should lie between $\bar{x} - 2\sigma$ and $\bar{x} + 2\sigma$, which is $58 - 4$ and $58 + 4$. Therefore, 95% of the scores should lie between 54 and 62.

PROBLEM 7: TRIGONOMETRY

The sine of an angle is positive in quadrants I *and* II.

The cosine of an angle is negative in quadrants II *and* III, so the secant of an angle is also negative in quadrants II and III.

Therefore, the terminal side of $\angle A$ must be in quadrant II.

PROBLEM 8: COMPLEX NUMBERS

To find the sum of $3 + 2i$ and $-4 - 5i$ we add the real components and the imaginary components separately. This gives us:

$$\left(3 + (-4)\right) + \left(2 + (-5)\right)i = -1 - 3i.$$

Next, to determine the quadrant, we look at the signs of the components. *The real component corresponds to the x-coordinate of the graph, and the imaginary component corresponds to the y-coordinate of the graph.* Because both components are negative, the graph has negative x- and y-coordinates. Therefore, the sum of $3 + 2i$ and $-4 - 5i$ lies in quadrant III.

PROBLEM 9: EXPONENTIAL EQUATIONS

$$2^{x+2} = 4^{x-1}$$

In order to solve this equation, we need to have a common base for the two sides of the equation. Because $4 = 2^2$, we can rewrite the right side as $2^{x+2} = \left(2^2\right)^{x-1}$.

The rules of exponents say that *when a number raised to a power is itself raised to a power, we multiply the exponents.* Thus, we get:

$$2^{x+2} = 2^{2x-2}.$$

Now, we can set the powers equal to each other and solve.

$$x + 2 = 2x - 2$$
$$x = 4.$$

PROBLEM 10: LAW OF COSINES

The Law of Cosines says that, in any triangle, with angles $A, B,$ and $C,$ and opposite sides of $a, b,$ and $c,$ respectively, $c^2 = a^2 + b^2 - 2ab \cos C.$ (This is on the formula sheet.) Don't let the different letters confuse you. If we have a triangle and we are given SAS (side, angle, side), then the side opposite the included angle is c, and the included angle is C. In this case, the included angle is T, and its opposite side is t. If we substitute the values we get: $t^2 = 4^2 + 5^2 - 2(4)(5)\left(\dfrac{1}{8}\right)$.

We can solve this easily.

$$t^2 = 16 + 25 - 5 = 36$$
$$t = 6.$$

PROBLEM 11: SUMMATIONS

To evaluate a sum, we plug each of the consecutive values for k into the expression, starting at the bottom value and ending at the top value and sum the results. We get:

$$\sum_{k=3}^{7}(3k + 2) = [3(3) + 2] + [3(4) + 2] + [3(5) + 2] + [3(6) + 2] + [3(7) + 2] =$$
$$11 + 14 + 17 + 20 + 23 = 85.$$

Another way to do this is to use your calculator. Push **2nd STAT** (**LIST**) and under the **MATH** menu, choose **SUM**. Then, again push **2nd STAT** and under the **OPS** menu, choose **SEQ**.

We put in the following: *(expression, variable, start, finish, step)*

For *expression*, we put in the formula, using x as the variable.

For *variable*, we ALWAYS put in x.

For *start*, we put in the bottom value

For *finish*, we put in the top value

For *step*, we ALWAYS put in 1.

Therefore, we put in the calculator: **SUM SEQ** $(3x + 2, x, 3, 7, 1)$. The calculator should return the value 85.

PROBLEM 12: COMPLEX NUMBERS

Whenever you are asked to simplify a fraction that has a complex number in the denominator, all that you have to do is to multiply the top and bottom by the complex conjugate of the denominator. *The complex conjugate of a number $a + bi$ is the number $a - bi$.*

Here, we multiply $\dfrac{5}{2-i}$ by $\dfrac{2+i}{2+i}$. We get: $\dfrac{5}{2-i} \cdot \dfrac{(2+i)}{(2+i)}$.

We distribute the 5 in the numerator, and we FOIL the denominators and we get:

$$\frac{10+5i}{4+2i-2i-i^2} = \frac{10+5i}{4-i^2}.$$

Next, because $i^2 = -1$, we have

$$\frac{10+5i}{4-(-1)} = \frac{10+5i}{5} = \frac{10}{5} + \frac{5}{5}i = 2+i.$$

PROBLEM 13: EXPONENTS

A number raised to a negative power is the same as the reciprocal of that number raised to the corresponding positive power. Here

$$y^{-\frac{1}{2}} = \frac{1}{y^{\frac{1}{2}}}.$$

A number raised to $\dfrac{1}{n}$ is the same as the nth root of that number.
Here $\dfrac{1}{y^{\frac{1}{2}}} = \dfrac{1}{\sqrt{y}}$.

Now we can solve the equation.

$$\frac{1}{\sqrt{y}} = \frac{1}{3}$$
$$\sqrt{y} = 3$$
$$y = 9.$$

PROBLEM 14: LAW OF SINES

The Law of Sines says that, in any triangle, with angles $A, B,$ and $C,$ and opposite sides of a, b, and c, respectively, $\dfrac{a}{\sin A} = \dfrac{b}{\sin B}$. (This is on the formula sheet.)

Here, $\dfrac{6}{\sin A} = \dfrac{7}{\sin 30°}$

We can easily solve this. $\dfrac{6}{\sin A} = \dfrac{7}{\frac{1}{2}} = 14$.

$$\dfrac{6}{14} = \sin A = \dfrac{3}{7}.$$

PROBLEM 15: ABSOLUTE VALUE EQUATIONS

When we have an absolute value equation we break it into two equations. First, rewrite the left side without the absolute value and leave the right side alone. Second, rewrite the left side without the absolute value, and change the sign of the right side.

Here, we have $|3x - 2| = 16$, so we rewrite it as: $3x - 2 = 16$ or $3x - 2 = -16$. If we solve each of these independently, we get $x = 6$ or $x = -\dfrac{14}{3}$.

Multiple-Choice Problems

For problems 16–35, you will find that you can sometimes **plug-in** or **backsolve** to get the right answer. You will also find that a calculator will simplify the arithmetic. In addition, most of the problems only require knowing which formula to use.

PROBLEM 16: CIRCLE RULES

The measure of an inscribed angle is always half of the measure of the arc that it subtends (intercepts). Here, the inscribed angle is 38°, so the subtended arc, $\overset{\frown}{AC}$, has measure 76.

The measure of a central angle is equal to the measure of the arc that it subtends.

Because the subtended arc has measure 76, so too does angle AOC. Therefore, $m\angle AOC = 76$.

The answer is (3).

PROBLEM 17: CIRCLE RULES

A *diameter of a circle that intersects a chord at right angles* **bisects** *the chord.* So here, the diameter cuts the chord in half at L, and thus $CL = LD$.

The answer is (1).

PROBLEM 18: TRANSFORMATIONS

You should know that the size and shape of an object is preserved under translation, rotation, and reflection. Therefore, the area is preserved under those transformations. **But**, dilation changes the size of an object, and therefore, the area is not preserved.

The answer is (4).

PROBLEM 19: TRIGONOMETRY FORMULAS

Notice that this problem has the form $\sin A \cos B + \cos A \sin B$. If you look on the formula sheet, under <u>Functions of the Sum of Two Angles</u>, you will find the formula $\sin (A + B) = \sin A \cos B + \cos A \sin B$. Thus we can rewrite the problem as

$$\sin 50° \cos 40° + \cos 50° \sin 40° = \sin(50° + 40°) = \sin 90°.$$

The answer is (3).

Another way to answer this question was to use your calculator to find the value of $\sin 50° \cos 40° + \cos 50° \sin 40°$. **MAKE SURE THAT YOUR CALCULATOR IS IN DEGREE MODE.** You should get 1. Now find the value of each of the answer choices with your calculator.

$\sin 10° = .1736$, so (1) is not the answer.

$\cos 10° = .9848$, so (2) is not the answer.

$\sin 90° = 1$, **so (3) is the answer**. Just in case, check the last choice.

$\cos 90° = 0$, so (4) is not the answer.

PROBLEM 20: LOGARITHMS

You should know the rule $\log_b x^a = a \log_b x$.

Here, $\log_b x^2 = 2 \log_b x = 2y$.

The answer is (2).

PROBLEM 21: INVERSE VARIATION

An *inverse variation* between x and y means that $xy = k$, where k is a positive constant. The graph of $xy = k$ looks like this:

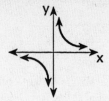

The answer is (4).

PROBLEM 22: INVERSES

The graph of the inverse of a function $f(x)$ is the reflection of the graph in the line $y = x$. In other words, the graphs are symmetric with respect to the line $y = x$.

The answer is (4).

PROBLEM 23: TRIGONOMETRIC AREA

The area of a triangle, if we are given the lengths of the two sides a and b, and their included angle θ, is $A = \frac{1}{2} ab \sin \theta$. In other words, if we are given SAS (side, angle, side), we find the area by multiplying $\frac{1}{2}$ by the product of the two sides by the sine of the included angle. Here, the two sides are a and c and the included angle is B. Plugging in, we get: $A = \frac{1}{2}(6)(10) \sin 60° = 30 \frac{\sqrt{3}}{2} = 15\sqrt{3}$.

The answer is (3).

PROBLEM 24: TRIGONOMETRY GRAPHS

A graph of the form $y = a \sin bx$ or $y = a \cos bx$ has an amplitude of $|a|$ and a period of $\frac{2\pi}{b}$.

Here, the period is $\frac{2\pi}{2} = \pi$.

The answer is (4).

PROBLEM 25: TRIGONOMETRIC IDENTITIES

You can find the identities for double angles on the formula sheet under *Functions of the Double Angle*. It says $\cos 2A = 2\cos^2 A - 1$ and $\sin 2A = 2\sin A \cos A$. Substituting this into the expression, we get:

$$\frac{1 + \cos 2x}{\sin 2x} = \frac{1 + 2\cos^2 x - 1}{2\sin x \cos x} = \frac{2\cos^2 x}{2\sin x \cos x}.$$

If we cancel like terms, we get: $\frac{2\cos^2 x}{2\sin x \cos x} = \frac{\cos x}{\sin x} = \cot x$.

The answer is (2).

PROBLEM 26: TRIGONOMETRY

If we substitute $\frac{\pi}{4}$ for x, we get $4\cos\left(\frac{3\pi}{4}\right) = 4\left(-\frac{\sqrt{2}}{2}\right) = -2\sqrt{2}$

The answer is (2).

You should know your trig functions at the special angles, but if you don't, put the calculator in radian mode and find the value of $4\cos\left(\frac{3\pi}{4}\right) = -2.828$. Then find the value of the answer choice that matches.

PROBLEM 27: TRIGONOMETRY

Where the denominator of a fraction is zero, the fraction is undefined. Here we need to know where $\cos \theta = 0$. Cosine is zero at $90°$

The answer is (4).

Another way to get this right is to **backsolve**. Plug each of the values of the answer choices into $f(x)$ and see which one gives you an *error*. **MAKE SURE THAT THE CALCULATOR IS IN DEGREE MODE.**

For choice (1), we get $\dfrac{6}{\cos 0} = 1$.

For choice (2), we get $\dfrac{6}{\cos 30} = 6.9282$.

For choice (3), we get $\dfrac{6}{\cos 60} = 12$.

For choice (4), we get $\dfrac{6}{\cos 90} = err$.

PROBLEM 28: QUADRATIC EQUATIONS

The quadratic formula says that, given a quadratic equation of the form $ax^2 + bx + c = 0$, the roots of the equation are:

$$x = \frac{-b \pm \sqrt{b^2 - 4ac}}{2a}.$$

We have to try each of the answer choices in the formula to see which one works.

Choice (1):

$$x = \frac{-6 \pm \sqrt{6^2 - 4(1)(7)}}{2(1)} = \frac{-6 \pm \sqrt{8}}{2} = \frac{-6 \pm 2\sqrt{2}}{2} = -3 \pm \sqrt{2}$$

Choice (2):

$$x = \frac{-(-6) \pm \sqrt{(-6)^2 - 4(1)(7)}}{2(1)} = \frac{6 \pm \sqrt{8}}{2} = \frac{6 \pm 2\sqrt{2}}{2} = 3 \pm \sqrt{2}$$

We don't have to bother with the other choices because each is going to have different roots.

The answer is (2).

Another way to get this right was to find the value of $3 + \sqrt{2} \approx 4.414$ on your calculator and plug it into each answer choice to see if it gave you zero. Then repeat the process with $3 - \sqrt{2} \approx 1.586$.

PROBLEM 29: PROBABILITY

If the probability of a particular outcome is **p**, then the probability of that outcome occurring **r** times out of a possible **n** times is $_nC_r(p)^r(1-p)^{n-r}$. This is known as *binomial probability*.

Here, we are told that the probability of Nicole's making a basket is $\frac{4}{10} = 0.4$. The probability of her making exactly 2 baskets out of a possible 3 is $_3C_2(0.4)^2(1-0.4)^{3-2}$. This can be simplified to $_3C_2(0.4)^2(0.6)^1$

We can then evaluate $_3C_2$ according to the formula $_nC_r = \dfrac{n!}{(n-r)!\,r!}$

If we plug in, we get $_3C_2 = \dfrac{3!}{(3-2)!\,2!} = \dfrac{3!}{1!\,2!} = \dfrac{3 \cdot 2 \cdot 1}{(1)(2 \cdot 1)} = 3$

Therefore, the probability of making exactly 2 out of 3 baskets is $3(0.4)^2(0.6)^1 = 0.288$

The answer is (1).

PROBLEM 30: RATIONAL EXPRESSIONS

First we find a common denominator for the bottom of the fraction

$$\frac{1}{R_1} + \frac{1}{R_2} = \frac{R_1 + R_2}{R_1 R_2}$$

Now, because we have one over a fraction, we invert the fraction

$$\frac{1}{\dfrac{R_1 + R_2}{R_1 R_2}} = \frac{R_1 R_2}{R_1 + R_2}$$

The answer is (4).

Another way to get this right is by plugging in. First make up values for R_1 and R_2. For example, let $R_1 = 2$ and $R_2 = 3$. If we plug them into the problem, we get $\dfrac{1}{\dfrac{1}{2} + \dfrac{1}{3}} = 1.2$. Then, plug $R_1 = 2$ and $R_2 = 3$ into each of the answer choices and pick the one that matches.

Choice (1): $2 + 3 = 5$; so answer (1) is wrong.

Choice (2): $\dfrac{2 + 3}{2 \cdot 3} = 0.833$; so answer (2) is wrong.

Choice (3): $2 \cdot 3 = 6$; so answer (3) is wrong.

Choice (4): $\dfrac{2 \cdot 3}{2 + 3} = 1.2$; **so answer (4) is correct.**

PROBLEM 31: QUADRATIC INEQUALITIES

When we are given a quadratic inequality, the first thing that we do is factor it. We get: $(x - 2)(x + 2) > 0$. The roots of the quadratic are $x = 2$, $x = -2$. If we plot the roots on a number line

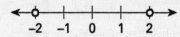

we can see that there are three regions: $x < -2$, $-2 < x < 2$, and $x > 2$. Now we try a point in each region to see whether the point satisfies the inequality.

First, we try a number less than -2. Let's try -3. Is $(-3 - 2)(-3 + 2) > 0$? Yes.

Now, we try a number between -2 and 2. Let's try 0. Is $(0 - 2)(0 + 2) > 0$? No.

Last, we try a number greater than 2. Let's try 3. Is $(3 - 2)(3 + 2) > 0$? Yes.

Therefore, the regions that satisfy the inequality are $x < -2$ or $x > 2$.

The answer is (3).

PROBLEM 32: QUADRATIC EQUATIONS

We can determine the nature of the roots of a quadratic equation of the form $ax^2 + bx + c = 0$ by using the discriminant $b^2 - 4ac$.

You should know the following rule:

> - If $b^2 - 4ac < 0$, the equation has two imaginary roots.
> - If $b^2 - 4ac = 0$, the equation has one rational root.
> - If $b^2 - 4ac > 0$, and $b^2 - 4ac$ is a perfect square, then the equation has two rational roots.
> - If $b^2 - 4ac > 0$, and $b^2 - 4ac$ is not a perfect square, then the equation has two irrational roots.

So to find where this equation has real roots, we need to find where

$$b^2 - 4ac \geq 0 .$$

Plugging into the discriminant, we get:

$$b^2 - 4ac = 16 - 4(4)c = 16 - 16c \geq 0 .$$
$$16c \leq 16$$
$$c \leq 1$$

The answer is (1).

PROBLEM 33: TRIGONOMETRY

You should know that $\sin x = \cos(90° - x)$ and that $\cos x = \sin(90° - x)$. We could use either substitution, so if we use the latter substitution in the equation we get:

$$\sin(90° - (2x - 1)) = \sin(3x + 6).$$

Therefore $(90 - (2x - 1)) = (3x + 6)$. If we solve for x, we get·

$$90 - 2x + 1 = 3x + 6$$
$$91 - 2x = 3x + 6$$
$$85 = 5x$$
$$17 = x.$$

The answer is (2).

Another way to get the answer is to **backsolve**. Try each of the values for x in the problem and see which one works. Make sure that your calculator is in Degree mode.

Choice (1): Does $\cos(2(-7) - 1)° = \sin(3(-7) + 6)°$? No.

Choice (2): Does $\cos(2(17) - 1)° = \sin(3(17) + 6)°$? **Yes.**

Choice (3): Does $\cos(2(35) - 1)^\circ = \sin(3(35) + 6)^\circ$? No.

Choice (4): Does $\cos(2(71) - 1)^\circ = \sin(3(71) + 6)^\circ$? No.

PROBLEM 34: TRANSFORMATIONS

Here we are asked to find the results of a composite transformation $(R_{-180^\circ} \circ R_{90^\circ})(B)$. *Whenever we have a composite transformation, always do the **right** one first.*

R_{90° means a rotation counterclockwise of 90°. Therefore, $R_{90^\circ}(B) \rightarrow A$.

R_{-180° means a rotation clockwise of 180°. Therefore, $R_{-180^\circ}(A) \rightarrow C$

The answer is (3).

PROBLEM 35: CONIC SECTIONS

A graph of the form $\dfrac{x^2}{a^2} + \dfrac{y^2}{b^2} = 1$, *where* **a** *and* **b** *have the same value, is the equation of a circle. If* **a** *and* **b** *have different values, the graph is that of an ellipse.*

Here we have $\dfrac{x^2 + y^2}{2} = 5$. We can divide both sides by 5 and we get.

$$\frac{x^2 + y^2}{10} = 1$$

We can then break this into two fractions and we get.

$\dfrac{x^2}{10} + \dfrac{y^2}{10} = 1$, so the graph is that of a circle.

The answer is (1).

Part II

PROBLEM 36: CIRCLE RULES

(a) We are given that AC is a diameter of the circle; therefore, the measure of arc $\overset{\frown}{ADC}$ is 180. We also know that the ratio of $\dfrac{m\overset{\frown}{CD}}{m\overset{\frown}{DA}} = \dfrac{1}{8}$. If we let $m\overset{\frown}{CD} = x$, then $m\overset{\frown}{DA} = 8x$; and, because $m\overset{\frown}{CD} + m\overset{\frown}{DA} = 180$, $x + 8x = 180°$. If we solve this for x, we get $x = 20°$. Therefore, $m\overset{\frown}{CD} = 20$ and $m\overset{\frown}{DA} = 160$.

(b) *The measure of an angle formed by a pair of secants, or a secant and a tangent, is equal to half of the difference between the larger and the smaller arcs that are formed by the secants, or the secant and the tangent.*

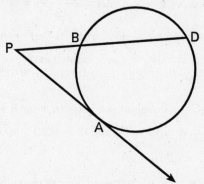

Here, the larger and smaller arcs formed by angle P are $\overset{\frown}{DA}$ and $\overset{\frown}{AB}$, respectively. Thus, $m\angle P = \dfrac{m\overset{\frown}{DA} - m\overset{\frown}{AB}}{2}$. Because we know that $m\angle P = 40$, we have $40 = \dfrac{160 - m\overset{\frown}{AB}}{2}$.

If we solve for $m\overset{\frown}{AB}$, we get that $m\overset{\frown}{AB} = 80$.

(c) *If two chords intersect within a circle, then the angle between the two chords is the average of their intercepted arcs.*

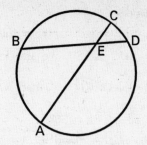

Therefore, $m\angle BEA = \dfrac{m\widehat{CD} + m\widehat{AB}}{2}$.

Substituting, we get: $m\angle BEA = \dfrac{20 + 80}{2} = 50$.

(d) *The measure of an inscribed angle is half of the arc it subtends (intercepts).* Thus, if we can find the measure of arc $\widehat{BC}$, we know that the measure of $\angle BAC$ is half of that.

Because AC is a diameter, we know that $m\widehat{AB} + m\widehat{BC} = 180$. We also know that $m\widehat{AB} = 80$, and thus $m\widehat{BC} = 100$.

Therefore, $m\angle BAC = 50$.

(e) *A tangent to a circle is perpendicular to the radius of the circle at the point of tangency.* In other words, $m\angle PAC = 90$. Because we know that $m\angle BAC = 50$, we know that $m\angle PAB = 40$. We also know that $m\angle P = 40$; therefore, $m\angle PBA = 180 - (40 + 40) = 100$.

PROBLEM 37:

(a) Radical Equations

The first thing that we do is to isolate the radical term by subtracting x from both sides. We get: $\sqrt{2x-1} = 8 - x$. Next, we square both sides.

$$2x - 1 = (8 - x)^2$$

$$2x - 1 = 64 - 16x + x^2.$$

Now we combine terms and we get:

$$x^2 - 18x + 65 = 0.$$

If we factor this we get:

$$(x - 5)(x - 13) = 0.$$

The roots of this equation are $x = 5$ and $x = 13$.

However, whenever we have a radical in an equation, we have to check the roots to see if they satisfy the original equation. Sometimes, in the process of squaring both sides, we will get answers that are invalid. So let's check the answers.

Does $5 + \sqrt{2(5) - 1} = 8$? **Yes.**

Does $13 + \sqrt{2(13) - 1} = 8$? **NO!**

Therefore, the answer is only $x = 5$.

(b) Rational Expressions

First the easy way.

Combine the terms with like denominators. We get:

$$\frac{y}{y - 1} - \frac{1}{y - 1} = \frac{8}{y}$$

$$\frac{y - 1}{y - 1} = \frac{8}{y}$$

Now, simplify the left side. $1 = \dfrac{8}{y}$. Therefore, $y = 8$.

Some of you may have done this the hard way.

The first thing that we do is get a common denominator. In this case, we want all terms to have a denominator of $y(y - 1)$. If we multiply the top and bottom of the first term by y, we get $\dfrac{y^2}{y(y - 1)}$. If we multiply the top and bottom of the second term by $y - 1$, we get $\dfrac{8(y - 1)}{y(y - 1)}$. If

we multiply the top and bottom of the last term by y, we get $\dfrac{y}{y(y - 1)}$.

Then our equation becomes: $\dfrac{y^2}{y(y - 1)} = \dfrac{8(y - 1)}{y(y - 1)} + \dfrac{y}{y(y - 1)}$.

We can now combine the fractions on the right side to get:

$$\dfrac{8(y - 1) + y}{y(y - 1)} = \dfrac{9y - 8}{y(y - 1)}.$$

Now, we can ignore the denominators and set the numerators equal to each other. (Note that if we get the answers $y = 0$ or $y = 1$ we must throw them out because either will make the denominator undefined.) $y^2 = 9y - 8$. Let's solve this equation.

$$y^2 = 9y - 8$$
$$y^2 - 9y + 8 = 0$$
$$(y - 1)(y - 8) = 0$$
$$y = 1, 8$$

We throw out the answer $y = 1$ (see the note above) so the answer is $y = 8$.

PROBLEM 38:

(a) Trigonometric Equations

First, put all of the terms on the left side of the equals sign. We get $2 \sin^2 x - \sin x - 1 = 0$.

Note that the trig equation $2 \sin^2 x - \sin x - 1 = 0$ has the same form as a quadratic equation $2x^2 - x - 1 = 0$. Just as we could factor the quadratic equation, so too can we factor the trig equation. We get: $(2 \sin x + 1)(\sin x - 1) = 0$.

This gives us $2 \sin x + 1 = 0$ and $\sin x - 1 = 0$. If we solve each of these equations, we get: $\sin x = -\dfrac{1}{2}$ and $\sin x = 1$.

Therefore $x = \sin^{-1}\left(-\dfrac{1}{2}\right)$ and $x = \sin^{-1}(1)$. The solutions to both equations are easy because they are special angles.

You should know that $\sin 210° = -\dfrac{1}{2}$ and that $\sin 330° = -\dfrac{1}{2}$. You should also know that $\sin 90° = 1$, so the solutions are $x = 90°, 210°, 330°$.

(b) Trigonometric Identities

All of the identities that you will need to know are contained on the formula sheet.

First, let's get rid of the double angle terms.

Using $\cos 2x = 1 - 2\sin^2 x$ and $\sin 2x = 2\sin x \cos x$, we can rewrite the right side of the equation as: $\dfrac{2\sin x \cos x}{1 - \left(1 - 2\sin^2 x\right)}$.

Next, we cancel like terms and simplify:

$$\frac{2\sin x \cos x}{1 - \left(1 - 2\sin^2 x\right)} = \frac{2\sin x \cos x}{2\sin^2 x} = \frac{\cos x}{\sin x}.$$

Finally, we can rewrite the right side as $\cot x$ and set the two sides equal, and we have proved the identity.

PROBLEM 39: TRIGONOMETRIC GRAPHS

(a) You should know what the basic tangent graph looks like.

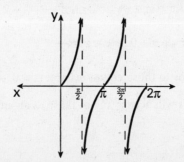

(b) A graph of the form $y = a \sin bx$ or $y = a \cos bx$ has an amplitude of $|a|$ and a period of $\frac{2\pi}{b}$. Therefore, the equation $y = 2 \sin x$ has an amplitude of 2 and a period of 2π.

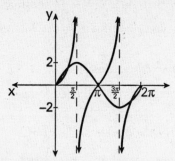

Remember, in graphing sines and cosines, the shape of the graph doesn't change, but it can be stretched or shrunk depending on the amplitude or period.

(c) We can find the intersections two ways: First, we can find the intersections from our graphs.

As you can see, the graphs clearly intersect at 0, π, and 2π.

The second way is to solve the equation algebraically. Replace $\tan x$ with $\frac{\sin x}{\cos x}$. Now we have: $\frac{\sin x}{\cos x} = 2 \sin x$.

Now set the equation equal to zero. We get: $\frac{\sin x}{\cos x} - 2 \sin x = 0$.

Factor out $\sin x$ and we get: $\sin x \left(\frac{1}{\cos x} - 2 \right) = 0$.

Thus, either $\sin x = 0$ or $\cos x = \frac{1}{2}$.

The solutions to these equations are: $x = 0, \frac{\pi}{3}, \pi, \frac{5\pi}{3}, 2\pi$.

The Regents will accept any of these as an answer.

Problem 40: Vectors

(a) Vector problems on the Regents exams can be solved with the Law of Cosines or the Law of Sines.

First, let's draw a picture of the situation.

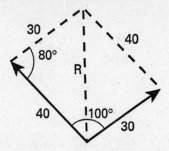

We find the resultant force by making a parallelogram and finding the length of the diagonal marked R. We know that the other angle is 80°, so now we have SAS and we can use the Law of Cosines.

The Law of Cosines says that, in any triangle, with angles A, B and C, and opposite sides of a, b, and c, respectively, $c^2 = a^2 + b^2 - 2ab \cos C$ (This is on the formula sheet). If we have a triangle and we are given SAS (side, angle, side), then the side opposite the included angle is c, and the included angle is C. In this case, the included angle is 80° and its opposite side is R

$$R^2 = 30^2 + 40^2 - 2(30)(40) \cos 80°$$
$$R^2 = 900 + 1600 - 2400 \cos 80°$$
$$R^2 \approx 2083.244$$
$$R \approx 45.6$$

(b) Law of Sines

Now we want to find the angle labeled A.

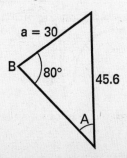

The Law of Sines says that, given a triangle with angles A B and C and opposite sides of a b and c.

$$\frac{a}{\sin A} = \frac{b}{\sin B} = \frac{c}{\sin C}$$

(This is on the formula sheet)

Substituting we have $\dfrac{45.6}{\sin 80°} = \dfrac{30}{\sin A}$

Now we solve this for A.

$$46.3 = \frac{30}{\sin A}$$

$$\sin A = \frac{30}{46.3} = 0.65$$

$$A = \sin^{-1}(0.65) = 40.4 \approx 40$$

PROBLEM 41: GRAPHING

(a) The graph of $\log_2 x = y$ goes through the point $(1, 0)$ the point $(2, 1)$ the point $(4, 2)$ and the point $(8, 3)$

The graph $xy = 8$ is a hyperbola in quadrants I and III, that goes through the points $(1, 8)$ and $(8, 1)$ in quadrant I and the points $(-1, -8)$ and $(-8, -1)$ in quadrant III

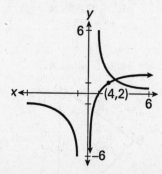

(b) If you look at the graphs, you can see that 5 or 6 will work

PROBLEM 42:

(a) Probability

(1) This problem requires that you know something called *binomial probability*. The rule is. *If the probability of a particular outcome is* **p**, *then the probability of that outcome occurring* **r** *times out of a possible* **n** *times is* $_nC_r(p)^r(1-p)^{n-r}$

The probability that the coin will come up heads at least two times out of a possible three is the sum of the probability that it will be heads two times and the probability that it will be heads all three times

The probability that the coin will come up heads on any given flip is $\frac{3}{4}$

Therefore, the probability that the coin will come up heads two out of three times is $_3C_2\left(\frac{3}{4}\right)^2\left(1-\frac{3}{4}\right)^1$ and the probability of it coming up heads three times is $_3C_3\left(\frac{3}{4}\right)^3\left(1-\frac{3}{4}\right)^0$

The rule for finding $_nC_r$ is $_nC_r = \frac{n!}{(n-r)!\,r!}$ Therefore,

$$_3C_2 = \frac{3!}{1!\,2!} = \frac{3\cdot 2\cdot 1}{(1)(2\cdot 1)} = 3$$

and

$$_3C_3 = \frac{3!}{0!\,3!} = \frac{3\cdot 2\cdot 1}{(1)(3\cdot 2\cdot 1)} = 1$$

(By the way, 0! = 1). Thus, the probability that the coin will come up heads at least two times is:

$$3\left(\frac{3}{4}\right)^2\left(\frac{1}{4}\right)^1 + 1\left(\frac{3}{4}\right)^3\left(\frac{1}{4}\right)^0 = \frac{27}{64} + \frac{27}{64} = \frac{54}{64}$$

(2) The probability that the coin will come up heads at most one time out of a possible four is the sum of the probability that it will be heads one time and the probability that it will be heads zero times.

The probability that the coin will come up heads one out of four flips is $_4C_1\left(\frac{3}{4}\right)^1\left(1-\frac{3}{4}\right)^3$, and the probability of it coming up heads zero times is $_4C_0\left(\frac{3}{4}\right)^0\left(1-\frac{3}{4}\right)^4$.

$$_4C_1 = \frac{4!}{3!\,1!} = \frac{4\cdot 3\cdot 2\cdot 1}{(3\cdot 2\cdot 1)(1)} = 4$$

and

$$_4C_0 = \frac{4!}{4!\,0!} = \frac{4\cdot 3\cdot 2\cdot 1}{(4\cdot 3\cdot 2\cdot 1)(1)} = 1$$

Thus, the probability that the coin will come up heads at most one time is:

$$4\left(\frac{3}{4}\right)^1\left(\frac{1}{4}\right)^3 + 1\left(\frac{3}{4}\right)^0\left(\frac{1}{4}\right)^4 = \frac{12}{256} + \frac{1}{256} = \frac{13}{256}.$$

(b) Statistics

The method for finding a standard deviation is simple, but time-consuming.

First, find the average of the scores.

$$\overline{x} = \frac{100 + 99 + 99 + 97 + 96 + 96 + 95 + 94 + 93 + 91}{10} = 96.$$

Second, subtract the average from each actual score.

$100 - 96 = 4$	$96 - 96 = 0$
$99 - 96 = 3$	$95 - 96 = -1$
$99 - 96 = 3$	$94 - 96 = -2$
$97 - 96 = 1$	$93 - 96 = -3$
$96 - 96 = 0$	$91 - 96 = -5$

Third, square each difference

$(4)^2 = 16$	$(0)^2 = 0$
$(3)^2 = 9$	$(-1)^2 = 1$
$(3)^2 = 9$	$(-2)^2 = 4$
$(1)^2 = 1$	$(-3)^2 = 9$
$(0)^2 = 0$	$(-5)^2 = 25$

Fourth find the sum of these squares.

$$16 + 9 + 9 + 1 + 0 + 0 + 1 + 4 + 9 + 25 = 74$$

Fifth, divide this sum by the total number of people.

$$\frac{74}{10} = 7.4$$

Last, take the square root of this number. This is the standard deviation.

$$\sigma = \sqrt{7.4} \approx 2.72 \text{ (rounded to the nearest hundredth).}$$

Formulas

Pythagorean and Quotient Identities

$$\sin^2 A + \cos^2 A = 1 \qquad \tan A = \frac{\sin A}{\cos A}$$

$$\tan^2 A + 1 = \sec^2 A \qquad \cot A = \frac{\cos A}{\sin A}$$

$$\cot^2 A + 1 = \csc^2 A$$

Functions of the Sum of Two Angles

$$\sin (A + B) = \sin A \cos B + \cos A \sin B$$

$$\cos (A + B) = \cos A \cos B - \sin A \sin B$$

$$\tan (A + B) = \frac{\tan A + \tan B}{1 - \tan A \tan B}$$

Functions of the Difference of Two Angles

$$\sin (A - B) = \sin A \cos B - \cos A \sin B$$

$$\cos (A - B) = \cos A \cos B + \sin A \sin B$$

$$\tan (A - B) = \frac{\tan A - \tan B}{1 + \tan A \tan B}$$

Law of Sines

$$\frac{a}{\sin A} = \frac{b}{\sin B} = \frac{c}{\sin C}$$

Law of Cosines

$$a^2 = b^2 + c^2 - 2bc \cos A$$

Functions of the Double Angle

$$\sin 2A = 2 \sin A \cos A$$

$$\cos 2A = \cos^2 A - \sin^2 A$$

$$\cos 2A = 2 \cos^2 A - 1$$

$$\cos 2A = 1 - 2 \sin^2 A$$

$$\tan 2A = \frac{2 \tan A}{1 - \tan^2 A}$$

Functions of the Half Angle

$$\sin \frac{1}{2} A = \pm \sqrt{\frac{1 - \cos A}{2}}$$

$$\cos \frac{1}{2} A = \pm \sqrt{\frac{1 + \cos A}{2}}$$

$$\tan \frac{1}{2} A = \pm \sqrt{\frac{1 - \cos A}{1 + \cos A}}$$

Area of Triangle

$$K = \frac{1}{2} ab \sin C$$

Standard Deviation

$$\text{S.D.} = \sqrt{\frac{1}{n} \sum_{i=1}^{n} \left(x_i - \bar{x} \right)^2}$$

EXAMINATION
JANUARY 1997

Part 1

Answer 30 questions from this part. Each correct answer will receive 2 credits. No partial credit will be allowed. Write your answers in the spaces on your answer sheet. Where applicable, answers may be left in terms of π or in radical form.

1 In the accompanying diagram, $\overline{AB}$ is tangent to circle O at B and $\overline{ACD}$ is a secant. If $AB = 9$ and $AD = 27$, find AC.

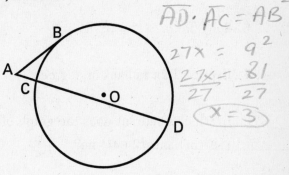

$$\overline{AD} \cdot \overline{AC} = AB^2$$
$$27x = 9^2$$
$$\frac{27x}{27} = \frac{81}{27}$$
$$\boxed{x = 3}$$

2 In terms of i, express in simplest form $\sqrt{-64} - 3\sqrt{-4}$.

$8i - 3 \cdot 2i$

$8i - 6i = 2i$

3 In $\triangle ABC$, $\sin A = \dfrac{2}{3}$, $\sin B = \dfrac{4}{5}$, and side $a = 20$. Find side b.

$\dfrac{20}{2/3} = \dfrac{b}{4/5}$

$120 = \overset{b}{5b}$

$24 = b$

$20 \cdot \dfrac{3}{2} = b \cdot \dfrac{5}{4}$

$30 = b$

[OVER]

4 In the accompanying diagram, $\overline{CD}$ is tangent to circle O at B, AO and BO are radii, and chord $\overline{AB}$ is drawn. If m $\angle AOB = 108$, find m $\angle ABD$.

$m\angle B = \frac{1}{2} \overparen{AB}$
$m\angle B = \frac{1}{2} 108$
$m\angle B = 54$

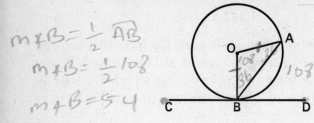

5 Solve for y: $\dfrac{4}{5y-3} = \dfrac{2}{3y+4}$. $y = -11$

$\dfrac{4}{-58} = \dfrac{2}{-29}$

6 Solve for y: $3^{y+1} = 9^{y-1}$.

$27 = $ $y = 2$

7 Express 1.2π radians in degrees.

216

8 In which quadrant does the graph of the sum of $(-3-5i)$ and $(2+4i)$ lie? III

9 Evaluate: $\displaystyle\sum_{n=1}^{3}(2n-1)$. 9

10 If $f(x) = x - 3$ and $g(x) = x^2$, what is the value of $(f \circ g)(2)$? $F(4) = 1$

11 In $\triangle ABC$, $a = 8$, $b = 9$, and $\cos C = \dfrac{2}{3}$. Find c

$c = 7$

12 In a normal distribution, 68% of the scores fall between 72 and 86 and the mean is 79. What is the standard deviation?

7

13 In a circle with a radius of 4 centimeters, what is the number of radians in the central angle that intercepts an arc of 8 centimeters?

2

14 If x varies inversely as y, and $x = 9$ when $y = 8$, find x when $y = 12$.

6

Directions (15–35): For *each* question chosen, write the *numeral* preceding the word or expression that best completes the statement or answers the question.

15 If $f(x) = 3^x$, then $f(-2)$ equals

   (1) $\dfrac{1}{9}$ $F(-2) = 3^{-2}$

   (2) 9 $F(-2) = 1/3^2$

   (3) −6 $F(-2) = 1/9$

   (4) −9

16 The expression $\dfrac{\cot\theta}{\csc\theta}$ is equivalent to

   (1) $\dfrac{\cos\theta}{\sin^2\theta}$ $\cos\theta$

   (2) $\sin\theta$ $\dfrac{\sin\theta}{}$

   (3) $\tan\theta$

   (4) $\cos\theta$ $\dfrac{1}{\sin\theta}$

$\dfrac{\cos\theta}{\sin\theta}$, $\dfrac{\sin\theta}{1}$

[OVER]

17 In the accompaning diagram of circle O, the measure of RS is 64°.

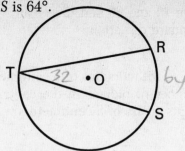

$m \angle T = \frac{1}{2} RS$

$x = \frac{1}{2} \cdot 64$

What is m $\angle RTS$?

(1) 32   (3) 96
(2) 64   (4) 128

I, II

18 If $\sin A > 0$ and $(\sin A) < 0$, in which quadrant does $\angle A$ terminate?

(1) I   (3) III
(2) II   (4) IV

19 The solution set of the inequality $|x - 3| < 5$ is

(1) $\{x < 8 \text{ and } x < -2\}$

(2) $\{x < 8 \text{ or } x < -2\}$

(3) $\{x < 8 \text{ and } x > -2\}$

(4) $\{x > 8 \text{ or } x < -2\}$

$-(x - 3| < 5$ $x - 3 < 5$
$-x + 3 < 5$ $x < 8$
$-x < 2$
$-x$
$x > -2$

20 For which value of x is $f(x) = \dfrac{\sin x}{\cos x}$ undefined?

(1) 0   (3) $\dfrac{\pi}{2}$

(2) $\dfrac{\pi}{4}$ (4) π

21 The expression $\cos 80° \cos 20° - \sin 80° \sin 20°$ is equivalent to

(1) $\cos 60°$ (3) $\sin 100°$

(2) $\cos 100°$ (4) $\sin 60°$

22 Which statement is true about the roots of the equation $\sqrt{x^2 - 5x + 5} = 1$?

(1) The only root is 1.
(2) The only root is 4.
(3) Both 1 and 4 are the roots.
(4) Neither 1 nor 4 is a root.

23 If the coordinates of P are $(-2,7)$, what are the coordinates of $\left(D_2 \circ r_{y=x}\right)(P)$?

(1) $(4,-14)$ (3) $(-4,14)$

(2) $(-14,4)$ (4) $(14,-4)$

[OVER]

24 Which graph represents the equation $\frac{x^2}{4} + \frac{y^2}{4} = 1$?

(1)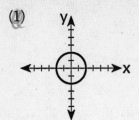

(3)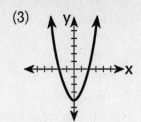

(2)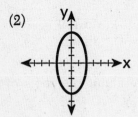

(4)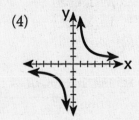

25 What is the period of the graph of the equation $y = a \sin bx$?

(1) $\frac{2\pi}{a}$

(3) a

(2) $\frac{2\pi}{b}$

(4) b

26 What is the product of the roots of the equation $2x^2 - x - 2 = 0$?

(1) 1

(3) −1

(2) 2

(4) −2

27 What is the image of $A(5,2)$ under $R_{90°}$?

(1) (−5,2)

(3) (2,5)

(2) (5,−2)

(4) (−2,5)

28 The set $\{0,1,-1\}$ is closed under the operation of

(1) addition   (3) subtraction
(2) multiplication   (4) division

29 If $\sin\left(x+20°\right) = \cos x$, the value of x is

(1) 35°   (3) 55°
(2) 45°   (4) 70°

30 The probability of Gordon's team winning any given game in a 5-game series is 0.3. What is the probability that Gordon's team will win *exactly* 2 games in the series?

(1) $\left(0.3\right)^{2}\left(0.7\right)^{3}$ (3) $10\left(0.3\right)^{2}\left(0.7\right)^{3}$

(2) $5\left(0.3\right)^{3}\left(0.7\right)^{2}$ (4) $5\left(0.3\right)^{2}\left(0.7\right)$

31 What is the solution set of the inequality $x^{2} - 3x - 10 > 0$?

(1) $\{x|-2 < x < 5\}$ (3) $\{x|x < -5 \text{ or } x > 2\}$
(2) $\{x|-5 < x < 2\}$ (4) $\{x|x < -2 \text{ or } x > 5\}$

32 What is the middle term in the expansion of $\left(3x - 2y\right)^{4}$?

(1) $-6x^{2}y^{2}$ (3) $-216x^{2}y^{2}$
(2) $36x^{2}y^{2}$ (4) $216x^{2}y^{2}$

[OVER]

33 If the graphs of the equations $y = \log_3 x$ and $y = 2$ are drawn on the same set of axes, they will intersect where x is equal to

(1) 1   (3) 3

(2) 2   (4) 9

34 An obtuse angle of a parallelogram has a measure of 150°. If the sides of the parallelogram measure 10 and 12 centimeters, what is the area of the parallelogram?

(1) 30 cm²   (3) $60\sqrt{2}$ cm²

(2) 60 cm²   (4) $60\sqrt{3}$ cm²

35 The accompanying diagram shows a sketch of a quadratic function, $f(x)$

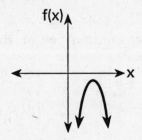

What is the nature of the roots of the quadratic equation $f(x) = 0$?

(1) imaginary

(2) real, rational, and equal

(3) real, rational, and unequal

(4) real, irrational, and unequal

Part II

Answer four questions from this part. Clearly indicate the necessary steps, including appropriate formula substitutions, diagrams, graphs, charts, etc. Calculations that may be obtained by mental arithmetic or the calculator do not need to be shown. [40]

36 *a* Solve for x and express the roots in terms of i:

$$\frac{x+3}{3} + \frac{x+3}{x} = 2. \quad [4]$$

   b Solve for x and express the roots in simplest $a + bi$ form:

$$x^2 = 6x - 10. \quad [6]$$

37 *a* Find, to the *nearest degree*, all values of θ in the interval $0° \le \theta < 360°$ that satisfy the equation $3 \sin^2 \theta - \sin \theta - 2 = 0$. [8]

   b Solve for x to the *nearest tenth*: $5^x = 30$. [2]

38 *a* On the same set of axes, sketch and label the graphs of the equations $y = -2 \sin x$ and and $y = \cos 2x$ as x varies from 0 to 2π radians. [8]

   b Using the graphs sketched in part *a*, determine the number of points in the interval $0 \le x \le 2\pi$ that satisfy the equation $\cos 2x = -2 \sin x$. [2]

[OVER]

39 *a* The table below shows the set of score data for an English examination.

x_i	f_i
100	2
90	3
80	6
70	5
60	4

Find the standard deviation of these scores to the *nearest tenth.* [4]

b In the accompanying diagram, a regular hexagon with a spinner is divided in six equal areas labeled with a letter or a number. [3]

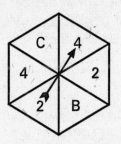

If the spinner is spun four times, find the probability that it will land in a

(1) numbered area *at most* one time [3]
(2) lettered area *at least* three times [3]

40 *a* Express in simplest form:

$$\frac{1-\dfrac{1}{x}}{\dfrac{1}{x^2}-\dfrac{1}{x}} \quad \text{[5]}$$

b For all values of x for which the expressions are defined, prove the following is an identity:

$$\frac{\cos 2x}{\sin x}+\frac{\sin 2x}{\cos x}=\csc x \quad \text{[5]}$$

41 In the accompanying diagram of circle O, $\triangle ABC$ is formed by tangent $\overline{AB}$, secant $\overline{BDC}$, and chord $\overline{AC}$; $\overline{CA}\cong\overline{CD}$; $m\widehat{AC}=140$; and $AC=10$.

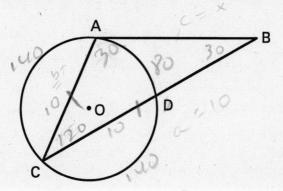

Find:

a $m\widehat{AD}$ [2]   80
b $m\angle B$ [2]   = 30
c AB to the *nearest tenth*   [6]

$m\widehat{AC} + m\widehat{CD} + \widehat{AD} = 360$
$140 + 140 + x = 360$
$280 + x = 360$
$x = 80$

$m\angle B = \tfrac{1}{2}(140 - 80)$

[OVER]

42   *a*   *On your answer paper*, copy and complete the table for the values of y for the equation $y = \log_2 x$.   [4]

     b   *On graph paper*, using the completed table, draw the graph of the equation $y = \log_2 x$ for the interval $\frac{1}{4} \leq x \leq 4$. Label the graph b.   [2]

     c   On the same set of axes, reflect the graph drawn in part *b* in the y-axis and label it *c*.   [2]

     d   On the same set of axes, reflect the graph drawn in part *b* in the line $y = x$ and label it *d*.   [2]

ANSWER KEY

Part I

1. 3	13. 2	25. (2)
2. 2*i*	14. 6	26. (3)
3. 24	15. (1)	27. (4)
4. 54	16. (4)	28. (2)
5. −11	17. (1)	29. (1)
6. 3	18. (2)	30. (3)
7. 216	19. (3)	31. (4)
8. III	20. (3)	32. (4)
9. 9	21. (2)	33. (4)
10. 1	22. (3)	34. (2)
11. 7	23. (4)	35. (1)
12. 7	24. (1)	

ANSWERS AND EXPLANATIONS
JANUARY 1997

Part I

In problems 1–14, you must supply the answer. Because of this, you will not be able to **plug-in** or **backsolve**. But, you can use your calculator to simplify calculations. In addition, you will find that these problems tend to test your basic knowledge of Sequential Math III and are not tricky. Many can be solved just by knowing the right formula and by plugging in the numbers.

PROBLEM 1: CIRCLE RULES

Let x be the length of AC. We know from our circle rules that *the product of the secant and its external segment is equal to the square of the tangent segment*. Therefore $(AD)(AC) = (AB)^2$. Thus we have $27x = 9^2$. If we solve for x we find that $x = 3$.

PROBLEM 2: COMPLEX NUMBERS

$\sqrt{-64} - 3\sqrt{-4}$ can be rewritten as $\sqrt{64}\sqrt{-1} - 3\sqrt{4}\sqrt{-1}$. Replacing $\sqrt{-1}$ with i we get: $\sqrt{64}\,i - 3\sqrt{4}\,i$.

Simplifying the radicals, we get: $8i - 6i = 2i$.

PROBLEM 3: LAW OF SINES

From the Law of Sines, we know that $\dfrac{\sin A}{a} = \dfrac{\sin B}{b}$

If we substitute the values we are given, we get

$$\frac{\frac{2}{3}}{20} = \frac{\frac{4}{5}}{b}$$

Next, we cross-multiply to get:

$$\frac{2}{3}(b) = \frac{4}{5}(20)$$

Now we solve:

$$\frac{2}{3}(b) = 16$$

$$b = 16\left(\frac{3}{2}\right) = 24.$$

PROBLEM 4: CIRCLE RULES

The measure of a central angle is the same as the arc it subtends (intercepts).

If $m\angle AOB = 108$ then $m\widehat{AB} = 108$.

Then, because CD is tangent to the circle at B, $m\angle ABD$ is half of the arc $m\widehat{AB}$. Thus $m\angle ABD = 54$.

PROBLEM 5: RATIONAL EXPRESSIONS

$$\frac{4}{5y - 3} = \frac{2}{3y + 4}$$

Cross-multiply to obtain $4(3y + 4) = 2(5y - 3)$.

Distributing, we get:

$$12y + 16 = 10y - 6.$$

Now we can solve

$$2y = -22$$
$$y = -11.$$

PROBLEM 6: EXPONENTIAL EQUATIONS

$$3^{y+1} = 9^{y-1}.$$

In order to solve this equation, we need to have a common base for the two sides of the equation. Because $9 = 3^2$, we can rewrite the right side as $3^{y+1} = \left(3^2\right)^{y-1}$.

The rules of exponents say that *when a number raised to a power is itself raised to a power, we multiply the exponents.* Thus, we get

$$3^{y+1} = 3^{2y-2}.$$

Now, we can set the powers equal to each other and solve. $y + 1 = 2y - 2$

$$y = 3$$

PROBLEM 7: CONVERTING FROM RADIANS TO DEGREES

All we have to do is multiply 1.2π by $\dfrac{180°}{\pi}$

$$1.2\pi\left(\dfrac{180°}{\pi}\right) = 1.2(180°) = 216°$$

PROBLEM 8: COMPLEX NUMBERS

To find the sum of $-3 - 5i$ and $2 + 4i$ we add the real components and the imaginary components separately. This gives us:

$$(-3 + 2) + (-5 + 4)i = -1 - i$$

Next, to determine the quadrant, we look at the signs of the components. *The real component corresponds to the x-coordinate of the graph, and the imaginary component corresponds to the y-coordinate of the graph.* Because both components are negative, the graph has negative x- and y-coordinates. Therefore, the graph is in quadrant III.

PROBLEM 9: SUMMATIONS

To evaluate a sum, we plug each of the consecutive values for n into the expression, starting at the bottom value and ending at the top value and sum the results We get:

$$\sum_{n=1}^{3}(2n - 1) = [2(1) - 1] + [2(2) - 1] + [2(3) - 1] = 1 + 3 + 5 = 9.$$

Another way to do this is to use your calculator. Push **2nd STAT** (**LIST**) and under the **MATH** menu, choose **SUM**. Then, again push **2nd STAT** and under the **OPS** menu, choose **SEQ**.

We put in the following: (*expression, variable, start, finish, step*)

For *expression*, we put in the formula, using x as the variable.

For *variable*, we ALWAYS put in x.

For *start*, we put in the bottom value.

For *finish* we put in the top value

For *step* we ALWAYS put in 1.

Therefore, we put in the calculator: **SUM SEQ** $(2x - 1 \ x \ 1 \ 3, 1)$ The calculator should return the value 9

PROBLEM 10: COMPOSITE FUNCTIONS

When evaluating a composite function, start with the one on the right.

Here we first plug 2 into $g(x)$. We obtain $g(2) = 2^2 = 4$.

Next, we plug 4 into $f(x)$. We obtain $f(4) = 4 - 3 = 1$.

PROBLEM 11: LAW OF COSINES

The Law of Cosines says that, in any triangle, with angles A, B, and C, and opposite sides of a, b, and c, respectively, $c^2 = a^2 + b^2 - 2ab \cos C$ Here, we substitute the values and we get:

$$c^2 = 8^2 + 9^2 - 2(8)(9)\left(\frac{2}{3}\right)$$

If we solve this, we get.

$$c^2 = 64 + 81 - 96 = 49$$
$$c = 7$$

PROBLEM 12: STATISTICS

You should know the following rule about normal distributions

In a normal distribution, with a mean of $\bar{x}$ and a standard deviation of σ:
- approximately 68% of the outcomes will fall between $\bar{x} - \sigma$ and $\bar{x} + \sigma$.
- approximately 95% of the outcomes will fall between $\bar{x} - 2\sigma$ and $\bar{x} + 2\sigma$.
- approximately 99.5% of the outcomes will fall between $\bar{x} - 3\sigma$ and $\bar{x} + 3\sigma$.

Here, we are told that 68% of the scores lie between 72 and 86 with a mean of 79 Therefore, the standard deviation is $86 - 79 = 7$

Problem 13: Arc Length

In a circle with a central angle, θ, measured in radians; and a radius, r, the arc length; s is found by $s = r\theta$.

Here we have $r = 4$ and $s = 8$. Therefore, the central angle is 2 radians.

Problem 14: Inverse Variation

When two variables, x, and y, are said to vary inversely, their product is always a constant. In other words, $xy = k$, where k is a constant.

Here, we are given that $x = 9$ when $y = 8$. Thus, $(8)(9) = k$ and $k = 72$. So we now have $xy = 72$. If we plug in $y = 12$ and solve, we find that $x = 6$.

Multiple-Choice Problems

For problems 15–35, you will find that you can sometimes **plug-in** or **backsolve** to get the right answer. You will also find that a calculator will simplify the arithmetic. In addition, most of the problems only require knowing which formula to use.

Problem 15: Exponents

A number raised to a negative power is the same as the reciprocal of that number raised to the corresponding positive power Here

$$3^{-2} = \frac{1}{3^2} = \frac{1}{9}$$

The answer is (1).

Problem 16: Trigonometric Identities

Here you need to know that $\cot \theta = \dfrac{\cos \theta}{\sin \theta}$ and that $\csc \theta = \dfrac{1}{\sin \theta}$

Substituting into the expression, we get:

$$\frac{\cot \theta}{\csc \theta} = \frac{\dfrac{\cos \theta}{\sin \theta}}{\dfrac{1}{\sin \theta}}$$

Now we can simplify. Invert the denominator and multiply.

$$\frac{\dfrac{\cos\theta}{\sin\theta}}{\dfrac{1}{\sin\theta}} = \frac{\cos\theta}{\sin\theta} \cdot \frac{\sin\theta}{1} = \cos\theta.$$

The answer is (4).

Another way to get the right answer is to plug-in a value for θ. After you obtain a value for the expression, plug the same value for θ into each of the answer choices and find the one that matches.

For example, let $\theta = 30°$. Then $\dfrac{\cot\theta}{\csc\theta} = \dfrac{\cot 30°}{\csc 30°} = \dfrac{\sqrt{3}}{2}$.

Now plug 30° into each of the answers.

For choice (1) we get $2\sqrt{3}$.

For choice (2) we get $\dfrac{1}{2}$.

For choice (3) we get $\dfrac{\sqrt{3}}{3}$.

For choice (4) we get $\dfrac{\sqrt{3}}{2}$. This one matches so it's the right answer.

PROBLEM 17: CIRCLE RULES

The measure of an inscribed angle is always half of the measure of the arc that it subtends (intercepts). Here, the measure of the subtended arc is 64, so $m\angle RTS = 32$.

The answer is (1).

PROBLEM 18: TRIGONOMETRY

The sine of an angle is positive in quadrants I *and* II.

If the product of the sine and cosine is negative, then, because the sine is positive, the cosine must be negative.

The cosine of an angle is negative in quadrants II *and* III. Therefore the angle must be in quadrant II.

The answer is (2).

PROBLEM 19: ABSOLUTE VALUE EQUATIONS

When we have an absolute value inequality, we break it into two equations. First, we rewrite the left side without the absolute value and leave the right side alone. Second, we rewrite the left side without the absolute value, flip the inequality sign, and change the sign of the right side.

Here, we have $|x - 3| < 5$, so we rewrite it as: $x - 3 < 5$ and $x - 3 < -5$
If we solve each of these independently, we get $x < 8$ and $x > -2$.

The answer is (3).

PROBLEM 20: TRIGONOMETRY

Where the denominator of a fraction is zero, it is undefined. Here we need to know where $\cos x = 0$. Cosine is zero at $\dfrac{\pi}{2}$ radians

The answer is (3).

Another way to get this right is to **backsolve**. Plug each of the values of the answer choices into $f(x)$ and see which one gives you an error **MAKE SURE THAT THE CALCULATOR IS IN RADIAN MODE.**

For choice (1), we get $\dfrac{\sin 0}{\cos 0} = 0$.

For choice (2), we get $\dfrac{\sin \dfrac{\pi}{4}}{\cos \dfrac{\pi}{4}} = 1$

For choice (3), we get $\dfrac{\sin \dfrac{\pi}{2}}{\cos \dfrac{\pi}{2}} = err.$

For choice (4), we get $\dfrac{\sin \pi}{\cos \pi} = 0$.

Note that, even though we found the answer we wanted on choice (3) we tried all four choices. You never know when you might be making a mistake!

PROBLEM 21: TRIGONOMETRY FORMULAS

Notice that this problem has the form cos A cos B – sin A sin B. Each Regents exam comes with a formula sheet that contains all of the trig formulas that you will need to know. If you look under <u>Functions of the Sum of Two Angles</u> , you will find the formula

$$\cos(A + B) = \cos A \cos B - \sin A \sin B$$

Thus we can rewrite the problem as:

$$\cos 80° \cos 20° - \sin 80° \sin 20° = \cos(80° + 20°) = \cos 100°$$

The answer is (2).

Another way to answer this question is to use your calculator to find the value of cos 80° cos 20° – sin80° sin 20°. **MAKE SURE THAT YOUR CALCULATOR IS IN DEGREE MODE.** You should get –0.1736. Now find the value of each of the answer choices with your calculator.

cos 60° = 0.5, so (1) is not the answer.

cos 100° = –0.1736, so (2) is the answer.

Just in case, check the other two values.

sin 100° = –0.9848, so (3) is not the answer

sin 60° = 0.8660, so (4) is not the answer

PROBLEM 22: RADICAL EQUATIONS

If we square both sides of the equation, we get $x^2 - 5x + 5 = 1$. If we subtract 1 from both sides, we get $x^2 - 5x + 4 = 0$. Now we can factor. $(x - 4)(x - 1) = 0$. This means that the roots of the equation are 1 and 4. **BUT**, whenever we have a radical in an equation, we have to check the roots to see if they satisfy the original equation. Sometimes, in the process of squaring both sides, we will get answers that are invalid So let's check the answers.

Does $\sqrt{(1)^2 - 5(1) + 5} = 1$? $\sqrt{1} = 1$: so, yes, it does.

Does $\sqrt{(4)^2 - 5(4) + 5} = 1$? $\sqrt{1} = 1$: so, yes, it does.

Therefore, both are roots and **the answer is (3).**

Another way to do this is to **backsolve**. Try each answer choice and see if it works. We plug-in 1 and it works, so the answer cannot be (2) or (4). We then plug-in 4 and it too works, so the answer cannot be (1) and the answer must be (3).

PROBLEM 23: TRANSFORMATIONS

Here we are asked to find the results of a composite transformation $(D_2 \circ r_{y=x})(P)$, where P is $(-2, 7)$. *Whenever we have a composite transformation, always do the **right** one first.* The transformation $r_{y=x}$ means to reflect the point in the line $y = x$. This means that you simply switch the x- and y-coordinates. This gives us $r_{y=x}(-2, 7) \rightarrow (7, -2)$. The transformation D_2 means to dilate the point by 2. This means that you multiply each of the coordinates by 2. This gives us $D_2(7, -2) \rightarrow (14, -4)$.

The answer is (4).

PROBLEM 24: CONIC SECTIONS

A graph of the form $\dfrac{x^2}{a^2} + \dfrac{y^2}{b^2} = 1$, *where **a** and **b** have the same value, is the equation of a circle   If **a** and **b** have different values, the graph is that of an ellipse*

Here we have $\dfrac{x^2}{4} + \dfrac{y^2}{4} = 1$, so the graph is that of a circle

The answer is (1).

PROBLEM 25: TRIGONOMETRY GRAPHS

A graph of the form $y = a \sin bx$ *or* $y = a \cos bx$ *has an amplitude of* $|a|$ *and a period of* $\dfrac{2\pi}{b}$

The answer is (2).

PROBLEM 26: QUADRATIC EQUATIONS

Given an equation of the form $ax^2 + bx + c = 0$, the product of the roots is $\frac{c}{a}$, and the sum of the roots is $-\frac{b}{a}$.

Here, we have the equation $2x^2 - x - 2 = 0$, so the product of the roots is $\frac{c}{a} = \frac{-2}{2} = -1$.

The answer is (3).

Another way to get the answer is to actually find the roots and multiply them. The equation doesn't factor, so we use the quadratic formula

$$x = \frac{-b \pm \sqrt{b^2 - 4ac}}{2a}$$

Plugging in, we get $x = \frac{1 \pm \sqrt{(-1)^2 - 4(2)(-2)}}{2(2)} = \frac{1 \pm \sqrt{17}}{4}$. The roots are thus $\frac{1 + \sqrt{17}}{4} \approx 1.2808$ and $\frac{1 - \sqrt{17}}{4} \approx -.7808$. If we multiply them, we get -1.00004864, which is closest to answer (3).

PROBLEM 27: TRANSFORMATIONS

$R_{90°}$ means a rotation counterclockwise of 90°. You should know the following rotation rules:

$$R_{90°}(x, y) = (-y, x) \quad R_{180°}(x, y) = (-x, -y)$$
$$R_{270°}(x, y) = (y, -x)$$

Therefore, $R_{90°}(5, 2) \rightarrow (-2, 5)$.

The answer is (4).

PROBLEM 28: ALGEBRA THEORY

The Regents always asks an Algebra theory question that is usually very basic. Here, they are testing to see if you know what it means to say that a set is closed

A set is closed under an operation if the result of performing that operation on any element of the set yields an answer that is an element of the set.

In other words, if you perform the operation on any of the numbers in the set, do you always get an answer that is also a number in the set? If the answer is yes, then the set is closed. If not, then the set is not closed.

Let's try each of the operations.

Addition: $0 + 1 = 1$; $0 + (1) = -1$; $-1 + 1 = 0$; $1 + 1 = 2$. 2 is not an element of the set. Therefore, the set is not closed under addition.

Multiplication: $(0)(1) = 0$; $(0)(-1) = 0$; $(1)(-1) = -1$; $(0)(0) = 0$; $(-1)(-1) = 0$; $(1)(1) = 1$. Notice how each of the answers is an element of the set. Therefore, the set is closed under multiplication.

Subtraction: $0 - 1 = -1$; $0 - (-1) = 1$; $1 - (-1) = 2$. 2 is not an element of the set. Therefore, the set is not closed under subtraction.

Division: $\dfrac{1}{0} = undefined$; therefore, the set is not closed under division.

The answer is (2).

PROBLEM 29: TRIGONOMETRY

You should know that $\sin x = \cos(90° - x)$ and that $\cos x = \sin(90° - x)$. If we use the latter substitution in the equation we get:

$$\sin(x + 20°) = \sin(90° - x).$$

Therefore $(x + 20°) = (90° - x)$. If we solve for x, we get $x = 35°$.

The answer is (1).

Another way to get the answer is to **backsolve**. Try each of the values for x in the problem and see which one works.

Does $\sin(35° + 20°) = \cos 35°$? If you plug both into your calculator you will see that they are equal

Just in case, try the other choices. You will see that none of them give equal values.

PROBLEM 30: PROBABILITY

*If the probability of a particular outcome is **p**, then the probability of that outcome occurring **r** times out of a possible **n** times is*

$_nC_r(p)^r(1-p)^{n-r}$. This is known as *binomial probability*.

Here, we are told that the probability of Gordon's team winning is 0.3. The probability of the team winning exactly 2 games out of a possible 5 is $_5C_2(0.3)^2(1-0.3)^{5-2}$. This can be simplified to $_5C_2(0.3)^2(0.7)^3$.

We can then evaluate $_5C_2$ according to the formula $_nC_r = \dfrac{n!}{(n-r)!\,r!}$. If we plug in, we get:

$$_5C_2 = \frac{5!}{(5-2)!\,2!} = \frac{5!}{3!\,2!} = \frac{5 \cdot 4 \cdot 3 \cdot 2 \cdot 1}{(3 \cdot 2 \cdot 1)(2 \cdot 1)} = 10$$

which means that the answer is $10(0.3)^2(0.7)^3$.

The answer is (3).

PROBLEM 31: QUADRATIC INEQUALITIES

When we are given a quadratic inequality, the first thing that we do is factor it. We get:

$$(x-5)(x+2) > 0.$$

The roots of the quadratic are $x = 5$, $x = -2$. If we plot the roots on a number line

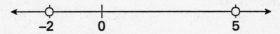

we can see that there are three regions: $x < -2$, $-2 < x < 5$, and $x > 5$. Now we try a point in each region to see whether the point satisfies the inequality.

First, we try a number less than –2. Let's try –3. Is $(-3-5)(-3+2) > 0$? Yes.

Now, we try a number between –2 and 5. Let's try 0. Is $(0-5)(0+2) > 0$? No.

Last, we try a number greater than 5. Let's try 6. Is $(6-5)(6+2) > 0$? Yes.

Therefore, the regions that satisfy the inequality are $x < -2$ or $x > 5$.

The answer is (4).

PROBLEM 32: BINOMIAL EXPANSIONS

The binomial theorem says that if you expand $(a + b)^n$, you get the following terms:

$$_nC_0a^n + _nC_1a^{n-1}b^1 + _nC_2a^{n-2}b^2 + \ldots + _nC_{n-2}a^2b^{n-2} + _nC_{n-1}a^1b^{n-1} + _nC_nb^n.$$

Therefore, if we expand $(3x - 2y)^4$, we get:

$$_4C_0(3x)^4 + _4C_1(3x)^3(-2y)^1 + _4C_2(3x)^2(-2y)^2 + _4C_3(3x)^1(-2y)^3 + _4C_4(-2y)^4.$$

Next, we use the rule that $_nC_r = \dfrac{n!}{(n-r)!\,r!}$. This gives us:

$$(3x - 2y)^4 = (3x)^4 + 4(3x)^3(-2y)^1 + 6(3x)^2(-2y)^2 + 4(3x)^1(-2y)^3 + 1(-2y)^4,$$

which simplifies to $(3x - 2y)^4 = 81x^4 - 216x^3y + 216x^2y^2 - 96xy^3 + 16y^4$.

The middle term is $216x^2y^2$.

The answer is (4).

PROBLEM 33: LOGARITHMS

This question is asking us where $\log_3 x = 2$. The definition of a logarithm says that $\log_b x = a$ means that $b^a = x$. Therefore, $\log_3 x = 2$ means that $3^2 = x = 9$.

The answer is (4).

PROBLEM 34: TRIGONOMETRIC AREA

The area of a parallelogram, if we are given the lengths of the two different sides a and b, and their included angle θ, is $A = ab\sin\theta$. Plugging in, we get:

$$A = (10)(12)\sin 150° = 60.$$

The answer is (2).

PROBLEM 35: QUADRATIC EQUATIONS

You should know the following rule:

If the graph of a quadratic equation does not intersect the x-axis, its roots are imaginary.

If the graph of a quadratic equation intersects the x-axis once (is tangent to it), its roots are real, rational, and equal.

If the graph of a quadratic equation intersects the x-axis twice, its roots are real and unequal.

Here, the graph does not intersect the x-axis, so its roots are imaginary.

The answer is (1).

Part II

PROBLEM 36:

(a) Rational Expressions

The first thing that we do is get a common denominator. In this case, we want all terms to have a denominator of $3x$. If we multiply the top and bottom of the first term by x, we get:

$$\frac{x(x+3)}{3x}.$$

If we multiply the top and bottom of the second term by 3, we get:

$$\frac{3(x+3)}{3x}.$$

If we multiply the top and bottom of the last term by $3x$, we get:

$$\frac{6x}{3x}.$$

Then our equation becomes:

$$\frac{x(x+3)}{3x} + \frac{3(x+3)}{3x} = \frac{6x}{3x}.$$

We can now combine the fractions on the left side to get:

$$\frac{x(x+3) + 3(x+3)}{3x} = \frac{6x}{3x}.$$

Now, we can ignore the denominators and set the numerators equal to each other. (Note that if we get the answer $x = 0$ we must throw it out because it will make the denominator zero).

$$x(x+3) + 3(x+3) = 6x.$$

Let's solve this equation.

$$x^2 + 3x + 3x + 9 = 6x$$
$$x^2 + 6x + 9 = 6x$$
$$x^2 + 9 = 0$$
$$x^2 = -9$$
$$x = \pm\sqrt{-9}$$
$$x = 3i, -3i.$$

(b) Quadratic Equations

If we move all of the terms to the left side, we get $x^2 - 6x + 10 = 0$. We can now use the quadratic formula to solve for x.

The formula says that, given a quadratic equation of the form $ax^2 + bx + c = 0$, the roots of the equation are:

$$x = \frac{-b \pm \sqrt{b^2 - 4ac}}{2a}.$$

Plugging in to the formula we get:

$$x = \frac{-(-6) \pm \sqrt{(-6)^2 - 4(1)(10)}}{2(1)} = \frac{6 \pm \sqrt{-4}}{2} =$$

$$\frac{6 \pm 2i}{2} = \frac{6}{2} \pm \frac{2}{2}i = 3 \pm i$$

PROBLEM 37:

(a) Trigonometric Equations

Note that the trig equation $3\sin^2\theta - \sin\theta - 2 = 0$ has the same form as a quadratic equation $3x^2 - x - 2 = 0$. Just as we could factor the quadratic equation, so too can we factor the trig equation. We get:

$$(3\sin\theta + 2)(\sin\theta - 1) = 0.$$

This gives us $3\sin\theta + 2 = 0$ and $\sin\theta - 1 = 0$. If we solve each of these equations, we get:

$$\sin\theta = -\frac{2}{3} \text{ and } \sin\theta = 1.$$

Therefore $\theta = \sin^{-1}\left(-\frac{2}{3}\right)$ and $\theta = \sin^{-1}(1)$. The answer to the second equation is easy. This is a special angle and you should know that sin 90° is 1, so one answer is $\theta = 90°$.

There is not a special angle whose sine is $-\frac{2}{3}$, so we need to use the calculator to find the value. Using the calculator, push **2nd SIN (–2/3) ENTER. (MAKE SURE THAT THE CALCULATOR IS IN DEGREE MODE!)** This gives us –41.8° ≈ –42° as a reference angle, but, we are asked to find values of θ between 0° and 360°.

We can find the required values of θ by making a small graph.

Notice that –42° is the same as 318°. The other place where sine is negative and has a reference angle of 42° is in quadrant III. The other angle then is 180° + 42° = 222°.

Therefore, the answers are θ = 90°, 222°, and 318°.

(b) Logarithms

If we take the log of both sides, we get:

$$x \log 5 = \log 30.$$

Now, divide both sides by log 5 and we get:

$$x = \frac{\log 30}{\log 5} \approx 2.1133.$$

Rounded to the nearest tenth, we get: $x \approx 2.1$

PROBLEM 38: TRIGONOMETRIC GRAPHS

(a) A graph of the form $y = a \sin bx$ or $y = a \cos bx$ has an amplitude of $|a|$ and a period of $\frac{2\pi}{b}$. Therefore, the equation $y = -2 \sin x$ has an amplitude of 2 and a period of 2π. The negative sign means that you should reflect the graph about the x-axis (turn it upside down). The second graph has an amplitude of 1 and a period of π. The two graphs look like this:

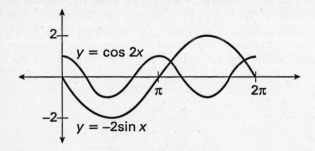

Remember, in graphing sines and cosines, the shape of the graph doesn't change, but it can be stretched or shrunk depending on the amplitude or period.

(b) We can find the intersections two ways: First, we can look at the number of intersections that we graphed on the paper. **The graphs intersect twice.**

The second way is to solve the equation algebraically. However, because you are only asked for the number of intersections, not the values of the intersections, the easiest thing to do is to look at the graphs.

PROBLEM 39:

(a) Statistics

The method for finding a standard deviation is simple, but time-consuming.

First, find the average of the scores. We do this by multiplying each score by the number of people who obtained that score, and adding them up.

$$(2)(100) + (3)(90) + (6)(80) + (5)(70) + (4)(60) = 1540.$$

Next, we divide by the total number of people $(2 + 3 + 6 + 5 + 4) = 20$, to obtain the average $\bar{x} = 77$.

Second, subtract the average from each actual score.

$$100 - 77 = 23$$
$$90 - 77 = 13$$
$$80 - 77 = 3$$
$$70 - 77 = -7$$
$$60 - 77 = -17.$$

Third, square each difference.

$$(23)^2 = 529$$
$$(13)^2 = 169$$
$$(3)^2 = 9$$
$$(-7)^2 = 49$$
$$(-17)^2 = 289.$$

Fourth, multiply the square of the difference by the corresponding number of people and sum.

$$(529)(2) = 1058$$
$$(169)(3) = 507$$
$$(9)(6) = 54$$
$$(49)(5) = 245$$
$$(289)(4) = 1156$$
$$1058 + 507 + 54 + 245 + 1156 = 3020.$$

Fifth, divide this sum by the total number of people.

$$\frac{3020}{20} = 151.$$

Last, take the square root of this number. This is the standard deviation. $\sigma = \sqrt{151} \approx 12.3$.

(b) Probability

(1) This problem requires that you know something called *binomial probability*. The rule is: *If the probability of a particular outcome is* p, *then the probability of that outcome occurring* r *times out of a possible* n *times is* $_nC_r(p)^r(1-p)^{n-r}$.

The probability that the spinner will land in a numbered area at most one time out of a possible four is the sum of the probability that it will land in a numbered area zero times and the probability that it will land in a numbered area one time.

The probability that the spinner will land in a numbered area on any given spin is $\dfrac{4}{6} = \dfrac{2}{3}$. Therefore, the probability that the spinner will land in a numbered area once out of four spins is $_4C_1\left(\dfrac{2}{3}\right)^1\left(1-\dfrac{2}{3}\right)^3$, and the probability of it landing zero times is $_4C_0\left(\dfrac{2}{3}\right)^0\left(1-\dfrac{2}{3}\right)^4$.

The rule for finding $_nC_r$ is $_nC_r = \dfrac{n!}{(n-r)!\,r!}$. Therefore,

$$_4C_1 = \frac{4!}{3!\,1!} = \frac{4\cdot3\cdot2\cdot1}{(3\cdot2\cdot1)(1)} = 4$$

and

$$_4C_0 = \frac{4!}{4!\,0!} = \frac{4\cdot3\cdot2\cdot1}{(4\cdot3\cdot2\cdot1)(1)} = 1$$

(By the way, $0! = 1$) Thus, the probability that the spinner will land at most one time is:

$$4\left(\frac{2}{3}\right)\left(\frac{1}{3}\right)^3 + 1\left(\frac{2}{3}\right)^0\left(\frac{1}{3}\right)^4 = \frac{8}{81} + \frac{1}{81} = \frac{9}{81} = \frac{1}{9}.$$

(2) The probability that the spinner will land in a lettered area at least three times out of a possible four is the sum of the

probability that it will land in a lettered area three times and the probability that it will land in a lettered area four times

The probability that the spinner will land in a lettered area on any given spin is $\frac{2}{6} = \frac{1}{3}$

Using our binomial probability formula, the probability that the spinner will land in a lettered area three out of four spins is:

$$_4C_3\left(\frac{1}{3}\right)^3\left(1-\frac{1}{3}\right)^1$$

and the probability of it landing four times is

$$_4C_4\left(\frac{1}{3}\right)^4\left(1-\frac{1}{3}\right)^0.$$

$$_4C_3 = \frac{4!}{1!\,3!} = \frac{4 \cdot 3 \cdot 2 \cdot 1}{(1)(3 \cdot 2 \cdot 1)} = 4$$

and $_4C_4 = \frac{4!}{0!\,4!} = \frac{4 \cdot 3 \cdot 2 \cdot 1}{(1)(4 \cdot 3 \cdot 2 \cdot 1)} = 1.$

Thus, the probability that the spinner will land at least three times is:

$$4\left(\frac{1}{3}\right)^3\left(\frac{2}{3}\right) + 1\left(\frac{1}{3}\right)^4\left(\frac{2}{3}\right)^0 = \frac{8}{81} + \frac{1}{81} = \frac{9}{81} = \frac{1}{9}.$$

Did you notice that the probability that the spinner will land in a numbered area zero or one times is the same as the probability that the spinner will land in a lettered area four or three times? Can you see why?

PROBLEM 40:

(a) Rational Expressions

The first thing we do is get common denominators independently for the top and the bottom. For the top, the common denominator is x and for the bottom, the common denominator is x^2. Thus, we can rewrite this expression as:

$$\frac{1 - \dfrac{1}{x}}{\dfrac{1}{x^2} - \dfrac{1}{x}} = \frac{\dfrac{x}{x} - \dfrac{1}{x}}{\dfrac{1}{x^2} - \dfrac{x}{x^2}} = \frac{\dfrac{x-1}{x}}{\dfrac{1-x}{x^2}}.$$

Next, we can invert the bottom fraction and multiply it by the top fraction.

$$\frac{\dfrac{x-1}{x}}{\dfrac{1-x}{x^2}} = \frac{x-1}{x} \cdot \frac{x^2}{1-x}.$$

Now we can cancel and simplify and we get:

$$\frac{x-1}{x} \cdot \frac{x^2}{1-x} = -x.$$

(b) Trigonometric Identities

Almost all of the identities that you will need to know are contained on the formula sheet.

First, let's get rid of the double angle terms.

Using $\cos 2x = 1 - 2\sin^2 x$ and $\sin 2x = 2 \sin x \cos x$, we can rewrite the left side of the equation as:

$$\frac{1 - 2\sin^2 x}{\sin x} + \frac{2 \sin x \cos x}{\cos x}.$$

Next, we break the first fraction into two fractions:

$$\frac{1}{\sin x} - \frac{2 \sin^2 x}{\sin x} + \frac{2 \sin x \cos x}{\cos x}.$$

Next, we cancel terms:

$$\frac{1}{\sin x} - 2 \sin x + 2 \sin x = \frac{1}{\sin x}.$$

Finally, we can rewrite the left side as csc x and set the two sides equal, and we have proved the identity.

(a) *If two chords are congruent, they cut off equal arcs.*

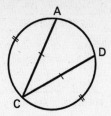

Because arc $\overset{\frown}{AC}$ and arc $\overset{\frown}{CD}$ are congruent, they both measure 140. Furthermore, arc $\overset{\frown}{AC}$, arc $\overset{\frown}{CD}$, and arc $\overset{\frown}{AD}$ form a circle. Therefore, arc $\overset{\frown}{AD}$ must have a measure equal to 360 minus the sum of the measures of arcs $\overset{\frown}{AC}$ and $\overset{\frown}{CD}$. $m\overset{\frown}{AD} = 360 - (140 + 140) = 80$.

(b) *The measure of an angle formed by a pair of secants or a secant and a tangent is equal to half of the difference between the larger and the smaller arcs that are formed by the secants, or the secant and the tangent.*

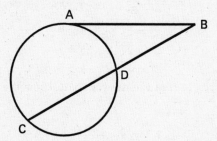

Here, the larger and smaller arcs formed by angle B are $\overset{\frown}{AC}$ and $\overset{\frown}{AD}$, respectively.

$$m\angle B = \frac{m\overset{\frown}{AC} - m\overset{\frown}{AD}}{2} = \frac{140 - 80}{2} = 30$$

(c) Law of Sines

If we look at triangle ABC, we know that angle B is 30° (from part b), and we know that angle C is 40° (because it is half of $\overset{\frown}{AD}$). So, we can use the Law of Sines to find the length of AB.

The Law of Sines says that, given a triangle with angles A, B, and C and opposite sides of a, b, and c

$$\frac{a}{\sin A} = \frac{b}{\sin B} .$$

(This is on the formula sheet.)

Substituting, we get:

$$\frac{AB}{\sin 40°} = \frac{10}{\sin 30°}$$

We solve this for AB and get: $AB = \dfrac{10\left(\sin 40°\right)}{\sin 30°} \approx 12.9$

PROBLEM 42: LOGARITHMS

(a) From the definition of logarithm, we know that $\log_2 x = y$ means that $2^y = x$. So we can complete the table as follows:

$$2^y = \frac{1}{4} \qquad y = -2$$

$$2^y = \frac{1}{2} \qquad y = -1$$

$$2^y = 1 \qquad y = 0$$

$$2^y = 2 \qquad y = 1$$

$$2^y = 4 \qquad y = 2$$

Another way to find the answers is to know the change of base rule.

$$\log_b x = \frac{\log x}{\log b} .$$

Using this rule, we get:

$$\log_2 \frac{1}{4} = \frac{\log \frac{1}{4}}{\log 2} = -2$$

$$\log_2 \frac{1}{2} = \frac{\log \frac{1}{2}}{\log 2} = -1$$

$$\log_2 1 = \frac{\log 1}{\log 2} = 0$$

(Note: The log of one is **ALWAYS** zero.)

$$\log_2 2 = \frac{\log 2}{\log 2} = 1$$

$$\log_2 4 = \frac{\log 4}{\log 2} = 2$$

(b) Just plot the points. Don't forget to label the graph *b.*

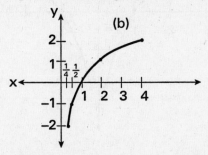

(c) When you reflect a graph in the *y*-axis, you flip the left and the right sides. Don't forget to label the graph *c.*

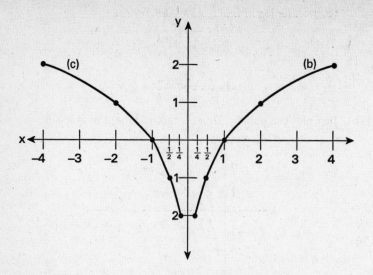

(d) When you reflect the graph from part (b) in the line $y = x$, it looks like this:

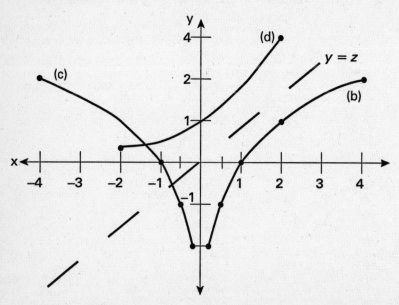

Don't forget to label the graph d.

Formulas

Pythagorean and Quotient Identities

$$\sin^2 A + \cos^2 A = 1 \qquad \tan A = \frac{\sin A}{\cos A}$$

$$\tan^2 A + 1 = \sec^2 A \qquad \cot A = \frac{\cos A}{\sin A}$$

$$\cot^2 A + 1 = \csc^2 A$$

Functions of the Sum of Two Angles

$$\sin (A + B) = \sin A \cos B + \cos A \sin B$$

$$\cos (A + B) = \cos A \cos B - \sin A \sin B$$

$$\tan (A + B) = \frac{\tan A + \tan B}{1 - \tan A \tan B}$$

Functions of the Difference of Two Angles

$$\sin (A - B) = \sin A \cos B - \cos A \sin B$$

$$\cos (A - B) = \cos A \cos B + \sin A \sin B$$

$$\tan (A - B) = \frac{\tan A - \tan B}{1 + \tan A \tan B}$$

Law of Sines

$$\frac{a}{\sin A} = \frac{b}{\sin B} = \frac{c}{\sin C}$$

Law of Cosines

$$a^2 = b^2 + c^2 - 2bc \cos A$$

Functions of the Double Angle

$$\sin 2A = 2 \sin A \cos A$$

$$\cos 2A = \cos^2 A - \sin^2 A$$

$$\cos 2A = 2 \cos^2 A - 1$$

$$\cos 2A = 1 - 2 \sin^2 A$$

$$\tan 2A = \frac{2 \tan A}{1 - \tan^2 A}$$

Functions of the Half Angle

$$\sin \frac{1}{2} A = \pm\sqrt{\frac{1 - \cos A}{2}}$$

$$\cos \frac{1}{2} A = \pm\sqrt{\frac{1 + \cos A}{2}}$$

$$\tan \frac{1}{2} A = \pm\sqrt{\frac{1 - \cos A}{1 + \cos A}}$$

Area of Triangle

$$K = \frac{1}{2} ab \sin C$$

Standard Deviation

$$\text{S.D.} = \sqrt{\frac{1}{n} \sum_{i=1}^{n} \left(x_i - \bar{x}\right)^2}$$

EXAMINATION
JUNE 1997

Part I

Answer 30 questions from this part. Each correct answer will receive 2 credits. No partial credit will be allowed. Write your answers in the spaces provided. Where applicable, answers may be left in terms of π or in radical form.   [60]

1 Express 240° in the radian measure

$$\frac{240}{1} \cdot \frac{\pi}{180} = \frac{4\pi}{3}$$

2 In $\triangle ABC$, $a = 12$, $\sin A = 0.45$, and $\sin B = 0.15$
   Find b.

$$\frac{12}{0.45} = \frac{b}{0.15} \qquad 0.45b = 12 \cdot 0.15$$
$$0.45b = 1.8$$
$$\boxed{b = 4}$$

3 Find the value of $\displaystyle\sum_{k=1}^{3}(3k-5)$

4 Solve for x: $4^{(3x+5)} = 16$.

$$2^{2(3x+5)} = 2^4 \qquad 6x+10 = 4$$
$$6x = -6$$
$$\boxed{x = -1}$$

5 Express the sum of $\sqrt{-64}$ and $3\sqrt{-4}$ as a monomial in terms of i.

$$8i + 3 \cdot 2i$$
$$8i + 6i = 14i$$

6 Solve for all values of x: $|2x+5| = 7$

$$2x+5=7 \qquad\qquad 2x+5=-7$$
$$2x = 2 \qquad\qquad 2x = -12$$
$$x = 1 \qquad\qquad x = -6$$

7 What will be the amplitude of the image of the curve $y = 2\sin 3x$ after a dilation of scale factor 2?

8 What is the solution of the equation $\sqrt{5x-9} - 3 = 1$?

$$5x = 16 + 9$$
$$5x - 9 = 16 \qquad 5x = 25$$
$$\boxed{x = 5}$$

9 In the interval $90° \leq \theta \leq 180°$, find the value of
   θ that satisfies the equation $2 \sin \theta - 1 = 0$.

   $\theta = 150$ $O = O$

10 Express in simplest form: $\dfrac{1}{\dfrac{1}{a} + \dfrac{1}{b}}$.

   $i = \dfrac{b+a}{ab}$ $1 \cdot \dfrac{ab}{b+a}$

11 If $f(x) = x^0 + x^{\frac{2}{3}} + x^{-\frac{2}{3}}$, find $f(8)$.

   $F(8) = 8^0 + 8^{2/3} + 8^{-2/3}$ $F(8) = \dfrac{4 + 16 + 1}{4} = \dfrac{21}{4}$

   $1 + 4 + 1/4$

12 When the sum of $4 + 5i$ and $-3 - 7i$ is represented
   graphically, in which quadrant does the sum lie?

13 In the accompanying diagram, $\overline{AP}$ is a tangent and
   $\overline{PBC}$ is a secant to circle O. If $PC = 12$ and $BC = 9$,
   find the length of $\overline{AP}$.

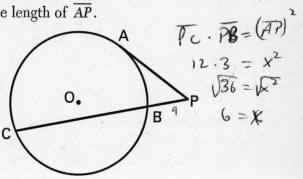

   $PC \cdot PB = (AP)^2$

   $12 \cdot 3 = x^2$

   $\sqrt{36} = \sqrt{x^2}$

   $6 = x$

14 Circle O has a radius of 10. Find the length of an
   arc subtended by a central angle measuring 1.5
   radians.

   $S = r\theta$

   $S = 10 \cdot 1.5 = 15$

15 If $f(x) = 5x - 2$ and $g(x) = \sqrt[3]{x}$, evaluate $(f \circ g)(-8)$

   $g(8) = \sqrt[3]{8}$ $F(2) = 5 \cdot 2 - 2$
   $g(8) = 2$ $F(2) = 10 - 2$
                            $F(2) = 8$

Directions (16–35): For *each* question chosen, write on the separate answer sheet the *numeral* preceding the word or expression that best completes the statement or answers the question.

16 For which value of x is the expression $\dfrac{1}{1-\cos x}$ undefined?

(1) 90°   (3) 270°
(2) 180°   (4) 360°

17 The expression $\log \sqrt{\dfrac{x}{y}}$ is equivalent to

(1) $\dfrac{1}{2}(\log x - \log y)$ (3) $\dfrac{1}{2}\log x - \log y$

(2) $\log \dfrac{1}{2}x - \log \dfrac{1}{2}y$ (4) $\log \dfrac{1}{2}x - \log y$

18 If $f(x) = \cos 3x + \sin x$, then $f\left(\dfrac{\pi}{2}\right)$ equals

(1) 1   (3) –1 $F(\pi/2)=\cos 3 \cdot \pi/2 +$
(2) 2   (4) 0 $\sin \pi/2$

$F(\pi/2) = 0 + 1 = 1$

19 Expessed in $a + bi$ form, $(1 + 3i)^2$ is equivalent to

(1) 10 + 6i   (3) 10 – 6i
(2) –8 + 6i   (4) –8 – 6i

$(1+3i)(1+3i)$ $1+6i+9i^2$ $-8+6i$
$1+3i+3i+9i^2$ $1+6i+9(-1)$

20 The expression $\dfrac{\tan \theta}{\sec \theta}$ is equivalent to

(1) cot θ   (3) cos θ
(2) csc θ   (4) sin θ

$\dfrac{\frac{\sin\theta}{\cos\theta}}{\frac{1}{\cos\theta}}$ $\dfrac{\sin\theta}{\cos\theta} \div \dfrac{1}{\cos\theta}$

$\dfrac{\sin\theta \cdot \cos\theta}{\cos\theta}$

21 Which equation is represented by the graph below?

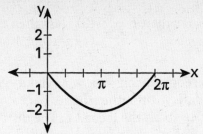

(1) $y = -2 \sin \frac{1}{2} x$ (3) $y = \frac{1}{2} \sin 2x$

(2) $y = -\frac{1}{2} \sin 2x$ (4) $y = 2 \sin \frac{1}{2} x$

22 Expressed in $a + bi$ form, $\dfrac{5}{3 + i}$ is equivalent to

(1) $\dfrac{15}{8} - \dfrac{5}{8} i$ (3) $\dfrac{3}{2} - \dfrac{1}{2} i$ $\dfrac{5}{3+i} \cdot \dfrac{(3-i)}{(3-i)}$

(2) $\dfrac{5}{3} - 5i$ (4) $15 - 5i$ $\dfrac{15 - 15i}{9 - 3i + 3i - i^2}$

$\dfrac{15 - 5i}{10} = \dfrac{3}{2} - \dfrac{5i}{2}$

23 Gordon tosses a fair die six times. What is the probability that he will toss *exactly* two 5s?

$n = 6$
$r = 2$

(1) $_6C_5\left(\dfrac{5}{6}\right)^2\left(\dfrac{1}{6}\right)^4$ (3) $_6C_5\left(\dfrac{1}{6}\right)^2\left(\dfrac{5}{6}\right)^4$ $p = \frac{1}{6}$

$q = \frac{5}{6}$

(2) $_6C_2\left(\dfrac{5}{6}\right)^2\left(\dfrac{1}{6}\right)^4$ (4) $_6C_2\left(\dfrac{1}{6}\right)^2\left(\dfrac{5}{6}\right)^4$ $_6C_2\left(\frac{1}{6}\right)^2\left(\frac{5}{6}\right)^4$

24 If sin θ is negative and cot θ is positive, in which quadrant does θ terminate?

(1) I   (3) III
(2) II   (4) IV

$\dfrac{\sin \theta}{\text{II, III}}$ $\dfrac{1}{\sin \theta}$

25 The domain of the equation $y = \dfrac{1}{(x-1)^2}$ is all real numbers

(1) greater than 1   (3) less than 1
(2) except 1   (4) except 1 and –1

26 In the accompanying diagram, about 68% of the scores fall within the shaded area, which is symmetric about the mean, $\bar{x}$. The distribution is normal and the scores in the shaded area range from 50 to 80.

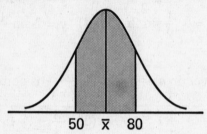

50    $\bar{x}$ 80

What is the standard deviation of the scores in this distribution?

(1) $7\dfrac{1}{2}$ (3) 30
(2) 15   (4) 65

27 The expression 2 sin 30° cos 30° has the same value as

(1) sin 15°   (3) sin 60°
(2) cos 60°   (4) cos 15°

28 In the accompanying diagram of a unit circle, the ordered pair (x,y) represents the point where the terminal side of θ intersects the unit circle.

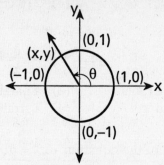

If m ∠θ = 120, what is the value of x in simplest form?

(1) $-\dfrac{\sqrt{3}}{2}$ (3) $-\dfrac{1}{2}$

(2) $\dfrac{\sqrt{3}}{2}$ (4) $\dfrac{1}{2}$

29 In △ABC, side a is twice as long as side b and m∠C = 30. In terms of b, the area of △ABC is

(1) $0.25\,b^2$ (3) $0.866\,b^2$
(2) $0.5\,b^2$ (4) b^2

30 Which quadratic equation has roots $3 + i$ and $3 - i$?

(1) $x^2 - 6x + 10 = 0$ (3) $x^2 - 6x + 8 = 0$
(2) $x^2 + 6x - 10 = 0$ (4) $x^2 + 6x - 8 = 0$

31 Which is the fourth term in the expansion of $(\cos x + 3)^5$?

(1) $90 \cos^2 x$ (3) $90 \cos^3 x$

(2) $270 \cos^2 x$ (4) $270 \cos^3 x$

32 The graph of the equation $y = \dfrac{6}{x}$ forms

(1) a hyperbola   (3) a parabola

(2) an ellipse   (4) a straight line

33 The roots of the equation $-3x^2 = 5x + 4$ are

(1) real, rational, and unequal

(2) real, irrational, and unequal

(3) real, rational, and equal

(4) imaginary

34 Which equation does *not* represent a function?

(1) $y = 4$ (3) $y = x - 4$

(2) $y = x^2 - 4$ (4) $x^2 + y^2 = 4$

35 If the point $(2, -5)$ is reflected in the line $y = x$, then the image is

(1) $(5, -2)$ (3) $(-5, 2)$

(2) $(-2, 5)$ (4) $(-5, -2)$

Part II

Answer four questions from this part. Clearly indicate the necessary steps, including appropriate formula substitutions, diagrams, graphs, charts, etc. Calculations that may be obtained by mental arithmetic or the calculator do not need to be shown. [40]

36 *a* On the same set of axes, sketch and label the graphs of the equations $y = 2 \cos x$ and $y = \sin \frac{1}{2} x$ in the interval $-\pi \le x \le \pi$. [8]

  b Using the graphs drawn in part *a*, determine the number of values in the interval $-\pi \le x \le \pi$ that satisfy the equation $\sin \frac{1}{2} x = 2 \cos x$. [2]

37 In the accompanying diagram, isosceles triangle
    ABC is inscribed in circle *O*, and vertex angle *BAC*
    measures 40°. Tangent $\overline{PC}$, secant $\overline{PBA}$, and
    diameters $\overline{BD}$ and $\overline{AE}$ are drawn.

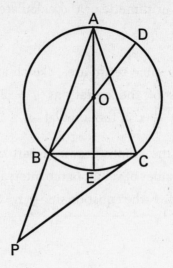

Find:

a	m $\overset{\frown}{BC}$	[2]
b	m∠*ABD*	[2]
c	m∠*DOE*	[2]
d	m∠*P*	[2]
e	m∠*ACP*	[2]

38 Find, to the *nearest ten minutes* or *nearest tenth
    of a degree,* all values of *x* in the interval
    $0° \leq x < 360°$ that satisfy the equation
    $2 \sin 2x + \cos x = 0$.   [10]

39 *a* Find the standard deviation, to the *nearest hundredth,* for the following measurements:

24, 28, 29, 30, 30, 31, 32, 32, 32, 33, 35, 36.   [4]

b A circle that is partitioned into five equal sectors has a spinner. The colors of the sectors are red, orange, yellow, blue, and green. If four spins are made, find the probability that the spinner will land in the green sector

(1) on *exactly* two spins   [2]
(2) on *at least* three spins   [4]

40 *a* Express in simplest form:

$$\frac{3y + 15}{25 - y^2} + \frac{2}{y - 5}.$$ [5]

b Solve for *x* and express the roots in simplest *a* + *bi* form:

$$2 + \frac{5}{x^2} = \frac{6}{x}.$$ [5]

41 In $\triangle ABC$, $AB = 14$, $AC = 20$, and m$\angle CAB = 49$.

a Find the length of $\overline{BC}$ to the *nearest tenth.*   [6]

b Using the results from part *a,* find m $\angle C$ to the *nearest degree.*   [4]

42 Given: $f = \{(x,y) \mid y = \log_2 x\}$

a On graph paper, sketch and label the graph of the function f. [3]

b Write a mathematical explanation of how to form the inverse of function f. [3]

c On the same set of axes, sketch and label the graph of the function f^{-1}, the inverse of f. [3]

d Write an equation for f^{-1}. [2]

ANSWER KEY

Part I

1. $\dfrac{4\pi}{3}$

2. 4

3. 3

4. −1

5. 14i

6. −6, 1

7. 4

8. 5

9. 150°

10. $\dfrac{ab}{b+a}$

11. $5\dfrac{1}{4}$

12. IV

13. 6

14. 15

15. −12

16. (4)

17. (1)

18. (1)

19. (2)

20. (4)

21. (1)

22. (3)

23. (4)

24. (3)

25. (2)

26. (2)

27. (3)

28. (3)

29. (2)

30. (1)

31. (2)

32. (1)

33. (4)

34. (4)

35. (3)

ANSWERS AND EXPLANATIONS
JUNE 1997

Part I

In problems 1–15, you must supply the answer. Because of this, you will not be able to **plug-in** or **backsolve**. But, you can use your calculator to simplify calculations. In addition, you will find that these problems tend to test your basic knowledge of Sequential Math III and are not tricky. Many can be solved just by knowing the right formula and by plugging in the numbers.

PROBLEM 1: CONVERTING FROM DEGREES TO RADIANS

We simply multiply the angle by $\dfrac{\pi}{180°}$. We get: $240° \dfrac{\pi}{180°} = \dfrac{4\pi}{3}$.

PROBLEM 2: LAW OF SINES

The Law of Sines says that, in any triangle, with angles $A, B,$ and C, and opposite sides of $a, b,$ and c, respectively, $\dfrac{a}{\sin A} = \dfrac{b}{\sin B}$ (This is on the formula sheet.)

Here, $\dfrac{12}{0.45} = \dfrac{b}{0.15}$.

We can easily solve this.

$$0.15 \cdot \frac{12}{0.45} = 4 = b.$$

PROBLEM 3: SUMMATIONS

To evaluate a sum, we plug each of the consecutive values for k into the expression, starting at the bottom value and ending at the top value and sum the results. We get:

$$\sum_{k=1}^{3} (3k - 5) = [(3(1) - 5)] + [(3(2) - 5)] + [(3(3) - 5)] = -2 + 1 + 4 = 3$$

Another way to do this is to use your calculator. Push **2nd STAT** (**LIST**) and under the **MATH** menu, choose **SUM**. Then, again push **2nd STAT** and under the **OPS** menu, choose **SEQ**.

We put in the following: (*expression, variable, start, finish, step*).

For *expression*, we put in the formula, using x as the variable.

For *variable*, we ALWAYS put in x.

For *start*, we put in the bottom value.

For *finish*, we put in the top value.

For *step*, we ALWAYS put in 1.

Therefore, we put in the calculator: **SUM SEQ** $(3x - 5, x, 1, 3, 1)$. The calculator should return the value 3.

PROBLEM 4: EXPONENTIAL EQUATIONS

$$4^{3x+5} = 16.$$

In order to solve this equation, we need to have a common base for the two sides of the equation. Because $4^2 = 16$, we can rewrite the right side as

$$4^{3x+5} = 4^2.$$

Now, we can set the powers equal to each other and solve.

$$3x + 5 = 2$$
$$x = -1$$

PROBLEM 5: COMPLEX NUMBERS

We can rewrite $\sqrt{-64}$ and $3\sqrt{-4}$ as $\sqrt{64}\sqrt{-1}$ and $3\sqrt{4}\sqrt{-1}$, respectively.

Next, we use the rule that $i = \sqrt{-1}$ to get:

$$\sqrt{64}\sqrt{-1} = 8i \text{ and } 3\sqrt{4}\sqrt{-1} = 6i.$$

Now we can add the two expressions to obtain $14i$.

PROBLEM 6: ABSOLUTE VALUE EQUATIONS

When we have an absolute value equation we break it into two equations. First, rewrite the left side without the absolute value and leave the right side alone. Second, rewrite the left side without the absolute value, and change the sign of the right side.

Here, we have $|2x + 5| = 7$, so we rewrite it as: $2x + 5 = 7$ or $2x + 5 = -7$. If we solve each of these independently, we get $x = 1$ or $x = -6$.

PROBLEM 7: TRANSFORMATIONS

A *dilation* means that we multiply the equation by the scale factor. In this case, we get $y = 4\sin 6x$.

Next, if an equation has the form $y = a \sin bx$ or $y = a \cos bx$, then its amplitude is $|a|$ and its period is $\dfrac{2\pi}{b}$. So, here, the amplitude is 4.

PROBLEM 8: RADICAL EQUATIONS

First, we add 3 to both sides   We get: $\sqrt{5x - 9} = 4$.

Next, square both sides.

$$5x - 9 = 16$$

Now, solve for x.

$$5x = 25$$
$$x = 5.$$

BUT, whenever we have a radical in an equation, we have to check the root to see if it satisfies the original equation. Sometimes, in the process of squaring both sides, we will get answers that are invalid. So let's check the answer.

Does $\sqrt{25 - 9} - 3 = 1$? $4 - 3 = 1$; so, yes, it does. Therefore the answer is $x = 5$.

PROBLEM 9: TRIGONOMETRIC EQUATIONS

Note that the trig equation $2\sin\theta - 1 = 0$ has the same form as the equation $2x - 1 = 0$. Just as we can solve the latter equation, so too can we solve the former.

First, add 1 to both sides: $2\sin\theta = 1$.

Next, divide both sides by 2: $\sin\theta = \dfrac{1}{2}$.

You should know the values of θ between $0°$ and $360°$ for which $\sin\theta = \dfrac{1}{2}$ because they are special angles. They are $\theta = 30°$ and $\theta = 150°$. Because we are asked for the answer in the region $90° \le \theta \le 180°$, the answer is $\theta = 150°$.

PROBLEM 10: RATIONAL EXPRESSIONS

First, get a common denominator for the bottom:

$$\frac{1}{\dfrac{1}{a} + \dfrac{1}{b}} = \frac{1}{\dfrac{b}{ab} + \dfrac{a}{ab}} = \frac{1}{\dfrac{b+a}{ab}}.$$

Now, invert the bottom fraction and multiply it by the top one.

We get: $(1)\left(\dfrac{ab}{b+a}\right) = \dfrac{ab}{b+a}$.

PROBLEM 11: FUNCTIONS

First, we plug in 8 for x: $f(8) = (8)^0 + (8)^{\frac{2}{3}} + (8)^{-\frac{2}{3}}$.

Any number raised to the power zero is 1. This gives us:

$$f(8) = 1 + (8)^{\frac{2}{3}} + (8)^{-\frac{2}{3}}$$

A number raised to a negative power is the same as the reciprocal of that number raised to the corresponding positive power. In other words,

$$x^{-a} = \frac{1}{x^a}.$$

This gives us: $f(8) = 1 + (8)^{\frac{2}{3}} + \dfrac{1}{(8)^{\frac{2}{3}}}$

The rule for raising a number to a fractional power is $x^{\frac{b}{a}} = \left(\sqrt[a]{x}\right)^{b}$.

This gives us:

$$f(8) = 1 + \frac{1}{\left(\sqrt[3]{8}\right)^2} = 1 + 2^2 + \frac{1}{2^2} = 5\frac{1}{4}.$$

PROBLEM 12: COMPLEX NUMBERS

To find the sum of $4 + 5i$ and $-3 - 7i$ we add the real components and the imaginary components separately. This gives us:

$$\left(4 + (-3)\right) + \left(5 + (-7)\right)i = 1 - 2i.$$

Next, to determine the quadrant, we look at the signs of the components. *The real component corresponds to the x-coordinate of the graph, and the imaginary component corresponds to the y-coordinate of the graph.* Because the real component is positive and the imaginary component is negative, the graph has a positive x-coordinate and a negative y-coordinate. Therefore, the graph is in quadrant IV.

PROBLEM 13: CIRCLE RULES

Here's a good rule to know: Given a point exterior to a circle, *the square of the tangent segment to the circle is equal to the product of the lengths of the secant and its external segment.*

In other words, $(\overline{PC})(\overline{PB}) = (\overline{AP})^2$. We are given that $\overline{BC} = 9$, so $\overline{PB} = 12 - 9 = 3$.

Now, we can plug into the equation and solve:

$$(12)(3) = (\overline{AP})^2$$

$$(\overline{AP})^2 = 36$$

$$\overline{AP} = \pm 6$$

Since $\overline{AP}$ can't have a negative length, the answer is $\overline{AP} = 6$

PROBLEM 14: ARC LENGTH

In a circle with a central angle, θ, measured in radians, and a radius, r; the arc length, s, is found by $s = r\theta$

Here we have $r = 10$ and $\theta = 1.5$. Therefore, the arc has length $s = (10)(1.5) = 15$.

PROBLEM 15: COMPOSITE FUNCTIONS

With composite functions, first we work with the function on the right and then the function on the left.

Here, we first plug -8 into $g(x)$. We obtain $g(-8) = \sqrt[3]{-8} = -2$

Next, we plug -2 into $f(x)$. We obtain $f(-2) = 5(-2) - 2 = -12$

Multiple-Choice Problems

For problems 16–35, you will find that you can sometimes **plug-in** or **backsolve** to get the right answer. You will also find that a calculator will simplify the arithmetic, In addition, most of the problems only require knowing which formula to use.

PROBLEM 16: RATIONAL EXPRESSIONS

A function is undefined where its denominator equals zero. If we set the denominator equal to zero we get: $\cos x = 1$

You should know the value of x for which $\cos x = 1$ because it is a special angle. It is $x = 0°$. Notice that $x = 0°$ is not one of the answer choices. But, $360°$ has the same reference angle as $0°$. Therefore $x = 360°$ also works.

The answer is (4).

Another way to get the right answer is to **backsolve**. Plug each of the answer choices into the problem and see which one gives you an error on your calculator

Choice (1): $\dfrac{1}{1 - \cos 90°} = 1$. This is the wrong answer.

Choice (2): $\dfrac{1}{1 - \cos 180°} = \dfrac{1}{2}$. This is the wrong answer.

Choice (3): $\dfrac{1}{1 - \cos 270°} = 1$. This is the wrong answer, which leaves . . .

Choice (4): $\dfrac{1}{1 - \cos 360°} = err$. This is the **right** answer.

PROBLEM 17: LOGARITHMS

You should know the following log rules:

    (i) $\log A + \log B = \log(AB)$

    (ii) $\log A - \log B = \log\left(\dfrac{A}{B}\right)$

    (iii) $\log A^B = B \log A$

First, using rule (iii), we can rewrite $\log \sqrt{\dfrac{x}{y}}$ as $\dfrac{1}{2}\left(\log \dfrac{x}{y}\right)$.

Next, using rule (ii), we can rewrite $\dfrac{1}{2}\left(\log \dfrac{x}{y}\right)$ as $\dfrac{1}{2}\left(\log x - \log y\right)$

The answer is (1).

Another way to get the right answer is to **plug-in**. First, let's make up values for x and y. For example, let $x = 36$ and $y = 10$.

Then we use our calculator to find $\log \sqrt{\dfrac{36}{10}} \approx 0.278$.

Next, we plug $x = 36$ and $y = 10$ into each of the answer choices to see which one matches.

Choice (1): $\dfrac{1}{2}\left(\log 36 - \log 10\right) \approx 0.278$, **so choice (1) is the answer.**

Choice (2): $\log 18 - \log 5 \approx 0.556$, so choice (2) is not the answer.

Choice (3): $\frac{1}{2}\log 36 - \log 10 \approx -0.222$, so choice (3) is not the answer

Choice (4): $\log 18 - \log 10 \approx 0.255$, so choice (4) is not the answer.

PROBLEM 18: TRIGONOMETRY

First, we plug in $\frac{\pi}{2}$ for x and we get: $f\left(\frac{\pi}{2}\right) = \cos\frac{3\pi}{2} + \sin\frac{\pi}{2}$.

You should know both of these values because they are special angles.

$\cos\frac{3\pi}{2} = 0$ and $\sin\frac{\pi}{2} = 1$, so $\cos\frac{3\pi}{2} + \sin\frac{\pi}{2} = 0 + 1 = 1$.

The answer is (1).

Of course, you could always just find the values on your calculator. Just be sure that you are in **radian** mode. Also, **pay attention to order of operations** when you put the angles in your calculator. If you put in $\cos 3\pi / 2$, the calculator will evaluate $\cos 3\pi$ first and then divide it by 2. You will get $-\frac{1}{2}$, which is NOT what you want. Therefore, you have to put in $\cos(3\pi / 2)$ so that the calculator gives you the correct value of $\cos(3\pi / 2) = 0$. This is something that you should watch out for every time that you evaluate a trig function, a log, or a root.

PROBLEM 19: COMPLEX NUMBERS

When we want to find the product of two complex numbers, the first thing we do is FOIL:

$$(1 + 3i)(1 + 3i) = 1 + 3i + 3i + 9i^2.$$

Next we combine the two middle terms:

$$1 + 3i + 3i + 9i^2 = 1 + 6i + 9i^2.$$

Next, because $i^2 = -1$, the last term becomes $1 + 6i - 9$.

This can be simplified to $-8 + 6i$.

The answer is (2).

PROBLEM 20: TRIGONOMETRY IDENTITIES

First, convert $\tan \theta$ into $\dfrac{\sin \theta}{\cos \theta}$ and $\sec \theta$ into $\dfrac{1}{\cos \theta}$

This gives us: $\dfrac{\tan \theta}{\sec \theta} = \dfrac{\dfrac{\sin \theta}{\cos \theta}}{\dfrac{1}{\cos \theta}}$.

Now, get invert the bottom fraction and multiply:

$$\dfrac{\dfrac{\sin \theta}{\cos \theta}}{\dfrac{1}{\cos \theta}} = \dfrac{\sin \theta}{\cos \theta} \cdot \dfrac{\cos \theta}{1} = \sin \theta$$

The answer is (4).

Another way to get the right answer is to **plug-in**. For example, let's let $\theta = 30°$.

Now we plug $\theta = 30°$ into the problem and we get $\dfrac{\tan 30°}{\sec 30°} = 0.5$.

Next, plug $\theta = 30°$ into the answer choices and see which one matches.

Choice (1): $\cot 30° \approx 1.732$. This is the wrong answer.

Choice (2): $\csc 30° = 2$. This is the wrong answer.

Choice (3): $\cos 30° \approx 0.866$. This is the wrong answer.

Choice (4): $\sin 30° = 0.5$. This is the **right** answer

PROBLEM 21: TRIGONOMETRY GRAPHS

A graph of the form $y = a \sin bx$ or $y = a \cos bx$ has an amplitude of $|a|$ and a period of $\dfrac{2\pi}{b}$

This graph is upside down, so the value of a has to be negative. This eliminates choices (3) and (4). Furthermore, the amplitude of the graph is 2, so this eliminates choice (2).

Therefore, The answer is (1).

You should also have noticed that only half of this graph is represented in the region from 0 to 2π, which means that the period of the graph is 4π. This means that $b = \dfrac{1}{2}$ and we could have used this as well for the process of elimination.

PROBLEM 22: COMPLEX NUMBERS

Whenever you are asked to simplify a fraction that has a complex number in the denominator, all that you have to do is to multiply the top and bottom by the complex conjugate of the denominator. *The complex conjugate of a number $a + bi$ is the number $a - bi$.*

Here, we multiply $\dfrac{5}{3+i}$ by $\dfrac{3-i}{3-i}$. We get: $\dfrac{5}{3+i}\dfrac{3-i}{3-i}$.

We distribute the 5 in the numerator, and we FOIL the denominators and we get:

$$\frac{15 - 5i}{9 + 3i - 3i - i^2} = \frac{15 - 5i}{9 - i^2}.$$

Next, because $i^2 = -1$, we have:

$$\frac{15 - 5i}{9 - (-1)} = \frac{15 - 5i}{10} = \frac{15}{10} - \frac{5}{10}i = \frac{3}{2} - \frac{1}{2}i$$

The answer is (3).

PROBLEM 23: PROBABILITY

If the probability of a particular outcome is p, then the probability of that outcome occurring r times out of a possible n times is ${}_nC_r(p)^r(1-p)^{n-r}$. This is known as *binomial probability.*

The probability of getting any particular number on one toss of a fair die is $\dfrac{1}{6}$. Therefore, the probability of getting exactly two 5's out of six rolls is:

$$ {}_6C_2\left(\frac{1}{6}\right)^2\left(1 - \frac{1}{6}\right)^{6-2} = {}_6C_2\left(\frac{1}{6}\right)^2\left(\frac{5}{6}\right)^4. $$

The answer is (4).

PROBLEM 24: TRIGONOMETRY

The sine of an angle is negative in quadrants III *and* IV.

The tangent of an angle is positive in quadrants I *and* III. Because the cotangent of an angle is just the reciprocal of the tangent of the angle, the sign of the cotangent of an angle is **ALWAYS** the same as the sign of a tangent of the same angle. *Thus, the cotangent of an angle is positive in quadrants* I *and* III .

Therefore, the angle must be in quadrant III.

The answer is (3).

PROBLEM 25: FUNCTIONS

A fraction is defined as long as its denominator is not equal to zero. Therefore, the domain of a fraction is all real values except where the denominator equals zero. Here, the denominator is equal to zero at $x = 1$.

The answer is (2).

PROBLEM 26: STATISTICS

You should know the following rule about normal distributions.

> In a normal distribution, with a mean of $\bar{x}$ and a standard deviation of σ:
> - approximately 68% of the outcomes will fall between $\bar{x} - \sigma$ and $\bar{x} + \sigma$
> - approximately 95% of the outcomes will fall between $\bar{x} - 2\sigma$ and $\bar{x} + 2\sigma$
> - approximately 99.5% of the outcomes will fall between $\bar{x} - 3\sigma$ and $\bar{x} + 3\sigma$.

Here, $\bar{x} - \sigma = 50$ and $\bar{x} + \sigma = 80$, so $\bar{x}$ must equal 65 and $\sigma = 15$.

The answer is (2).

PROBLEM 27: TRIGONOMETRY FORMULAS

Notice that this problem has the form 2sin A cos A. You can find the identities for double angles on the formula sheet under *Functions of the Double Angle*. It says sin 2A = 2 sin A cos A.

Thus we can rewrite the problem as $2 \sin 30° \cos 30° = \sin 2(30°) = \sin 60°$.

The answer is (3).

Another way to answer this question was to use your calculator to find the value of $2 \sin 30° \cos 30°$. **MAKE SURE THAT YOUR CALCULATOR IS IN DEGREE MODE.** You should get ≈ 0.866. Now find the value of each of the answer choices with your calculator.

Choice (1): $\sin 15° \approx 0.259$, so (1) is not the answer

Choice (2): $\cos 60° = 0.5$, so (2) is not the answer.

Choice (3): $\sin 60° \approx 0.866$, **so (3) is the winner!**

Choice (4): $\cos 15° \approx 0.966$, so (4) is not the answer.

PROBLEM 28: TRIGONOMETRY

In a unit circle, a terminal side with an angle of θ has an x-coordinate of $\cos θ$ *and a y-coordinate of* $\sin θ$.

Therefore, the x-coordinate is $\cos 120°$. You should know this value because it is a special angle. $\cos 120° = -\dfrac{1}{2}$.

The answer is (3)

If you don't know that $\cos 120° = -\dfrac{1}{2}$, you can always find the value with your calculator. Then pick the answer choice that matches what you get.

PROBLEM 29: TRIGONOMETRIC AREA

The area of a triangle, if we are given the lengths of the two sides a and b, and their included angle $θ$, is $A = \dfrac{1}{2}ab \sin θ$. In other words, if we are given SAS (side, angle, side) we find the area by multiplying $\dfrac{1}{2}$ by the product of the two sides by the sine of the included angle. Here, the two sides are a and b and the included angle is C.

Plugging in, we get:

$$A = \frac{1}{2} ab \sin 30°.$$

Next, given that $a = 2b$, we get:

$$\frac{1}{2} ab \sin 30° = \frac{1}{2}(2b)b \sin 30° = b^2 \sin 30°.$$

Finally, you should know that $\sin 30° = \frac{1}{2}$ because it is a special angle.

This give us: $b^2 \sin 30° = b^2\left(\frac{1}{2}\right) = \frac{b^2}{2}$ or $0.5b^2$.

The answer is (2).

Another way to get the right answer is to **plug-in**. Let's let $b = 4$, which makes $a = 8$. If we plug $a = 8$ and $b = 4$ into the problem, we get:

$$A = \frac{1}{2}(8)(4) \sin 30° = 8.$$

Now, plug $b = 4$ into the answer choices and see which one gives us 8.

Choice (1): $0.25(4^2) = 4$. Wrong.

Choice (2): $0.5(4^2) = 8$. **Right!** But, just in case . . .

Choice (3): $0.866(4^2) = 13.856$. Wrong.

Choice (4): This obviously can't be 8.

PROBLEM 30: QUADRATIC EQUATIONS

If we are given the roots of a quadratic equation, we can figure out the equation by using the following rules.

Given an equation of the form $ax^2 + bx + c = 0$, the *sum of the roots is*

$-\dfrac{b}{a}$ and the *product of the roots is* $\dfrac{c}{a}$.

Here, the roots are $3 + i$ and $3 - i$, so the sum of the roots is 6, and the product of the roots is: $(3 + i)(3 - i) = 9 - 3i + 3i - i^2 = 9 - i^2$.

Using the identity $i^2 = -1$, we get that the product of the roots is 10.

Thus $-\dfrac{b}{a} = 6$ and $\dfrac{c}{a} = 10$.

We can let a be anything we want, but it's traditional to let $a = 1$.

This makes $b = -6$ and $c = 10$.

Therefore, our equation is $x^2 - 6x + 10 = 0$.

The answer is (1).

Another way to get the right answer is to write the two roots as factors and multiply.

$$(x - (3 + i))(x - (3 - i)) = 0$$
$$(x - 3 - i)(x - 3 + i) = 0$$

Because this is the difference of two squares, we get:

$$(x - 3)^2 - i^2 = x^2 - 6x + 9 - i^2 = 0$$
$$x^2 - 6x + 10 = 0$$

As you can see, this way is a little faster, but you have to be more adept at multiplying trinomials.

Still another way to get the right answer is to use the quadratic formula to find the roots of each answer choice. Using the quadratic formula on Choice (1), we get:

$$x = \frac{-(-6) \pm \sqrt{(-6)^2 - 4(1)(10)}}{2(2)} = \frac{6 \pm \sqrt{-4}}{2} = \frac{6 \pm 2i}{2} = 3 \pm i.$$

Fortunately, we got the right answer on the first try. We obviously don't have to test the other three choices because they can't give the same answer. (If they did, they would be the same equation!)

PROBLEM 31: BINOMIAL EXPANSIONS

The binomial theorem says that if you expand $(a + b)^n$, you get the following terms:

$$_nC_0a^n +_n C_1a^{n-1}b^1 +_n C_2a^{n-2}b^2 + ... +_n C_{n-2}a^2b^{n-2} +_n C_{n-1}a^1b^{n-1} +_n C_nb^n.$$

Therefore, if we expand $(\cos x + 3)^5$, we get:

$$_5C_0(\cos x)^5 + {}_5C_1(\cos x)^4(3)^1 + {}_5C_2(\cos x)^3(3)^2 + {}_5C_3(\cos x)^2(3)^3 +$$
$$_5C_4(\cos x)^1(3)^4 + {}_5C_5(3)^5$$

Next, we use the rule that $_nC_r = \dfrac{n!}{(n-r)!\,r!}$. This gives us:

$$(\cos x)^5 + 5(\cos x)^4(3) + 10(\cos x)^3(9) + 10(\cos x)^2(27) + 5(\cos x)(81) + 243$$

which simplifies to

$$5\cos^5 x + 15\cos^4 x + 90\cos^3 x + 270\cos^2 x + 405\cos x + 243.$$

The fourth term is $270\cos^2 x$.

The answer is (2).

A shortcut is to know the rule that the *rth* term of the binomial expansion of $(a - b)^n$ is $_nC_{r-1}(a)^{n-r+1}(b)^{r-1}$. Thus the fourth term is:

$$_5C_{4-1}(\cos x)^{5-4+1}(3)^{4-1},$$

which can be simplified to $_5C_3(\cos x)^2(3)^3 = 270\cos^2 x$.

PROBLEM 32: CONIC SECTIONS

A graph of the form $y = \dfrac{k}{x}$ or $x = \dfrac{k}{y}$, where k is a constant, is a hyperbola.

The answer is (1).

You could always graph the equation $y = \dfrac{6}{x}$ and see what it looks like.

PROBLEM 33: QUADRATIC EQUATIONS

First, rewrite the equation in standard form as:

$$3x^2 + 5x + 4 = 0.$$

We can determine the nature of the roots of a quadratic equation of the form $ax^2 + bx + c = 0$ by using the discriminant $b^2 = -4ac$.

You should know the following rule:

If $b^2 - 4ac < 0$, the equation has two imaginary roots.

If $b^2 - 4ac = 0$, the equation has one rational root.

If $b^2 - 4ac > 0$, and $b^2 - 4ac$ is a perfect square, then the equation has two rational roots.

If $b^2 - 4ac > 0$, and $b^2 - 4ac$ is not a perfect square, then the equation has two irrational roots.

Here, the discriminant is $b^2 - 4ac = (5)^2 - 4(3)(4) = 25 - 48 = -23$.

The answer is (4)

PROBLEM 34: FUNCTIONS

For the purpose of the Regents exam, a function is defined as a relation between x and y where, for each x, there is one, and only one, y

This means that if you are given pairs of coordinates, you can never have two pairs with the same x-coordinate. It also means that *you can never have an equation where y is raised to an even power.*

Therefore, **choice (4) is not a function.**

PROBLEM 35: TRANSFORMATIONS

When an equation is reflected in the line $y = x$, all that you do is switch the x- and y-coordinates of each point. Therefore, when the point $(2, -5)$ is reflected in the line $y = x$, the image is $(-5, 2)$.

The answer is (3).

Part II

PROBLEM 36: TRIGONOMETRIC GRAPHS

(a) A graph of the form $y = a \sin bx$ or $y = a \cos bx$ has an amplitude of $|a|$ and a period of $\frac{2\pi}{b}$. The first equation $y = 2\cos x$ has an amplitude of 2 and a period of 2π. The second equation $y = \sin\frac{1}{2}x$ has an amplitude of 1 and a period of 4π. The two graphs look like this:

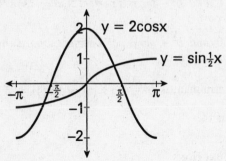

Remember, in graphing sines and cosines, the shape of the graph doesn't change, but it can be stretched or shrunk depending on the amplitude or period.

(b) We can find the intersections two ways: First, we can look at the number of intersections of our graphs. **The graphs intersect twice.**

The second way is to solve the equation algebraically. However, because you are only asked for the number of intersections, not the values of the intersections, the easiest thing to do is to look at the graphs

(a) *The measure of an inscribed angle is half of the arc it subtends (intercepts)*

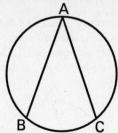

Here, we are given that $m\angle BAC = 40$. Therefore

$$m\overset{\frown}{BC} = 2(40) = 80$$

(b) We are given that triangle ABC is isosceles, so $m\angle ABC = m\angle ACB$. We are also given that $m\angle BAC = 40$, so the other two angles must add up to $180° - 40° = 140°$, and, since the angles are equal, $m\angle ABC = m\angle ACB = 70$.

Next, given the rule from part (a) above, we know that

$$m\overset{\frown}{AB} = m\overset{\frown}{ADC} = 2(70) = 140$$

Now, we are also given that $\overline{AE}$ is a diameter, so $m\overset{\frown}{ABE} = m\overset{\frown}{ACE} = 180$, so we can find $m\overset{\frown}{BE} = 180 - 140 = 40$ and $m\overset{\frown}{CE} = 180 - 140 = 40$.

Now, we know that $\angle BOE \cong \angle DOA$, because they are vertical angles, so their corresponding arcs are also congruent and thus, $m\overset{\frown}{AD} = 40$

Finally, using the rule from part (a) above, if $m\overset{\frown}{AD} = 40$, then

$$m\angle ABD = \frac{1}{2}(40) = 20.$$

(c) *A central angle has the same measure as the arc it subtends.*

We found in part (b) above that $m\overset{\frown}{AD} = 40$, so we know that

$m\angle AOD = 40$.

Next, because angles AOD and DOE form a line, their sum is $180°$. Therefore, we know that $m\angle DOE = 180 - 40 = 140$.

(d) Here, we can use another rule: *The measure of an angle formed by a pair of secants, or a secant and a tangent, is equal to half of the difference between the larger and the smaller arcs that are formed by the secants, or the secant and the tangent.*

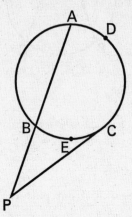

Here, the larger and smaller arcs formed by angle P are $\overgroup{ADC}$ and $\overgroup{CEB}$ respectively.

$$m\angle P = \frac{m\overgroup{ADC} - m\overgroup{CEB}}{2} = \frac{140 - 80}{2} = 30$$

(e) *An angle formed by a tangent and a chord is equal to half of its intercepted arc*

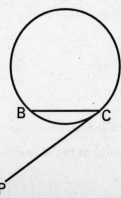

Therefore, $m\angle PCB = \dfrac{1}{2}\,m\widehat{BC} = \dfrac{1}{2}\,80 = 40$.

Next, we found in part (b) above that $m\angle ACB = 70$

This gives us.

$$m\angle ACP = m\angle PCB + m\angle ACB = 40 + 70 = 110$$

PROBLEM 38: TRIGONOMETRIC EQUATIONS

First, we use the trigonometric identity $\sin 2x = 2\sin x \cos x$ (it's on the formula sheet) and substitute it into the equation. We get:

$2(2\sin x \cos x) + \cos x$.

We then simplify and get:

$4\sin x \cos x + \cos x = 0$.

If we factor out $\cos x$, we get·

$(\cos x)(4\sin x + 1) = 0$.

This gives us $\cos x = 0$ and $4\sin x + 1 = 0$. If we solve each of these equations, we get: $\cos x = 0$ and $\cos x = -\dfrac{1}{4}$. Therefore:

$$x = \cos^{-1}(0) \text{ and } x = \cos^{-1}\left(-\dfrac{1}{4}\right).$$

The solution to the first equation is easy because it is a special angle. You should know that $\cos 90° = 0$ and that $\cos 270° = 0$.

We can find the solution to the second equation with the calculator. You should get $x \approx -14.478°$. We are asked to give the answer in the interval $0° \le x < 360°$, so we want to use $x \approx 360° - 14.478° \approx 345.522°$.

This is not the only value for x however. Sine is also negative in quadrant III, so there is another angle that satisfies the first equation It is: $x \approx 180° + 14.478° \approx 194.478°$

Rounding both of these to the nearest tenth of a degree, we get

$x = 194.5°, 345.5°$. Therefore, the solutions are $x = 90°, 194.5°, 270°, 345.5°$

PROBLEM 39:

(a) Statistics

The method for finding a standard deviation is simple, but time-consuming

First, find the sum of the measurements.

$24 + 28 + 29 + 30 + 30 + 31 + 32 + 32 + 32 + 33 + 35 + 36 = 372$.

Next, we divide by the total number of measurements, 12, to obtain the average $\bar{x} = \dfrac{372}{12} = 31$.

Second, subtract each actual score from the average.

$$24 - 31 = -7 \qquad\qquad 32 - 31 = 1$$
$$28 - 31 = -3 \qquad\qquad 32 - 31 = 1$$
$$29 - 31 = -2 \qquad\qquad 32 - 31 = 1$$
$$30 - 31 = -1 \qquad\qquad 33 - 31 = 2$$
$$30 - 31 = -1 \qquad\qquad 35 - 31 = 4$$
$$31 - 31 = 0 \qquad\qquad 36 - 31 = 5$$

Third, square each difference.

$$(-7)^2 = 49 \qquad\qquad (1)^2 = 1$$
$$(-3)^2 = 9 \qquad\qquad (1)^2 = 1$$
$$(-2)^2 = 4 \qquad\qquad (1)^2 = 1$$
$$(-1)^2 = 1 \qquad\qquad (2)^2 = 4$$
$$(-1)^2 = 1 \qquad\qquad (4)^2 = 16$$
$$(0)^2 = 0 \qquad\qquad (5)^2 = 25$$

Fourth, add up the squares of the differences.

$$49 + 9 + 4 + 1 + 1 + 0 + 1 + 1 + 1 + 4 + 16 + 25 = 112$$

Fifth, divide this sum by the total number of people

$$\frac{112}{12} = 9.33\overline{3}.$$

Last, take the square root of this number. This is the standard deviation.

$$\sigma = \sqrt{9.333} \approx 3.055$$

Rounded to the nearest hundredth, we get $\sigma \approx 3.06$.

(b) Probability

(1) This problem requires that you know something called *binomial probability*. The rule is: *If the probability of a particular outcome is **p**, then the probability of that outcome occurring **r** times out of a possible **n** times is* $_nC_r(p)^r(1-p)^{n-r}$

The probability that the spinner will land in the green sector on any given spin is $p = \dfrac{1}{5}$.

The probability that the spinner will land in the green sector exactly two times out of a possible four is:

$$_4C_2\left(\frac{1}{5}\right)^2\left(1-\frac{1}{5}\right)^{4-2} = {_4C_2}\left(\frac{1}{5}\right)^2\left(\frac{4}{5}\right)^2.$$

The rule for finding $_nC_r$ is $_nC_r = \dfrac{n!}{(n-r)!\,r!}$, so,

$$_4C_2 = \frac{4!}{2!\,2!} = \frac{4 \cdot 3 \cdot 2 \cdot 1}{(2 \cdot 1)(2 \cdot 1)} = 6.$$

Thus, the probability that spinner will land in the green sector exactly two out of four times is:

$$6\left(\frac{1}{5}\right)^2\left(\frac{4}{5}\right)^2 = \frac{96}{625}.$$

(2) The probability that the spinner will land in the green sector at least three times out of a possible four spins is the probability that it will land three times *plus* the probability that it will land all four times.

The probability that it will land three times is:

$$_4C_3\left(\frac{1}{5}\right)^3\left(\frac{4}{5}\right)^{4-3} = {_4C_3}\left(\frac{1}{5}\right)^3\left(\frac{4}{5}\right)^1.$$

The probability that it will land four times is:

$$_4C_4\left(\frac{1}{5}\right)^4\left(\frac{4}{5}\right)^{4-4} = {_4C_4}\left(\frac{1}{5}\right)^4\left(\frac{4}{5}\right)^0.$$

$$_4C_3 = \frac{4!}{1!\,3!} = \frac{4 \cdot 3 \cdot 2 \cdot 1}{(1)(3 \cdot 2 \cdot 1)} = 4$$

and $_4C_4 = \dfrac{4!}{0!\,4!} = \dfrac{4 \cdot 3 \cdot 2 \cdot 1}{(1)(4 \cdot 3 \cdot 2 \cdot 1)} = 1$

(By the way, $0! = 1$).

Therefore, the probability that it will land three times is·

$$4\left(\frac{1}{5}\right)^3\left(\frac{4}{5}\right)^1 = \frac{16}{625};$$

and the probability that it will land four times is:

$$1\left(\frac{1}{5}\right)^4\left(\frac{4}{5}\right)^0 = \frac{1}{625}.$$

The sum of the two probabilities is:

$$\frac{16}{625} + \frac{1}{625} = \frac{17}{625}.$$

PROBLEM 40:

(a) Rational Expressions

First, factor the numerator and denominator of the first expression

$$\frac{3y+15}{25-y^2} + \frac{2}{y-5} = \frac{3(y+5)}{(5+y)(5-y)} + \frac{2}{y-5}$$

Next, cancel like terms to simplify the first expression.

$$\frac{3(y+5)}{(5+y)(5-y)} + \frac{2}{y-5} = \frac{3}{5-y} + \frac{2}{y-5}$$

Next, multiply the numerator and denominator of the second expression by -1.

$$\frac{3}{5-y} + \frac{2}{y-5} \cdot \frac{-1}{-1} = \frac{3}{5-y} + \frac{-2}{5-y}$$

Finally, now that we have common denominators, we can combine terms

$$\frac{3}{5-y} + \frac{-2}{5-y} = \frac{1}{5-y}.$$

You also could have gotten $\dfrac{-1}{y-5}$.

(b) Quadratic Equations

If we move all of the terms to the left side, we get $2 + \dfrac{5}{x^2} - \dfrac{6}{x} = 0$.

Next, we multiply through by x^2 to get: $2x^2 - 6x + 5 = 0$. (Note that, because we had x and x^2 in the denominator of the original expression, if we get the answer $x = 0$, we have to throw it out.)

We can now use the quadratic formula to solve for x.

The formula says that, given a quadratic equation of the form $ax^2 + bx + c = 0$, the roots of the equation are:

$$x = \frac{-b \pm \sqrt{b^2 - 4ac}}{2a}.$$

Plugging in to the formula we get:

$$x = \frac{-(-6) \pm \sqrt{(-6)^2 - 4(2)(5)}}{2(2)} = \frac{6 \pm \sqrt{-4}}{4} = \frac{6 \pm 2i}{4} =$$

$$\frac{6}{4} \pm \frac{2}{4}i = \frac{3}{2} \pm \frac{1}{2}i$$

PROBLEM 41: TRIGONOMETRY

(a) We can find the measure of the angle using the Law of Cosines.

First, let's draw a picture of the situation.

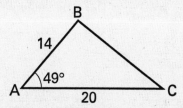

The Law of Cosines says that, in any triangle, with angles $A, B,$ and C, and opposite sides of a, b, and c, respectively, $c^2 = a^2 + b^2 - 2ab \cos C$ (This is on the formula sheet). If we have a triangle and we are given SAS (side, angle, side), then the angle opposite the side we are looking for is C, and that side is c. Here, we will use $\overline{BC}$ for side c, and $\angle CAB$ for angle C.

$$\left(\overline{BC}\right)^2 = (14)^2 + (20)^2 - 2(14)(20)\cos 49°$$

$$\left(\overline{BC}\right)^2 = 196 + 400 - 367.393$$

$$\left(\overline{BC}\right)^2 = 228.607$$

$$\overline{BC} = 15.120$$

Rounded to the nearest tenth, the answer is $\overline{BC} = 15.1$

(b) Now we can use the Law of Cosines again to find angle C. This time, we have SSS (side, side, side) and we are looking for the included angle. The side opposite the angle we are looking for is c, and the angle is C. Plugging in the information and answers from part (a), we get:

$$(14)^2 = (15.1)^2 + (20)^2 - 2(15.1)(20)\cos C$$

$$196 = 228.01 + 400 - 604\cos C$$

$$604\cos C = 228.01 + 400 - 196$$

$$604\cos C = 432.01$$

$$\cos C = \frac{432.01}{604}$$

$$C \approx 44.34°$$

Rounded to the nearest degree, we get $C \approx 44°$.

Another way to get the answer was to use the Law of Sines. The Law of Sines says that, in any triangle, with angles A, B, and C, and opposite sides of a, b, and c, respectively,

$$\frac{a}{\sin A} = \frac{b}{\sin B} = \frac{c}{\sin C} \quad \text{(This is on the formula sheet)}$$

So we have $\dfrac{15.1}{\sin 49°} = \dfrac{14}{\sin C}$.

If we solve for $\sin C$, we get: $\sin C = \dfrac{14\sin 49°}{15.1} \approx 0.6997$

This means that $C \approx \sin^{-1} 0.6997 \approx 44.405°$ Which, once again, rounds to $C \approx 44°$.

Notice how both methods give you the right answer and neither method is particularly faster

PROBLEM 42:

(a) Logarithms

All log graphs have the same shape. This one looks like:

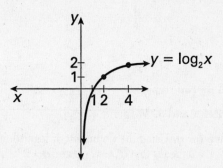

(b) Inverses

Either of the following is acceptable.

Switch the x and y terms and solve for y.

Reflect the function in the line $y = x$.

Your wording doesn't have to match ours, but you have to get close.

(c) Inverses

The graph of the inverse of $y = \log_2 x$ is the reflection in the line $y = x$.

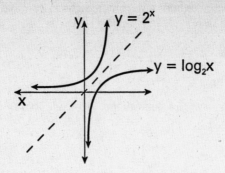

(d) Inverses

Just follow these two steps:

Step 1: **Switch x and y.** We get: $x = \log_2 y$.

Step 2: **Solve for y.** Using the definition of logarithm, which says that $\log_b x = a$ means that $b^a = x$, we get: $y = 2^x$.

Formulas

Pythagorean and Quotient Identities

$$\sin^2 A + \cos^2 A = 1 \qquad \tan A = \frac{\sin A}{\cos A}$$

$$\tan^2 A + 1 = \sec^2 A \qquad \cot A = \frac{\cos A}{\sin A}$$

$$\cot^2 A + 1 = \csc^2 A$$

Functions of the Sum of Two Angles

$$\sin (A + B) = \sin A \cos B + \cos A \sin B$$

$$\cos (A + B) = \cos A \cos B - \sin A \sin B$$

$$\tan (A + B) = \frac{\tan A + \tan B}{1 - \tan A \tan B}$$

Functions of the Difference of Two Angles

$$\sin (A - B) = \sin A \cos B - \cos A \sin B$$

$$\cos (A - B) = \cos A \cos B + \sin A \sin B$$

$$\tan (A - B) = \frac{\tan A - \tan B}{1 + \tan A \tan B}$$

Law of Sines

$$\frac{a}{\sin A} = \frac{b}{\sin B} = \frac{c}{\sin C}$$

Law of Cosines

$$a^2 = b^2 + c^2 - 2bc \cos A$$

Functions of the Double Angle

$$\sin 2A = 2 \sin A \cos A$$

$$\cos 2A = \cos^2 A - \sin^2 A$$

$$\cos 2A = 2 \cos^2 A - 1$$

$$\cos 2A = 1 - 2 \sin^2 A$$

$$\tan 2A = \frac{2 \tan A}{1 - \tan^2 A}$$

Functions of the Half Angle

$$\sin \frac{1}{2} A = \pm \sqrt{\frac{1 - \cos A}{2}}$$

$$\cos \frac{1}{2} A = \pm \sqrt{\frac{1 + \cos A}{2}}$$

$$\tan \frac{1}{2} A = \pm \sqrt{\frac{1 - \cos A}{1 + \cos A}}$$

Area of Triangle

$$K = \frac{1}{2} ab \sin C$$

Standard Deviation

$$\text{S.D.} = \sqrt{\frac{1}{n} \sum_{i=1}^{n} \left(x_i - \bar{x} \right)^2}$$

EXAMINATION
AUGUST 1997

Part I

Answer 30 questions from this part. Each correct answer will receive two credits. No partial credit will be allowed. Write your answers in the spaces provided on the separate answer sheet. Where applicable, answers may be left in terms of π or in radical form.

1 If $f(x) = \sqrt{3x} + \sqrt{12x}$, express f(−3) as a monomial in terms of i.

$$F(-3) = \sqrt{3(-3)} + \sqrt{12(-3)} \qquad \begin{vmatrix} F(-3) = 3i + 6i \\ F(-3) = 9i \end{vmatrix}$$
$$F(-3) = \sqrt{-9} + \sqrt{-36}$$

2 If $7^{(x^2+x)} = 49$, find the positive value of x.

$$7^{(x^2+x)} = 7^2 \qquad x^2 + x = \cancel{49} \qquad (x+2)(x+1) = 0$$
$$x^2 + x - \cancel{49} = 0 \qquad x = -2 \quad \boxed{x = 1}$$

3 Express 160° in radian measure.

$$160 \cdot \frac{\pi}{180} = \frac{8\pi}{9}$$

4 If cos 72° = sin x, find the number of degrees in the measure of acute angle x.

5 If $f(x) = x^{\frac{1}{2}} + x^{-2}$, what is the value of f(4)?

$$F(4) = 4^{1/2} + 4^{-2} \qquad F(4) = \frac{32+1}{16}$$
$$F(4) = 2 + \frac{1}{16} \qquad F(4) = \frac{33}{16}$$

6 Find the value of $\sum\limits_{x=0}^{2} 2^x$.

$$2^0 + 2^1 + 2^2$$
$$1 + 2 + 4 = 7$$

7 Perform the indicated operations and express in simplest form:

$$\frac{a+8}{7a^2} \cdot \frac{3a^2 - 24a}{a^2 - 64}$$

$$\frac{a+8}{7a^2} \cdot \frac{3a(a-8)}{(a+8)(a-8)}$$

$$\frac{3a}{7a^2}$$

8 Express sin 75° cos 15° – cos 75° sin 15° as a single trigonometric function of a positive acute angle.

9 If f(x) = x^2 and g(x) = x + 1, what is $(f \circ g)(2)$?

10 In which quadrant does the graph of the sum of 6 + 4i and 3 – 5i lie?

11 Two tangents are drawn to circle O from external point P. If the major arc formed has a measure of 280°, find m∠P.

12 If sin θ = $-\dfrac{8}{17}$ and tan θ is positive, what is the value of cos θ?

13 Find, to the *nearest tenth*, the positive value of x in the equation $\sqrt{x^2 + 21} = 2x$.

14 If the coordinates of A are (2, –3), what are the coordinates of A', the image of A after $R_{90°} \circ r_{y-axis}(A)$?

[OVER]

15 In the accompanying diagram, $\overline{AO}$ is tangent to circle O at D and $\overline{ABC}$ is a secant. If $AD = 6$ and $AC = 9$, find AB.

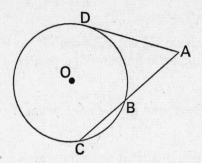

Directions (16–35): For *each* question chosen, write on the separate answer sheet the *numeral* preceding the word or expression that best completes the statement or answers the question.

16 In the accompanying table, y varies inversely as x.

x	3	6	12
y	8	4	z

What is the value of z?

(1) $\dfrac{1}{2}$ (3) 3

(2) 2   (4) $\dfrac{1}{4}$

17 What is the solution set of the equation $|2x + 1| = 9$?

(1) $\{-5\}$

(2) $\{-4, 5\}$

(3) $\{4, -5\}$

(4) $\{4\}$

18 The expression $\log\left(\dfrac{x^n}{\sqrt{y}}\right)$ is equivalent to

(1) $n \log x - \dfrac{1}{2} \log y$

(2) $n \log x - 2 \log y$

(3) $\log(nx) - \log\left(\dfrac{1}{2}y\right)$

(4) $\log(nx) - \log(2y)$

19 Which value is *not* in the range of the equation $y = \sin x$?

(1) 1

(2) 2

(3) $\dfrac{1}{2}$

(4) $-\dfrac{1}{2}$

[OVER]

20 Which trigonometric function is equivalent to the expression $\dfrac{\sin 2x}{2\sin x}$?

(1) $\tan x$
(2) $\cot x$
(3) $\sin x$
(4) $\cos x$

21 For which value of θ is the expression $\dfrac{2}{\tan\theta - 1}$ undefined?

(1) 0

(2) $\dfrac{3\pi}{4}$

(3) $\dfrac{\pi}{4}$

(4) $-\dfrac{\pi}{4}$

22 The value of $\sin\left(\dfrac{3\pi}{2}\right) - \cos\left(\dfrac{\pi}{3}\right)$ is

(1) $-1\dfrac{1}{2}$

(2) $1\dfrac{1}{2}$

(3) $\dfrac{1}{2}$

(4) $-\dfrac{1}{2}$

23 In which quadrants does the equation $xy = 10$ lie?

(1) I and II
(2) I and III
(3) II and IV
(4) III and IV

24 A set of scores with a normal distribution has a mean of 32 and a standard deviation of 3.7. Which score could be expected to occur the *least* often?

(1) 26
(2) 29
(3) 36
(4) 40

25 Which equation is represented in the graph below?

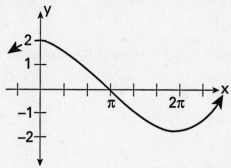

(1) $y = 2 \cos 2x$

(2) $y = \dfrac{1}{2} \cos 2x$

(3) $y = 2 \cos \dfrac{1}{2} x$

(4) $y = \dfrac{1}{2} \cos \dfrac{1}{2} x$

[OVER]

26 The product of $5 - 2i$ and i is

    (1) 7
    (2) $2 + 5i$
    (3) $5 - 2i$
    (4) $-2 + 5i$

27 Expressed in simplest form, $\dfrac{n - \dfrac{1}{n}}{1 + \dfrac{1}{n}}$ is equivalent to

    (1) $n - 1$
    (2) $n + 1$

    (3) $\dfrac{n - 1}{n + 1}$

    (4) n

28 Which is an equation of the inverse of $y = \dfrac{3}{2}x$?

    (1) $y = \dfrac{2}{3}x$

    (2) $y = -\dfrac{3}{2}x$

    (3) $y = 3x - 2$

    (4) $y = \dfrac{x + 3}{2}$

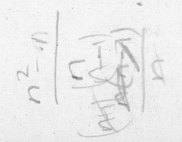

29 In a triangle, the sides measures 3, 5, and 7. What is the measure, in degrees, of the largest angle?

(1) 60
(2) 90
(3) 120
(4) 150

30 In $\triangle ABC$, m$\angle A = 45$, m$\angle B = 30$, and side $a = 10$. What is the length of side b?

(1) $5\sqrt{2}$
(2) $5\sqrt{3}$
(3) $10\sqrt{2}$
(4) $10\sqrt{3}$

31 Which relation is *not* a function?

(1) $\{(x,y) \mid y = \cos x\}$
(2) $\{(x,y) \mid x = y\}$
(3) $\{(x,y) \mid y = 3^x\}$
(4) $\{(x,y) \mid x = 3\}$

32 The roots of the quadratic equation $4x^2 = 2 + 7x$ are best described as

(1) real, equal, and rational
(2) real, unequal, and rational
(3) real, unequal, and irrational
(4) imaginary

[OVER]

33 What is the area of a parallelogram if two adjacent sides measure 4 and 5 and an included angle measures 60°?

(1) $5\sqrt{2}$

(2) $10\sqrt{2}$

(3) $5\sqrt{3}$

(4) $10\sqrt{3}$

34 Which quadratic equation has roots $2 + i$ and $2 - i$?

(1) $x^2 + 4x + 5 = 0$

(2) $x^2 - 4x - 5 = 0$

(3) $x^2 + 4x - 5 = 0$

(4) $x^2 - 4x + 5 = 0$

35 If three fair coins are tossed, what is the probability of getting at *least* two heads?

(1) $\dfrac{1}{8}$

(2) $\dfrac{3}{8}$

(3) $\dfrac{1}{2}$

(4) $\dfrac{2}{3}$

Part II

Answer four questions from this part. Clearly indicate the necessary steps, including appropriate formula substitutions, diagrams, graphs, charts, etc. Calculations that may be obtained by mental arithmetic or the calculator do not need to be shown   [40]

36 *a* Given the equation: $y = 2 \sin \dfrac{1}{2} x$

   (1) On graph paper, sketch and label the graph of this equation in the interval $0 \le x \le 2\pi$. [4]

   (2) On the same set of axes, sketch the image of the graph drawn in part *a*(1) after the transformation $r_{x\text{-}axis}$. Label the graph *T*. [2]

   (3) Write the equation of the graph drawn in part *a*(2).   [2]

   b The graph below *incorrectly* represents the equation $y = 2 \cos x$. Write a mathematical explanation of why this graph is incorrect. [2]

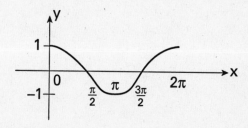

[OVER]

37 *a* On graph paper, sketch the graph of the equation $y = 2^x$ in the interval $-3 \le x \le 3$. [2]

b On the same set of axes, reflect the graph drawn in part *a* in the *y*-axis and label it *b*. [2]

c Write an equation of the function graphed in part *b*. [2]

d On the same set of axes, reflect the graph drawn in part *a* in the line $y = x$ and label the reflection *d*. [2]

e Write an equation of the function graphed in part *d*. [2]

38 *a* In the accompanying diagram, regular pentagon *ABCDE* is inscribed in circle *O*, chords $\overline{EC}$
and $\overline{DB}$ intersect at *F*, chord $\overline{DB}$ is extended
to *G*, and tangent $\overline{GA}$ is drawn.

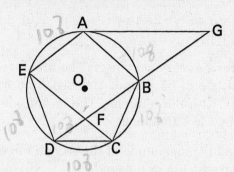

Find:

   (1) m∠*BDE*   [2]

   (2) m∠*BFC*   [2]

   (3) m∠*AGD*   [2]

b In the accompanying diagram of circle *O*, chords
$\overline{AB}$ and $\overline{CD}$ intersect at *E*, *AE* = *x*, *EB* = *x* + 1,
CE = *x* − 1, and *ED* = 2*x*. Find *AE*.   [4]

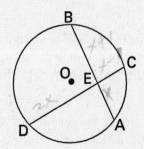

[OVER]

39 *a* Solve for *x* to the *nearest hundreth*:

$$6^x = 45 \qquad [4]$$

b Find, to the *nearest degree*, all values of *x* in the interval $0° \leq x \leq 360°$ that satisfy the equation $3\sin^2 x - 2\sin x = 1$. [6]

40 Two forces act on an object. The first force has a magnitude of 85 pounds and makes an angle of $31°\,30'$ with the resultant. The magnitude of the resultant is 130 pounds.

a Find the magnitude of the second force to *the nearest tenth of a pound*. [6]

b Using the results from part a, find, to the *nearest ten minutes* or *nearest tenth of a degree*, the angle that the second force makes with the resultant. [4]

41 *a* If a letter is selected at random from the name MARILYN in five separate trials, what is the probability that the M is chosen *exactly* three times? [2]

b If a letter is selected at random from the name DAPHNE in seven separate trials, what is the probability that a vowel is chosen *at least* six times? [4]

c If a letter is selected at random from the name NORMA in six separate trials, what is the probability that a consonant is chosen *at most* once? [4]

42 *a* For all values of *x* for which the expressions are defined, prove that the following is an identity:

$$\frac{\sin(A+B) + \sin(A-B)}{\sin(A+B) - \sin(A-B)} = \frac{\tan A}{\tan B} \quad [6]$$

b During a 10-game season, a high school football team scored the following number of points:

14, 17, 21, 10, 35, 27, 13, 7, 45, 21

Find the standard deviation of these scores to the nearest thousandth. [4]

ANSWER KEY

Part I

1. $9i$

2. 1

3. $\dfrac{8\pi}{9}$

4. 18

5. $2\dfrac{1}{16}$

6. 7

7. $\dfrac{3}{7a}$

8. $\sin 60°$

9. 9

10. IV

11. 100

12. $-\dfrac{15}{17}$

13. 2.6

14. $(3, -2)$

15. (4)

16. (2)

17. (3)

18. (1)

19. (2)

20. (4)

21. (3)

22. (1)

23. (2)

24. (4)

25. (3)

26. (2)

27. (1)

28. (1)

29. (3)

30. (1)

31. (4)

32. (2)

33. (4)

34. (4)

35. (3)

ANSWERS AND EXPLANATIONS
AUGUST 1997

Part I

In problems 1–15, you must supply the answer. Because of this, you will not be able to **plug-in** or **backsolve**. But, you can use your calculator to simplify calculations. In addition, you will find that these problems tend to test your basic knowledge of Sequential Math III and are not tricky. Many can be solved just by knowing the right formula and by plugging in the numbers.

PROBLEM 1: COMPLEX NUMBERS

First, substitute -3 for x in the equation. You should get $f(-3) = \sqrt{3(-3)} + \sqrt{12(-3)} = \sqrt{-9} + \sqrt{-36}$.

Next, you should know that $i = \sqrt{-1}$.

This means that $\sqrt{-9} = \sqrt{9}\sqrt{-1} = 3i$ and $\sqrt{-36} = \sqrt{36}\sqrt{-1} = 6i$.

Now combine the two monomials: $3i + 6i = 9i$.

PROBLEM 2: EXPONENTIAL EQUATIONS

$7^{(x^2+x)} = 49$.

In order to solve this equation, we need to have a common base for the two sides of the equation. Because $49 = 7^2$, we can rewrite the right side as $7^{(x^2+x)} = 7^2$

Now, we can set the powers equal to each other and solve.

$$x^2 + x = 2$$
$$x^2 + x - 2 = 0 \cdot$$
$$(x + 2)(x - 1) = 0$$

Thus, $x = -2$ or $x = 1$.

We are only asked for the positive value of x, so the answer is $x = 1$

Problem 3: Converting from Degrees to Radians

All we have to do is multiply $160°$ by $\dfrac{\pi}{180°}$.

$$160\left(\frac{\pi}{180}\right) = \frac{8\pi}{9}$$

Problem 4: Trigonometry

You should know that $\sin x = \cos(90° - x)$ and that $\cos x = \sin(90° - x)$. We could use either substitution, so if we use the latter substitution in the equation we get: $\sin(90° - 72°) = \sin(x)$. Therefore $(90° - 72°) = x$

If we solve for x, we get: $x = 18°$

Problem 5: Exponents

A number raised to a negative power is the same as the reciprocal of that number raised to the corresponding positive power. Here $x^{-2} = \dfrac{1}{x^2}$

A number raised to $\dfrac{1}{n}$ is the same as the nth root of that number Here $x^{\frac{1}{2}} = \sqrt{x}$

Now we can substitute 4 for x in the equation·

$$f(4) = \sqrt{4} + \frac{1}{4^2} = 2 + \frac{1}{16} = 2\frac{1}{16} \text{ or } \frac{33}{16}$$

Problem 6: Summations

To evaluate a sum, we plug each of the consecutive values for x into the expression, starting at the bottom value and ending at the top value and sum the results We get:

$$\sum_{x=0}^{2}\left(2^x\right) = \left[2^0\right] + \left[2^1\right] + \left[2^2\right] = 1 + 2 + 4 = 7$$

Another way to do this is to use your calculator. Push **2nd STAT** (**LIST**) and under the **MATH** menu, choose **SUM**. Then, again push **2nd STAT** and under the **OPS** menu, choose **SEQ**.

We put in the following: (*expression, variable, start, finish, step*).

For *expression*, we put in the formula, using x as the variable.

For *variable*, we ALWAYS put in x.

For *start*, we put in the bottom value.

For *finish*, we put in the top value.

For *step*, we ALWAYS put in 1.

Therefore, we put in the calculator: **SUM SEQ** ($2^\wedge x, x, 0, 2, 1$). The calculator should return the value 7.

PROBLEM 7: RATIONAL EXPRESSIONS

$$\frac{a+8}{7a^2} \cdot \frac{3a^2 - 24a}{a^2 - 64}$$

Factor the top and bottom of the right side to obtain

$$\frac{a+8}{7a^2} \cdot \frac{3a(a-8)}{(a-8)(a+8)}$$

Now, we cancel like terms and we get $\dfrac{3}{7a}$.

PROBLEM 8: TRIGONOMETRY FORMULAS

Notice that this problem has the form $\sin A \cos B - \cos A \sin B$. If you look on the formula sheet, under *Functions of the Difference of Two Angles*, you will find the formula $\sin(A - B) = \sin A \cos B - \cos A \sin B$. Thus we can rewrite the problem as

$$\sin 75° \cos 15° - \cos 75° \sin 15° = \sin(75° - 15°) = \sin 60°.$$

PROBLEM 9: COMPOSITE FUNCTIONS

When evaluating a composite function, start with the one on the right.

Here, we first plug 2 into $g(x)$. We obtain $g(2) = 2 + 1 = 3$.

Next, we plug 3 (the value that we found when we evaluated $g(2)$) into $f(x)$. We obtain $f(3) = 3^2 = 9$.

PROBLEM 10: COMPLEX NUMBERS

To find the sum of $6 + 4i$ and $3 - 5i$ we add the real components and the imaginary components separately. This gives us:

$$(6 + 3) + (4 + (-5))i = 9 - i.$$

Next, to determine the quadrant, we look at the signs of the components. The rule is: *The real component corresponds to the x-coordinate of the graph, and the imaginary component corresponds to the y-coordinate of the graph.* Thus, this number would have the coordinates $(9, -1)$. Because the x-coordinate is positive and the y-coordinate is negative, the sum of $3 + 2i$ and $-4 - 5i$ lies in quadrant IV.

PROBLEM 11: CIRCLE RULES

The measure of an angle formed by a pair of secants or tangents, or a secant and a tangent, is equal to half of the difference between the larger and the smaller arcs that are formed by the secants or tangents.

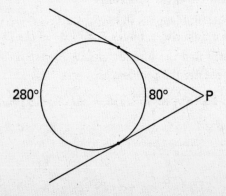

Here, the larger ("major") arc has a measure of $280°$, so the smaller ("minor") arc must have a measure of $360° - 280° = 80°$. Now we can find $m\angle P$:

$$m\angle P = \frac{280° - 80°}{2} = 100°.$$

PROBLEM 12: TRIGONOMETRY

The sine of an angle is negative in quadrants III and IV.

The tangent of an angle is positive in quadrants I and III.

This means that θ is in quadrant III.

You should know the identity $\sin^2\theta + \cos^2\theta = 1$. (This is on the formula sheet.)

We can use this identity to find $\cos\theta$:

$$\left(-\frac{8}{17}\right)^2 + \cos^2\theta = 1$$

$$\cos^2\theta = 1 - \frac{64}{289} = \frac{225}{289}$$

$$\cos\theta = \pm\frac{15}{17}$$

Because θ is in quadrant III, $\cos\theta$ is negative, so $\cos\theta = -\frac{15}{17}$.

PROBLEM 13: RADICAL EQUATIONS

First, square both sides of the equation to yield: $x^2 + 21 = 4x^2$.

Subtract x^2 from both sides: $3x^2 = 21$.

Take the square root of both sides: $x = \pm\sqrt{7} \approx \pm 2.6$.

The problem asks for the positive root, rounded to the nearest tenth, so the answer is $x = 2.6$.

HOWEVER, whenever we have a radical in an equation, we have to check the roots to see if they satisfy the original equation. Sometimes, in the process of squaring both sides, we will get answers that are invalid. So let's check the answer.

Does $\sqrt{2.6^2 + 21} \approx 2(2.6)$? $\sqrt{27.76} \approx 5.3$; close enough, so, yes, it does.

PROBLEM 14: TRANSFORMATIONS

Here we are asked to find the results of a composite transformation $\left(R_{90°} \circ r_{y-axis}\right)(A)$. *Whenever we have a composite transformation, always do the **right** one first.*

r_{y-axis} °means a reflection in the y-axis. We multiply the x-coordinate by -1, and leave the y-coordinate alone. That is, $(x, y) \xrightarrow{\ r_{y-axis}\ } (-x, y)$.

Therefore, $(2, -3) \xrightarrow{\ r_{y-axis}\ } (-2, -3)$.

$R_{90°}$ means a rotation counterclockwise of $90°$ about the origin. The rule is: If a point $P(x, y)$ is rotated $90°$ counterclockwise about the origin, its image is $P'(-y, x)$. That is, $(-2, -3) \xrightarrow{\ R_{90°}\ } P'(3, -2)$.

PROBLEM 15: CIRCLE RULES

Let x be the length of AB. We know from our circle rules that *the product of the secant and its external segment is equal to the square of the tangent segment.* Therefore $(AC)(AB) = (AD)^2$. Then we have $9x = 6^2$. If we solve for x we find that $x = 4$.

Multiple Choice Problems

For problems 16–35, you will find that you can sometimes *plug-in* and *backsolve* to get the right answer. You will also find that a calculator will simplify the arithmetic. In addition, most of the problems again only require knowing which formula to use.

PROBLEM 16: INVERSE VARIATION

An *inverse variation* between x and y means that $xy = k$, where k is a constant.

If we look at the first pair of points, we see that $(3)(8) = 24$. If we look at the second pair of points we see that $(6)(4) = 24$. Therefore, k must equal 24.

Now we have the equation $xy = 24$. If we plug in the third pair of points, we get $(12)(z) = 24$, and thus $z = 2$.

The answer is (2).

PROBLEM 17: ABSOLUTE VALUE EQUATIONS

When we have an absolute value equation we break it into two equations. First, rewrite the left side without the absolute value and leave the right side alone. Second, rewrite the left side without the absolute value, and change the sign of the right side.

Here, we have $|2x + 1| = 9$, so we rewrite it as: $2x + 1 = 9$ or $2x + 1 = -9$. If we solve each of these independently, we get $x = 4$ or $x = -5$. Thus, the solution set is $\{4, -5\}$.

The answer is (3).

PROBLEM 18: LOGARITHMS

(1) You should know the following log rules:

(i) $\log A + \log B = \log(AB)$

(ii) $\log A - \log B = \log\left(\dfrac{A}{B}\right)$

(iii) $\log A^B = B \log A$

First, using rule (ii), we get:

$$\log \frac{x^n}{\sqrt{y}} = \log x^n - \log \sqrt{y}\,.$$

Now, rewriting $\sqrt{y}$ as $y^{\frac{1}{2}}$, we get:

$$\log x^n - \log \sqrt{y} = \log x^n - \log y^{\frac{1}{2}}.$$

Finally, using rule (iii), we get:

$$\log x^n - \log y^{\frac{1}{2}} = n \log x - \frac{1}{2} \log y.$$

The answer is (1).

Another way to get the right answer is to *plug in* and use your calculator. Make up values for x, y, and n. For example, let $x = 2$, $y = 9$, and $n = 5$. Using a calculator, we get:

$$\log \frac{2^5}{\sqrt{9}} = \log \frac{32}{3} \approx 1.028\,.$$

Now, plug $x = 2$, $y = 9$, and $n = 5$ into the answers and pick the one that matches; or that is closest, because sometimes rounding will throw you off by a little bit.

Choice (1): $5 \log 2 - \dfrac{1}{2} \log 9 \approx 1.028$. Correct!

Just in case, we should check the other answers, but if you know your logs, then you know that they can't possibly be correct.

Choice (2): $5 \log 2 - 2 \log 9 \approx -0.403$. Wrong!

Choice (3): $\log 10 - \log \dfrac{9}{2} \approx 0.347$. Wrong!

Choice (4): $\log 10 - \log 18 \approx -0.255$. Wrong!

PROBLEM 19: TRIGONOMETRY GRAPHS

A graph of the form $y = a \sin bx$ *or* $y = a \cos bx$ *has an amplitude of* $|a|$

and a period of $\dfrac{2\pi}{|b|}$.

Here, the amplitude is 1 and thus its range is $-1 \le x \le 1$.

Therefore, 2 is not in the range.

The answer is (2).

PROBLEM 20: TRIGONOMETRIC IDENTITIES

You can find the identities for double angles on the formula sheet under *Functions of the Double Angle*. It says $\sin 2A = 2 \sin A \cos A$. Substituting this into the expression, we get:

$$\frac{\sin 2x}{2 \sin x} = \frac{2 \sin x \cos x}{2 \sin x}.$$

If we cancel like terms, we get:

$$\frac{2 \sin x \cos x}{2 \sin x} = \cos x.$$

The answer is (4).

Another way to get the problem right is to *plug in*. Make up a value for x and plug it into the problem. For example, let $x = 30°$. Using a calculator, we get:

$$\frac{\sin 60°}{2 \sin 30°} \approx 0.866.$$

Now, plug $x = 30°$ into the answer choices and pick the one that matches; or that is closest, because sometimes rounding will throw you off by a little bit.

Choice (1). $\tan 30° \approx 0.577$. Wrong

Choice (2): $\cot 30° \approx 1.732$. Wrong!

Choice (3): $\sin 30° = .05$. Wrong!

Choice (4): $\cos 30° \approx 0.866$. Correct!

PROBLEM 21: TRIGONOMETRY

Where the denominator of a fraction is zero, the fraction is undefined.

Here we need to know where $\tan \theta = 1$. Tangent is 1 at $45°$ or $\dfrac{\pi}{4}$ radians.

The answer is (3).

Another way to get this right is to *backsolve*. Plug each of the values of the answer choices into $f(x)$ and see which one gives you an error. **MAKE SURE THAT THE CALCULATOR IS IN RADIAN MODE.**

For choice (1), we get $\dfrac{2}{\tan 0 - 1} = -2$.

For choice (2), we get $\dfrac{2}{\tan \frac{3\pi}{4} - 1} = -1$.

For choice (3), we get $\dfrac{2}{\tan \frac{\pi}{4} - 1} = err$

For choice (4), we get $\dfrac{2}{\tan\left(-\frac{\pi}{4}\right) - 1} = -1$.

PROBLEM 22: TRIGONOMETRY

You should know your trigonometric functions at the special angles, but if you don't, put the calculator in radian mode and find the value of $\sin\left(\dfrac{3\pi}{2}\right) = -1$ and $\cos\left(\dfrac{\pi}{3}\right) = \dfrac{1}{2}$ This gives us:

$$\sin\left(\frac{3\pi}{2}\right) - \cos\left(\frac{\pi}{3}\right) = -1 - \frac{1}{2} = -1\frac{1}{2}$$

The answer is (1).

As mentioned above, you could always just find the values on your calculator. Just be sure that you are in **radian** mode. Also, **pay attention to order of operations** when you put the angles in your calculator. If you put in $\sin 3\pi / 2$, the calculator will evaluate $\sin 3\pi$ first and then divide it by 2. You will get 0, which is NOT what you want. Therefore, you have to put in $\sin(3\pi / 2)$ so that the calculator gives you the correct value of $\sin(3\pi / 2) = -1$. This is something that you should watch out for every time that you evaluate a trigonometric function, a log, or a root.

PROBLEM 23: INVERSE VARIATION

The graph of $xy = k$ is a hyperbola and looks like this:

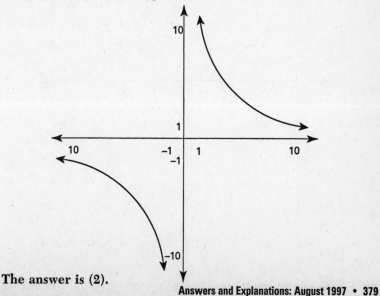

The answer is (2).

PROBLEM 24: STATISTICS

You should know the following rule about normal distributions:

*The **closer** a value is to the mean, the **more likely** it is to occur and the **farther** a value is from the mean, the **less likely** it is to occur.*

Just compare each score to the mean of 32 and choose the one with the greatest difference (if an answer is negative, use its absolute value).

Choice (1): $32 - 26 = 6$.

Choice (2): $32 - 29 = 3$.

Choice (3): $|32 - 36| = 4$.

Choice (4): $|32 - 40| = 8$. This one is the farthest.

The answer is (3).

PROBLEM 25: TRIGONOMETRY GRAPHS

You should know that the graph of $y = \cos x$ looks like this:

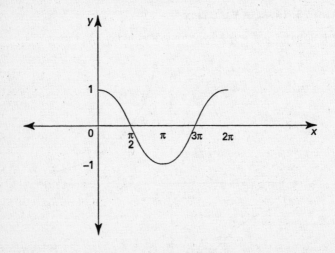

Notice that the highest point that the graph reaches is 2 and the lowest point that it reaches is −2. The amplitude of a cosine graph is found by subtracting the lowest point from the highest and dividing the result by 2. Here, we get:

$$\frac{2-(-2)}{2} = 2$$

Therefore, the amplitude of the graph is 2.

Next, you should note that only *half* of the graph is contained in the interval $0 \leq x \leq 2\pi$, so the period of the graph must be double this interval, or 4π.

A graph of the form $y = a\cos bx$ *has a period of* $\frac{2\pi}{|b|}$ *and an amplitude of* $|a|$

If we set $\frac{2\pi}{b} = 4\pi$, we get $b = \frac{1}{2}$. Therefore, the graph is

$$y = \cos\frac{1}{2}x$$

The answer is (3).

PROBLEM 26: COMPLEX NUMBERS

Here, we multiply. $i(5 - 2i) = 5i - 2i^2$

Next, because $i^2 = -1$, we have

$$5i - 2i^2 = 5i - 2(-1) = 2 + 5i$$

The answer is (2).

First we find a common denominator, n, for the numerator and the denominator of the fraction.

The numerator becomes·

$$\frac{n^2}{n} - \frac{1}{n} = \frac{n^2 - 1}{n}$$

and the denominator becomes

$$\frac{n}{n} + \frac{1}{n} = \frac{n+1}{n}$$

Next, invert the bottom fraction and multiply it by the top one

$$\frac{n^2 - 1}{n} \cdot \frac{n}{n+1}.$$

Now, factor the numerator of the left fraction to get·

$$\frac{(n-1)(n+1)}{n} \cdot \frac{n}{n+1}$$

Cancel like terms, and we get:

$$\frac{(n-1)(n+1)}{n} \cdot \frac{n}{n+1} = n - 1$$

The answer is (1).

Another way to get this right is by *plugging in*. First make up a values for n. For example, let $n = 3$. If we plug $n = 3$ into the problem, we get

$$\frac{3 - \dfrac{1}{3}}{1 + \dfrac{1}{3}} = 2$$

Then, plug $n = 3$ into each of the answer choices and pick the one that matches.

PROBLEM 28: INVERSES

In order to find the inverse of a function, follow these two steps:

(1) Switch x and y. Here, we get: $x = \dfrac{3}{2}y$.

(2) Solve for y. Multiply both sides by $\dfrac{2}{3}$ to get: $y = \dfrac{2}{3}x$

The answer is (1).

PROBLEM 29: LAW OF COSINES

The Law of Cosines says that, in any triangle, with angles A, B, and C, and opposite sides of a, b, and c, respectively, $c^2 = a^2 + b^2 - 2ab\cos C$ (This is on the formula sheet.) Don't let the different letters confuse you. If we have a triangle and we are given SAS (side, angle, side), then the side opposite the included angle is c, and the included angle is C. In a triangle, the largest angle is opposite the largest side, so we will call the largest angle C, and its opposite side is $c = 7$

If we substitute the values we get:

$$7^2 = 3^2 + 5^2 - 2(3)(5)(\cos C)$$

We can solve this for $\cos C$.

$49 = 9 + 25 - 30\cos C$

$30\cos C = 9 + 25 - 49$

$30\cos C = -15$

$\cos C = -\dfrac{1}{2}.$

You should know the answer to this because it is a special angle, but if you don't, use your calculator. You should get $120°$

The answer is (3).

PROBLEM 30: LAW OF SINES

The Law of Sines says that, in any triangle, with angles A, B, and C, and opposite sides of a, b, and c, respectively, $\dfrac{a}{\sin A} = \dfrac{b}{\sin B}$. (This is on the formula sheet.)

Here, $\dfrac{10}{\sin 45°} = \dfrac{b}{\sin 30°}$

Multiply both sides by $\sin 30°$: $\quad \dfrac{10 \sin 30°}{\sin 45°} = b$

Next, you should know that $\sin 30° = \dfrac{1}{2}$ and that $\sin 45° = \dfrac{1}{\sqrt{2}}$ because they are special angles. Otherwise use your calculator You should get:

$$\frac{10 \sin 30°}{\sin 45°} = \frac{5}{\dfrac{1}{\sqrt{2}}} = 5\sqrt{2} \approx 7.07.$$

The answer is (1).

PROBLEM 31: FUNCTIONS

A relation is a function if, for every x, there is only one y. The line $x = 3$ contains all possible values of y. Thus it will not be a function. Another way to tell is that the graph of a function must pass the *vertical line test*; that is, no vertical line can be drawn that intersects the graph more than once. The line $x = 3$ is a vertical line and so can be intersected by a vertical line an infinite number of times.

The answer is (4).

PROBLEM 32: QUADRATIC EQUATIONS

First, let's put this equation in standard quadratic form $ax^2 + bx + c = 0 \cdot$
$4x^2 - 7x - 2 = 0 \cdot$

We can determine the nature of the roots of a quadratic equation of the form $ax^2 + bx + c = 0$ by using the discriminant $b^2 - 4ac$.

You should know the following rule:

If $b^2 - 4ac < 0$, the equation has two imaginary roots.

If $b^2 - 4ac = 0$, the equation has one rational root.

If $b^2 - 4ac > 0$, and $b^2 - 4ac$ is a perfect square, then the equation has two rational roots.

If $b^2 - 4ac > 0$, and $b^2 - 4ac$ is not a perfect square, then the equation has two irrational roots.

Plugging into the discriminant, we get: $b^2 - 4ac = 49 - 4(4)(-2) = 81$.

Thus, the equation has two rational roots (which are unequal)

The answer is (2).

PROBLEM 33: TRIGONOMETRIC AREA

The area of a triangle, if we are given the lengths of the two sides a and b, and their included angle θ, is $A = \frac{1}{2}ab\sin\theta$. In other words, if we are given SAS (side, angle, side) we find the area by multiplying $\frac{1}{2}$ by the product of the two sides by the sine of the included angle.

Because a parallelogram can be thought of as the sum of two identical triangles, the area of a parallelogram, given SAS, is: $A = ab\sin\theta$

Here, the two sides are 4 and 5 and the included angle is 60°. Plugging in, we get: $A = (4)(5)\sin 60° = 20\frac{\sqrt{3}}{2} = 10\sqrt{3}$. (You should know that $\sin 60° = \frac{\sqrt{3}}{2}$, but if you don't, use your calculator.)

The answer is (4).

Problem 34: Quadratic Equations

The quadratic formula says that, given a quadratic equation of the form $ax^2 + bx + c = 0$, the roots of the equation are:

$$x = \frac{-b \pm \sqrt{b^2 - 4ac}}{2a}$$

We have to try each of the answer choices in the formula to see which one works.

Choice (1):

$$x = \frac{-4 \pm \sqrt{4^2 - 4(1)(5)}}{2(1)} = \frac{-4 \pm \sqrt{-4}}{2} = \frac{-4 \pm 2i}{2} = -2 \pm i$$

Choice (2):

$$x = \frac{-(-4) \pm \sqrt{(-4)^2 - 4(1)(-5)}}{2(1)} = \frac{4 \pm \sqrt{36}}{2} = \frac{4 \pm 6}{2} = 5 \ or \ -1$$

Choice (3):

$$x = \frac{-4 \pm \sqrt{4^2 - 4(1)(-5)}}{2(1)} = \frac{-4 \pm \sqrt{36}}{2} = \frac{-4 \pm 6}{2} = -5 \ or \ 1$$

Choice (4):

$$x = \frac{-(-4) \pm \sqrt{(-4)^2 - 4(1)(5)}}{2(1)} = \frac{4 \pm \sqrt{-4}}{2} = \frac{4 \pm 2i}{2} = 2 \pm i$$

This is the correct answer.

The answer is (4).

Problem 35: Probability

*If the probability of a particular outcome is **p**, then the probability of that outcome occurring **r** times out of a possible **n** times is*

$_nC_r(p)^r(1 - p)^{n-r}$ *This is known as binomial probability*

The probability of a fair coin coming up heads is $p = \frac{1}{2}$

In order to find the probability of getting heads *at least* two times out of three, we are going to need to find the probability of getting heads two times and the probability of getting heads three times, and then add the two probabilities.

The probability of 2 heads out of a possible 3 is $_3C_2\left(\dfrac{1}{2}\right)^2\left(1-\dfrac{1}{2}\right)^{3-2}$

This can be simplified to $_3C_2\left(\dfrac{1}{2}\right)^2\left(\dfrac{1}{2}\right)^1$.

We can then evaluate $_3C_2$ according to the formula $_nC_r = \dfrac{n!}{(n-r)!\,r!}$

If we plug in, we get:

$$_3C_2 = \frac{3!}{(3-2)!\,2!} = \frac{3!}{1!\,2!} = \frac{3\cdot2\cdot1}{(1)(2\cdot1)} = 3.$$

Therefore, the probability of 2 heads is

$$3\left(\frac{1}{2}\right)^2\left(\frac{1}{2}\right) = \frac{3}{8}.$$

The probability of 3 heads out of a possible 3 is

$$_3C_3\left(\frac{1}{2}\right)^3\left(1-\frac{1}{2}\right)^{3-3}$$

This can be simplified to

$$_3C_3\left(\frac{1}{2}\right)^3\left(\frac{1}{2}\right)^0 = {}_3C_3\left(\frac{1}{2}\right)^3.$$

We can then evaluate $_3C_3$ according to the formula $_nC_r = \dfrac{n!}{(n-r)!\,r!}$

If we plug in, we get:

$$_3C_3 = \frac{3!}{(3-3)!\,3!} = \frac{3!}{0!\,3!} = 1$$

(By the way, you should know that $_nC_n = {}_nC_0 = 1$)

Therefore, the probability of 3 heads is $\left(\dfrac{1}{2}\right)^3 = \dfrac{1}{8}$.

We add the two probabilities to get:

$$\frac{3}{8} + \frac{1}{8} = \frac{1}{2}.$$

The answer is (3).

Another way to get the right answer is to know that there are eight ways that three coins could come up: HHH, HHT, HTH, HTT, THH, THT, TTH, TTT. If we count, we find that half of these contain at least 2 heads.

Part II

PROBLEM 36: TRIGONOMETRIC GRAPHS

(a) (1) A graph of the form $y = a \sin bx$ or $y = a \cos bx$ has an amplitude of $|a|$ and a period of $\dfrac{2\pi}{b}$. Therefore, the graph of the equation $y = 2 \sin \dfrac{1}{2} x$ has an amplitude of 2 and a period of 4π.

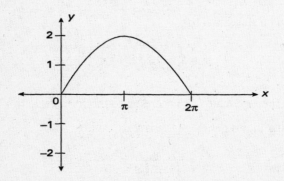

Remember, in graphing sines and cosines, the shape of the graph doesn't change, but it can be stretched or shrunk depending on the amplitude or period.

(2) The transformation r_{x-axis} means that we reflect the graph in the x-axis; that is, we turn it upside down. It looks like this:

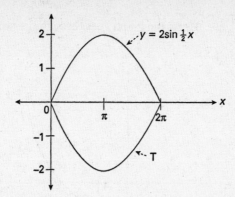

Don't forget to label the graph T.

(3) r_{x-axis} means a reflection in the x-axis. This means that we multiply each y-coordinate by -1, and leave the x-coordinates alone. The equation becomes: $-y = 2\sin\frac{1}{2}x$ or, multiplying through by -1, $y = -2\sin\frac{1}{2}x$.

(b) A graph of the form $y = a\sin bx$ or $y = a\cos bx$ has an amplitude of $|a|$ and a period of $\frac{2\pi}{b}$. Therefore, the graph of the equation $y = 2\cos x$ has an amplitude of 2 and a period of 2π. The graph shown incorrectly has an amplitude of 1.

PROBLEM 37:   TRANSFORMATIONS

(a) The graph of an equation of the form $y = a^x$ always has the same general shape. Here, the graph goes through the points $\left(-3, \frac{1}{8}\right)$, $(0,1)$, and $(3,8)$.

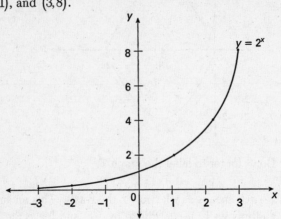

(b) The transformation $r_{y\text{-}axis}$ means that we reflect the graph in the y-axis; that is, we switch the left and right sides. It looks like this:

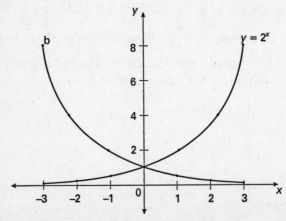

Don't forget to label the graph b.

(c) $r_{y\text{-}axis}$ means a reflection in the y-axis. This means that we multiply each x-coordinate by -1, and leave the y-coordinates alone. The equation becomes: $y = 2^{-x}$.

(d) When you reflect the graph from part (a) in the line $y = x$, it looks like this:

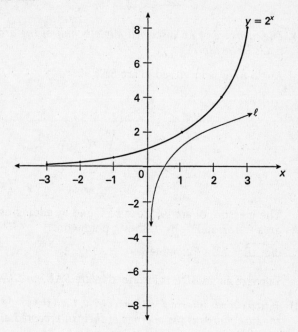

Don't forget to label the graph d.

(e) $r_{y\text{-}axis}$ means a reflection in the line $y = x$. This means that we switch the x- and y-coordinates. The equation becomes: $x = 2^y$, or if you are good with logarithms, the equation becomes $y = \log_2 x$.

PROBLEM 38: CIRCLE RULES

(a) We are given that $ABCDE$ is a regular pentagon. This means that the five sides of the pentagon are of equal length, and thus they cut the circle into five equal arcs. Therefore, the measure of each arc is:

$$\frac{360°}{5} = 72°.$$

(1) *The measure of an inscribed angle is half of the arc it subtends (intercepts).*

Angle BDE is inscribed in arc BAE.

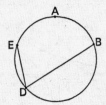

The measure of arc BAE can be found by adding two of the arcs formed by the pentagon. That is, $m\overset{\frown}{BAE} = 72° + 72° = 144°$.

Because angle BDE is inscribed in arc BAE, $m\angle BDE = 72°$.

(2) *If two chords intersect within a circle, then the angle between the two chords is the average of their intercepted arcs.*

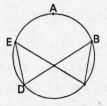

Each arc is formed by the pentagon, so each has measure $m\overset{\frown}{ED} = m\overset{\frown}{BC} = 72°$.

Therefore, $m\angle BFC = \dfrac{m\overset{\frown}{ED} + m\overset{\frown}{BC}}{2} = \dfrac{72° + 72°}{2} = 72°$.

(3) *The measure of an angle formed by a pair of secants or a secant and a tangent is equal to half of the difference between the larger and the smaller arcs that are formed by the secants or the secant and the tangent.*

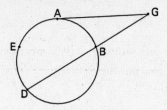

Here, the larger and smaller arcs formed by angle *AGD* are $\overset{\frown}{DEA}$ and $\overset{\frown}{AB}$ respectively. Thus, $m\angle AGD = \dfrac{m\overset{\frown}{DEA} - m\overset{\frown}{AB}}{2}$. We can easily find that $m\overset{\frown}{DEA} = 72° + 72° = 144°$ and $m\overset{\frown}{AB} = 72°$.

Therefore, $m\angle AGD = \dfrac{144° - 72°}{2} = 36°$.

(b) *If two chords intersect within a circle, then the product of the lengths of the segments of one chord is equal to the product of the lengths of the other.*

In other words, $\overline{AE} \cdot \overline{EB} = \overline{CE} \cdot \overline{ED}$.

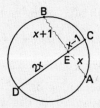

Substituting, we have:

$$(x) \cdot (x + 1) = (x - 1) \cdot (2x).$$

Distributing, we get:

$$x^2 + x = 2x^2 - 2x \cdot$$

This simplifies to:

$$x^2 - 3x = 0 \cdot$$

This factors to: $x(x - 3) = 0$, which means that $x = 0$ or $x = 3$. We can throw out the first answer (obviously), which leaves us with $AE = 3$.

PROBLEM 39:

(a) Logarithms

First, take the logarithm of both sides. We get:

$$\log 6^x = \log 45 \cdot$$

Next, use the log rule, $\log a^b = b \log a$, to rewrite the equation:

$$x \log 6 = \log 45 \cdot$$

Divide both sides by $\log 6$ and we get:

$$x = \frac{\log 45}{\log 6} \approx 2.12 \cdot$$

(b) Trigonometric Equations

First, subtract 1 from both sides of the equation to get:

$$3\sin^2 x - 2\sin x - 1 = 0 \cdot$$

Note that the trigonometric equation $3\sin^2 x - 2\sin x - 1 = 0$ has the same form as a quadratic equation $3x^2 - 2x - 1 = 0$. Just as we could factor the quadratic equation, so too can we factor the trigonometric equation. We get:

$$(3\sin x + 1)(\sin x - 1) = 0 \cdot$$

This gives us $3 \sin x + 1 = 0$ and $\sin x - 1 = 0$. If we solve each of these equations, we get·

$$\sin x = -\frac{1}{3} \text{ and } \sin x = 1.$$

Therefore $x = \sin^{-1}\left(-\frac{1}{3}\right)$ and $x = \sin^{-1}(1)$. The solutions to the

second equation is easy because it is a special angle, $x = 90°$.

We need to use a calculator to find the solution to the other equation.
Make sure that you are in **Degree** mode. You should get: $x = -19°$

In order to find the equivalent angles that are between $0°$ and $360°$
we draw a little picture and put in the answer to find the reference
angle.

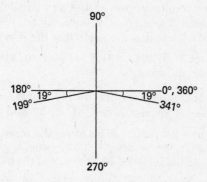

We can see that the reference angle is $19°$, thus the third quadrant
angle is $180° + 19° = 199°$ and the fourth quadrant angle is
$360° - 19° = 341°$

(a) Vector problems on the Regents exams can be solved with the Law of Cosines or the Law of Sines.

First, let's draw a picture of the situation.

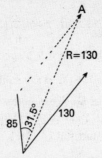

We find the resultant force by making a parallelogram and drawing a diagonal. The diagonal is the resultant. We label the second force F. We know that the angle that the first force makes with the resultant is $31°30'$, or $31.5°$, so now we have SAS and we can use the Law of Cosines.

The Law of Cosines says that, in any triangle, with angles A, B, and C, and opposite sides of a, b, and c, respectively, $c^2 = a^2 + b^2 - 2ab \cos C$. (This is on the formula sheet.) If we have a triangle and we are given SAS (side, angle, side), then the side opposite the included angle is c, and the included angle is C.

In this case, the included angle is $31.5°$, and its opposite side is F.

$F^2 = 85^2 + 130^2 - 2(85)(130) \cos 31.5°$

$F^2 = 7225 + 16900 - 22100 \cos 31.5°$

$R^2 \approx 5281.65$

$R \approx 72.7$ pounds

(b) Law of Sines

Now we want to find the angle labeled A.

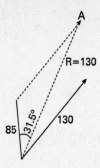

The Law of Sines says that, given a triangle with angles A, B, and C and opposite sides of a, b, and c:

$$\frac{a}{\sin A} = \frac{b}{\sin B} \quad \text{(This is on the formula sheet.)}$$

Substituting, we have:

$$\frac{72.7}{\sin 31.5°} = \frac{85}{\sin A}$$

Now we solve this for A.

First, we cross-multiply

$$72.7 \sin A = 85 \sin 31.5°$$

Next, we divide through by 72.7·

$$\sin A = \frac{85 \sin 31.5°}{72.7} \approx 0.611$$

Finally, use inverse sine to find the angle·

$$A = \sin^{-1}(0.611) \approx 37.66°$$

To decimal degrees to minutes multiply the decimal portion by 60. We get· 40′

Therefore, the answer is 37.7° or 37°40′

Problem 41:

(a) Probability

These problems require that you know something called *binomial probability*. The rule is: *If the probability of a particular outcome is **p**, then the probability of that outcome occurring **r** times out of a possible **n** times is* $_nC_r(p)^r(1-p)^{n-r}$

The probability of choosing the letter M is $\frac{1}{7}$ (because there are seven letters and one of them is M).

Therefore, the probability that M will be chosen 3 out of 5 times is

$$_5C_3\left(\frac{1}{7}\right)^3\left(1-\frac{1}{7}\right)^{5-3},$$

The rule for finding $_nC_r$ is·

$$_nC_r = \frac{n!}{(n-r)!\,r!}$$

Therefore, $_5C_3 = \frac{5!}{3!\,2!} = \frac{5\cdot4\cdot3\cdot2\cdot1}{(3\cdot2\cdot1)(2\cdot1)} = 10$

Thus, the probability of choosing M 3 out of 5 times is:

$$10\left(\frac{1}{7}\right)^3\left(\frac{6}{7}\right)^2 = \frac{360}{16807}$$

(b)

The probability of choosing a vowel is $\frac{2}{6} = \frac{1}{3}$. The probability of choosing a vowel *at least* six times is the sum of the probability that a vowel will be chosen six times and the probability that one will be chosen seven times.

We find the probability using the rule from part (a). The probability that a vowel will be chosen six times out of a possible seven is·

$$_7C_6\left(\frac{1}{6}\right)^6\left(1-\frac{1}{3}\right)^{7-6}$$

and the probability that one will be chosen seven times out of a possible seven is:

$$_7C_7\left(\frac{1}{3}\right)^7\left(1-\frac{1}{3}\right)^{7-7}.$$

$$_7C_6 = \frac{7!}{6!\,1!} = \frac{7\cdot6\cdot5\cdot4\cdot3\cdot2\cdot1}{(6\cdot5\cdot4\cdot3\cdot2\cdot1)(1)} = 7$$

(or use the rule that $_nC_{n-1} = n$)

and $$_7C_7 = \frac{7!}{7!\,0!} = \frac{7\cdot6\cdot5\cdot4\cdot3\cdot2\cdot1}{(7\cdot6\cdot5\cdot4\cdot3\cdot2\cdot1)(1)} = 1$$

(or use the rule that $_nC_n = 1$).

Thus, the probability of choosing a vowel at least six times out of a possible seven is:

$$7\left(\frac{1}{3}\right)^6\left(\frac{2}{3}\right)^1 + 1\left(\frac{1}{3}\right)^7\left(\frac{2}{3}\right)^0 = \frac{14}{2187} + \frac{1}{2187} = \frac{15}{2187}.$$

(c) The probability of choosing a consonant is $\frac{3}{5}$. The probability of choosing a consonant *at most* once is the sum of the probability that a consonant will be chosen one time and the probability that one will be chosen zero times.

We find the probability using the rule from part (a). The probability that a consonant will be chosen one time out of a possible six is:

$$_6C_1\left(\frac{3}{5}\right)^1\left(1-\frac{3}{5}\right)^{6-1}$$

and the probability that one will be chosen zero times out of possible six is:

$$_6C_0\left(\frac{3}{5}\right)^0\left(1-\frac{3}{5}\right)^{6-0}.$$

$$_6C_1 = \frac{6!}{5!\,1!} = \frac{6\cdot5\cdot4\cdot3\cdot2\cdot1}{(5\cdot4\cdot3\cdot2\cdot1)(1)} = 6$$

(or use the rule that $_nC_1 = n$)

and $_6C_0 = \dfrac{6!}{6!\,0!} = \dfrac{6 \cdot 5 \cdot 4 \cdot 3 \cdot 2 \cdot 1}{(6 \cdot 5 \cdot 4 \cdot 3 \cdot 2 \cdot 1)(1)} = 1$

(or use the rule that $_nC_0 = 1$).

Thus, the probability of choosing a consonant at most once out of a possible six is:

$$6\left(\dfrac{3}{5}\right)^1\left(\dfrac{2}{5}\right)^5 + 1\left(\dfrac{3}{5}\right)^0\left(\dfrac{2}{5}\right)^6 = \dfrac{576}{15625} + \dfrac{64}{15625} = \dfrac{640}{15625}.$$

PROBLEM 42:

(a) Trigonometric Identities

All of the identities that you will need to know are contained on the formula sheet.

Under *Functions of the Sum of Two Angles*, you will find the identity: $\sin(A + B) = \sin A \cos B + \cos A \sin B$, and under *Functions of the Difference of Two Angles*, you will find the identity: $\sin(A - B) = \sin A \cos B - \cos A \sin B$. Using these identities, we can rewrite the left side of the equation as:

$$\dfrac{(\sin A \cos B + \cos A \sin B) + (\sin A \cos B - \cos A \sin B)}{(\sin A \cos B + \cos A \sin B) - (\sin A \cos B - \cos A \sin B)}.$$

Next, we can distribute the minus sign in the denominator and remove the parentheses:

$$\dfrac{\sin A \cos B + \cos A \sin B + \sin A \cos B - \cos A \sin B}{\sin A \cos B + \cos A \sin B - \sin A \cos B + \cos A \sin B}$$

Simplify:

$$\dfrac{2 \sin A \cos B}{2 \cos A \sin B}$$

Cancel the 2's and use the identities $\tan A = \dfrac{\sin A}{\cos A}$ and $\cot A = \dfrac{\cos A}{\sin A}$ to simplify the left side some more: $\tan A \cot B$. Finally, because $\cot B = \dfrac{1}{\tan B}$, we can rewrite the left side as $\dfrac{\tan A}{\tan B}$ and set the two sides equal, and we have proved the identity.

(b) Statistics

The method for finding a standard deviation is simple, but time-consuming.

To simplify matters, let's put the scores in order:

7, 10, 13, 14, 17, 21, 21, 27, 35, 45.

First, find the average of the scores.

$$\bar{x} = \frac{7 + 10 + 13 + 14 + 17 + 21 + 21 + 27 + 35 + 45}{10} = 21.$$

Second, subtract the average from each actual score.

$7 - 21 = -14$

$10 - 21 = -11$

$13 - 21 = -8$

$14 - 21 = -7$

$17 - 21 = -4$

$21 - 21 = 0$

$21 - 21 = 0$

$27 - 21 = 6$

$35 - 21 = 14$

$45 - 21 = 24$

Third, square each difference.

$(-14)^2 = 196$

$(-11)^2 = 121$

$(-8)^2 = 64$

$(-7)^2 = 49$

$$(-4)^2 = 16$$

$$(0)^2 = 0$$

$$(0)^2 = 0$$

$$(6)^2 = 36$$

$$(14)^2 = 196$$

$$(24)^2 = 576$$

Fourth, find the sum of these squares.

$$196 + 121 + 64 + 49 + 16 + 0 + 0 + 36 + 196 + 576 = 1254$$

Fifth, divide this sum by the total number of games.

$$\frac{1254}{10} = 125.4$$

Last, take the square root of this number. This is the standard deviation.

$$\sigma = \sqrt{125.4} \approx 11.198$$

(Rounded to the nearest thousandth).

Formulas

Pythagorean and Quotient Identities

$$\sin^2 A + \cos^2 A = 1 \qquad \tan A = \frac{\sin A}{\cos A}$$

$$\tan^2 A + 1 = \sec^2 A \qquad \cot A = \frac{\cos A}{\sin A}$$

$$\cot^2 A + 1 = \csc^2 A$$

Functions of the Sum of Two Angles

$$\sin (A + B) = \sin A \cos B + \cos A \sin B$$

$$\cos (A + B) = \cos A \cos B - \sin A \sin B$$

$$\tan (A + B) = \frac{\tan A + \tan B}{1 - \tan A \tan B}$$

Functions of the Difference of Two Angles

$$\sin (A - B) = \sin A \cos B - \cos A \sin B$$

$$\cos (A - B) = \cos A \cos B + \sin A \sin B$$

$$\tan (A - B) = \frac{\tan A - \tan B}{1 + \tan A \tan B}$$

Law of Sines

$$\frac{a}{\sin A} = \frac{b}{\sin B} = \frac{c}{\sin C}$$

Law of Cosines

$$a^2 = b^2 + c^2 - 2bc \cos A$$

Functions of the Double Angle

$$\sin 2A = 2 \sin A \cos A$$

$$\cos 2A = \cos^2 A - \sin^2 A$$

$$\cos 2A = 2 \cos^2 A - 1$$

$$\cos 2A = 1 - 2 \sin^2 A$$

$$\tan 2A = \frac{2 \tan A}{1 - \tan^2 A}$$

Functions of the Half Angle

$$\sin \frac{1}{2} A = \pm \sqrt{\frac{1 - \cos A}{2}}$$

$$\cos \frac{1}{2} A = \pm \sqrt{\frac{1 + \cos A}{2}}$$

$$\tan \frac{1}{2} A = \pm \sqrt{\frac{1 - \cos A}{1 + \cos A}}$$

Area of Triangle

$$K = \frac{1}{2} ab \sin C$$

Standard Deviation

$$\text{S.D.} = \sqrt{\frac{1}{n} \sum_{i=1}^{n} \left(x_i - \bar{x} \right)^2}$$

EXAMINATION
JANUARY 1998

Part 1

Answer 30 questions from this part. Each correct answer will receive 2 credits. No partial credit will be allowed. Write you answers in the spaces provided on the separate answer sheet. Where applicable, answers may be left in terms of π or in radical form. [60]

1 Express the sum of $\sqrt{-25}$ and $2\sqrt{-9}$ as a monomial in terms of i.   5i + 2·3i
$$5i + 6i = 11i$$

2 In $\triangle ABC$, $b = 6$, $c = 3$, and $\sin B = 0.4$. Find the value of $\sin C$. $\dfrac{3}{c} = \dfrac{6}{0.4}$ 6c = 1.2   c = .2

3 An angle that measures $\dfrac{7\pi}{4}$ radians is in standard position. In which quadrant does its terminal side lie? $\dfrac{7\pi}{4}$, 180 π

4 In the accompanying diagram, segments $\overline{RS}, \overline{ST}$, and $\overline{TR}$ are tangent to circle O at A, B, and C respectively. If $SB = 3$, $BT = 5$, and $TR = 13$, what is the measure of $\overline{RS}$?

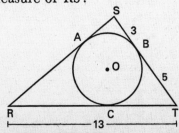

5 Solve for x: $\sqrt{5x+2}-3=0$

6 Evaluate: $\displaystyle\sum_{k=1}^{4}(k+2)^3$

7 If $f(x)=(16x)^0+x^{\frac{2}{3}}$, find f(64).

8 Express in simplest form: $\dfrac{\dfrac{x}{y}-\dfrac{y}{x}}{\dfrac{1}{y}+\dfrac{1}{x}}$

9 Find the image of A(4,–2) under the transformation $r_{y=x}$.

10 Solve for y: $2^{(y-3)}=\dfrac{1}{16}$

11 If a fair coin is tossed five times, what is the probability of tossing *exactly* three heads?

12 If $f(x)=2\sin^2 x+\sin x+1$, find the value of $f\left(\dfrac{\pi}{6}\right)$.

13 Find $m\angle\theta$ in the interval $180°\le\theta\le 270°$ that satisfies the equation $2\cos\theta+1=0$.

14 In the accompanying table, y varies inversely as x. What is the value of m?

x	2	4	m
y	18	9	3

15 In the accompanying diagram, tangent $\overline{CD}$ and secant $\overline{CBA}$ are drawn to circle O from external point C. If $DC = 4$ and $AB = 6$, find the length of $\overline{BC}$.

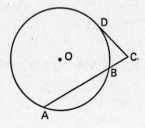

16 Write a single translation that is equivalent to $T_{3,-}$ followed by $T_{-5,5}$.

17 Express the product of $4 - 3i$ and $2 + i$ in simplest $a + bi$ form.

18 If $\log_x \dfrac{1}{4} = -2$, find x.

Directions (19–35): For *each* question chosen, write on the separate answer sheet the *numeral* preceding the word or expression that best completes the statement or answers the question.

19 Sin 50° cos 30° + cos 50° sin 30° is equivalent to
  (1) cos 80°   (3) cos 20°
  (2) sin 20°   (4) sin 80°

20 If $f(x) = 3$ and $g(x) = x^3$, then $f(g(3))$ is
  (1) 0   (3) 24
  (2) 6   (4) 30

21 Which curve has only one line of symmetry?
  (1) a circle   (3) a parabola
  (2) an ellipse   (4) a hyperbola

22 Which equation is represented in the accompanying graph?

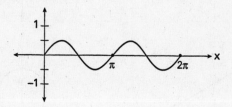

  (1) $y = 2 \sin 2x$ (3) $y = 2 \sin \dfrac{1}{2}x$

  (2) $y = \dfrac{1}{2} \sin \dfrac{1}{2}x$ (4) $y = \dfrac{1}{2} \sin 2x$

23 The solution set of $|x - 3| > 5$ is

(1) $\{x < 8 \text{ and } x < -2\}$ (3) $\{x < 8 \text{ and } x > -2\}$

(2) $\{x < 8 \text{ or } x < -2\}$ (4) $\{x > 8 \text{ or } x < -2\}$

24 The graph of the equation $y = 10^x$ lies entirely in Quadrants

(1) I and II   (3) I and IV

(2) II and III   (4) III and IV

25 Which graph of a relation is also a function?

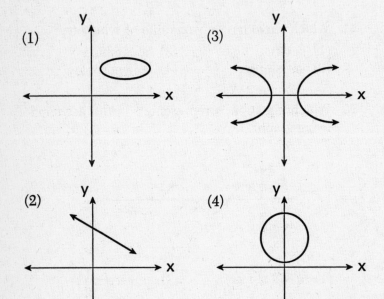

26 If $\theta = \text{Arc cos } \dfrac{\sqrt{2}}{2}$, what is the value of $\tan \theta$?

(1) 1

(3) $\dfrac{1}{\sqrt{2}}$

(2) $\sqrt{2}$

(4) $\dfrac{1}{2}$

27 If $\tan \theta < 0$ and $\text{CSC } \theta > 0$, in which quadrant does θ terminate?

(1) I

(3) III

(2) II

(4) IV

28 In the accompanying diagram of a unit circle, the ordered pair (x,y) represents the point where the terminal side of θ intersects the unit circle.

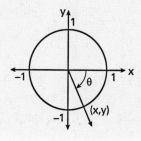

If $\theta = -\dfrac{\pi}{3}$, what is the value of y?

(1) $-\dfrac{\sqrt{3}}{2}$

(3) $-\sqrt{3}$

(2) $-\dfrac{\sqrt{2}}{2}$

(4) $-\dfrac{1}{2}$

29 What is the fourth term in the expansion of $(x - 2y)^5$?

(1) $80x^2y^3$ (3) $-80x^2y^3$

(2) $80xy^4$ (4) $-80xy^4$

30 The heights of a group of 1000 women are normally distibuted. The mean height of the group is 170 centimeters (cm) with a standard deviation of 10 cm. What is the best approximation of the number of women between 170 cm and 180 cm tall?

(1) 950   (3) 340

(2) 680   (4) 170

31 For which value of x is $f(x)$ undefined if
$$f(x) = \frac{\cos x}{1 - \cos 2x}?$$

(1) 1   (3) $\dfrac{\pi}{2}$

(2) $\dfrac{1}{2}$ (4) π

32 The graph of the equation $xy = -8$ is

(1) an ellipse   (3) a circle

(2) a hyperbola   (4) a parabola

33 The solution of $x^2 - 3x < 0$ is

(1) $0 < x < 3$ (3) $x < 0$ or $x > 3$

(2) $x > 3$ (4) $x < 0$

34 If m $\angle A$ = 45, AB = 10, and BC = 8, the greatest number of distinct triangles that can be constructed is

(1) 1   (3) 3

(2) 2   (4) 0

35 In $\triangle ABC$, $a = 8$, $b = 9$, and $m \angle C = 135$. What is the area of $\triangle ABC$?

(1) 18   (3) $18\sqrt{2}$

(2) 36   (4) $36\sqrt{2}$

Answers to the following questions are to be written on paper provided by the school.

Part II

Answer four questions from this part. Clearly indicate the necesary steps, including appropriate formula substitutions, diagrams, graphs, charts, etc Calculations that may be obtained by mental arithmetic or the calculator do not need to be shown   [40]

36 *a* On the same set of axes, sketch and label the graphs of the equations $y = \cos 2x$ and $y = -2 \sin x$ in the interval $0 \leq x \leq 2\pi$. [8]

   b Using the graphs sketched in part *a*, determine the number of values of x in the interval $0 \leq x \leq 2\pi$ that satisfy the equation $-2 \sin x - \cos 2x = 3$.   [2]

37 Find all positive values of θ less than 360° that satisfy the equation $2 \cos 2\theta - 3 \sin\theta = 1$. Express your answer to the *nearest ten minutes* or *nearest tenth of a degree*. [10]

38 *a* Given: $r = 2 - i$ and $s = 4 + 3i$

   (1) On graph paper, draw and label the graphs of these complex numbers.   [2]

   (2) On the same set of axes, graph the sum of *r* and *s* as drawn in part *a*(1) and label it *t*. [1]

   (3) On the same set of axes, draw the image of *t* after a counterclockwise rotation of 90° and label it *t′*.   [2]

b Express the roots of the equation $2x + \dfrac{5}{x} = 2$ in simplest $a + bi$ form. [5]

39 In the accompanying diagram of circle O, chord $\overline{AB}$ is parallel to diameter $\overline{EC}$, secant $\overline{PBD}$ intersects $\overline{EC}$ at F, tangent $\overline{PA}$ is drawn, $m\widehat{AB} = m\widehat{BC}$, and $m\widehat{CD} = 80$.

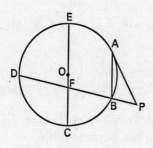

Find:

    a $m\widehat{AE}$ [2]

    b $m\angle ABD$ [2]

    c $m\angle DFC$ [2]

    d $m\angle P$ [2]

    e $m\angle PAB$ [2]

40 *a* Solve for all values of *x*:

$$\frac{2x}{x+3} + \frac{3}{x-3} = \frac{8}{x^2-9} \qquad [5]$$

b If $\log_b R = 0.75$ and $\log_b S = 0.25$, find the value of

   (1) $\log_b R^2 S$ [2]

   (2) $\log_b \dfrac{\sqrt[3]{R}}{RS}$ [3]

41 In $\triangle ABC$, $a = 6$, $b = 7$, and $c = 10$.

a Find the measure of angle *A* to the *nearest ten minutes* or *nearest tenth of a degree*. [6]

b Using the result from part *a*, find the area of $\triangle ABC$ to the *nearest tenth*. [4]

42 The table below shows the scores on a writing test in an English class.

x_i	f_i
95	4
85	13
75	11
70	6
65	2

a Using the accompanying set of data, find both the mean and the standard deviation to the *nearest tenth*. [4]

b What is the number of scores that fall within one standard deviation of the mean $(\bar{x} \pm 1\sigma)$? [2]

c Find, to the *nearest tenth*, the percentage of scores in this set of data that are within one standard deviation of the mean. [1]

d What is the number of scores that fall within two standard deviations of the mean $(\bar{x} \pm 2\sigma)$? [2]

e Find the percentage of scores in this set of data that are within two standard deviations of the mean. [1]

ANSWER KEY

Part I

1. $11i$

2. 0.2

3. IV

4. 11

5. $\dfrac{7}{5}$

6. 432

7. 17

8. $x - y$

9. $(-2,4)$

10. -1

11. $\dfrac{10}{32}$

12. 2

13. 240

14. 12

15. 2

16. $T_{-2,4}$

17. $11 - 2i$

18. 2

19. (4)

20. (3)

21. (3)

22. (4)

23. (4)

24. (1)

25. (2)

26. (1)

27. (2)

28. (1)

29. (3)

30. (3)

31. (4)

32. (2)

33. (1)

34. (2)

35. (3)

Part II

36. *b* 1   [2]

37. 14.5, 165.5, 270 *or*

     14°30′, 165°30′, 270°

38. $b \dfrac{1}{2} \pm \dfrac{3}{2} i$ [5]

39. *a* 60   [2]

    b 80   [2]

    c 100   [2]

    d 50   [2]

    e 30   [2]

40. *a* $\dfrac{1}{2}$, 1   [5]

    b   (1) 1.75   [2]

        (2) $\dfrac{1}{n}$ 0.75   [3]

41. *a* 36.2 *or* 36°10′   [6]

    b 20.7   [4]

42. *a* $\bar{x} = 79.4$, $\sigma = 8.4$ [4]

    b 24   [2]

    c 66.7   [1]

    d 36   [2]

    e 100   [1]

ANSWERS AND EXPLANATIONS
JANUARY 1998

Part I

In problems 1–18, you must supply the answer. Because of this, you will not be able to plug-in or backsolve. But, you can use your calculator to simplify calculations. In addition, you will find that these problems tend to test your basic knowledge of Sequential Math III and are not tricky. Many can be solved just by knowing the right formula and by plugging in the numbers.

PROBLEM 1: COMPLEX NUMBERS

We can rewrite $\sqrt{-25}$ as $\sqrt{25}\sqrt{-1}$ and $2\sqrt{-9}$ as $2\sqrt{9}\sqrt{-1}$

Next, we use the rule that $i = \sqrt{-1}$ to get:

$$\sqrt{25}\sqrt{-1} = 5i \text{ and } 2\sqrt{9}\sqrt{-1} = 6i.$$

Now we can add the two expressions to obtain $11i$

PROBLEM 2: LAW OF SINES

The Law of Sines says that, in any triangle, with angles A, B, and C, and opposite sides of a, b, and c, respectively:

$$\frac{a}{\sin A} = \frac{b}{\sin B} = \frac{c}{\sin C}.$$
(This is on the formula sheet.)

We substitute the values from the question to obtain

$$\frac{6}{0.4} = \frac{3}{\sin C}.$$

Next, we cross-multiply:

$$6\sin C = 1.2$$

Now we divide both sides by 6 and we get:

$$\sin C = 0.2$$

PROBLEM 3: CONVERTING FROM DEGREES TO RADIANS

If we want to convert $\dfrac{7\pi}{4}$ to degrees, we simply multiply the angle by

$\dfrac{180°}{\pi}$. We get·

$$\frac{7\pi}{4} \cdot \frac{180°}{\pi} = 315°$$

You should know that $315°$ lies in quadrant IV

PROBLEM 4: CIRCLE RULES

Here's a good rule to know: *Given a point exterior to a circle, the two tangents from that point to the circle are of equal length.*

In other words, $SA = SB$, $CT = BT$, and $RA = RC$

We are given that $SB = 3$ and $BT = 5$, so $SA = 3$ and $CT = 5$

Now, we can find RC:

$$RC = 13 - 5 = 8$$

This means that $RA = 8$ as well.

Therefore, $RS = 8 + 3 = 11$

PROBLEM 5: RADICAL EQUATIONS

First, we add 3 to both sides to isolate the radical. We get

$$\sqrt{5x + 2} = 3$$

Next, square both sides. $5x + 2 = 9$.

Now, solve for x

$5x = 7$

$x = \dfrac{7}{5}$.

BUT, whenever we have a radical in an equation, we have to check the root to see if it satisfies the original equation. Sometimes, in the process of squaring both sides, we will get answers that are invalid. So let's check the answer.

Does $\sqrt{5\left(\dfrac{7}{5}\right) + 2} - 3 = 0$? Yes. So $x = \dfrac{7}{5}$ is the answer.

PROBLEM 6: SUMMATIONS

To evaluate a sum, we plug each of the consecutive values for k into the expression, starting at the bottom value and ending at the top value and sum the results. We get:

$$\sum_{k=1}^{4}\left(k + 2\right)^{3} = \left(1 + 2\right)^{3} + \left(2 + 2\right)^{3} + \left(3 + 2\right)^{3} + \left(4 + 2\right)^{3} = 3^{3} + 4^{3} + 5^{3} + 6^{3} = 432$$

Another way to do this is to use your calculator. Push **2nd STAT (LIST)** and under the **MATH** menu, choose **SUM**. Then, again push **2nd STAT** and under the **OPS** menu, choose **SEQ**.

We put in the following: *(expression, variable, start, finish, step)*

For *expression*, we put in the formula, using x as the variable.

For *variable*, we ALWAYS put in x.

For *start*, we put in the bottom value.

For *finish*, we put in the top value.

For *step*, we ALWAYS put in 1.

Therefore, we put in the calculator: **SUM SEQ** $((x + 2)\wedge 3, x, 1, 4, 1)$. The calculator should return the value 432.

PROBLEM 7: FUNCTIONS

First, we plug in 64 for x:

$$f(64) = (16 \cdot 64)^0 + (64)^{\frac{2}{3}} .$$

Any number raised to the power zero is 1. This gives us.

$$f(64) = 1 + (64)^{\frac{2}{3}}$$

The rule for raising a number to a fractional power is:

$$x^{\frac{b}{a}} = \left(\sqrt[a]{x}\right)^b$$

This gives us:

$$f(64) = 1 + \left(\sqrt[3]{64}\right)^2 = 1 + 4^2 = 17$$

PROBLEM 8: RATIONAL EXPRESSIONS

First, get common denominators for the top and bottom.

$$\frac{\dfrac{x}{y} - \dfrac{y}{x}}{\dfrac{1}{y} + \dfrac{1}{x}} = \frac{\dfrac{x^2}{xy} - \dfrac{y^2}{xy}}{\dfrac{x}{xy} + \dfrac{y}{xy}} = \frac{\dfrac{x^2 - y^2}{xy}}{\dfrac{x + y}{xy}} .$$

Now, invert the bottom fraction and multiply it by the top one.
We get:

$$\left(\frac{x^2 - y^2}{xy}\right)\left(\frac{xy}{x + y}\right) = \frac{x^2 - y^2}{x + y}$$

Now factor the numerator into:

$$\frac{(x + y)(x - y)}{x + y}$$

Now we can cancel like terms, leaving us with·

$$x - y.$$

PROBLEM 9: TRANSFORMATIONS

$r_{y=x}$ means a *reflection* of the point in the line $y = x$. All that we have to do is switch the x-coordinate and the y-coordinate.

This gives us:

$$r_{y=x}(4, -2) \rightarrow (-2, 4).$$

PROBLEM 10: EXPONENTIAL EQUATIONS

In order to solve this equation, we need to have a common base for the two sides of the equation. Because $2^4 = 16$, we can rewrite $\dfrac{1}{16}$ as 2^{-4}

This gives us:

$$2^{y-3} = 2^{-4}.$$

Now, we can set the powers equal to each other and solve.

$y - 3 = -4$

$y = -1$

PROBLEM 11: PROBABILITY

*If the probability of a particular outcome is **p**, then the probability of that outcome occurring **r** times out of a possible **n** times is* $_nC_r(p)^r(1-p)^{n-r}$. This is known as *binomial probability*.

The probability of getting heads on one toss of a fair coin is $\dfrac{1}{2}$. Therefore, the probability of getting exactly three heads out of five tosses is:

$$_5C_3\left(\frac{1}{2}\right)^3\left(1 - \frac{1}{2}\right)^{5-3} = {}_5C_3\left(\frac{1}{2}\right)^3\left(\frac{1}{2}\right)^2.$$

We evaluate $_5C_3$ using the rule $_nC_r = \dfrac{n!}{(n-r)!\,r!}$.

This gives us:

$$_5C_3 = \frac{5!}{(5-3)!\,3!} = 10.$$

Thus, the probability of getting exactly three heads out of five tosses is:

$$_5C_3\left(\frac{1}{2}\right)^3\left(\frac{1}{2}\right)^2 = 10\left(\frac{1}{2}\right)^5 = \frac{10}{32}.$$

PROBLEM 12: TRIGONOMETRY

We plug in $\dfrac{\pi}{6}$ for x and we get:

$$f\left(\frac{\pi}{6}\right) = 2\sin^2\frac{\pi}{6} + \sin\frac{\pi}{6} + 1.$$

You should know that $\sin\dfrac{\pi}{6} = \dfrac{1}{2}$ because it is a special angle.

Therefore, $2\sin^2\dfrac{\pi}{6} + \sin\dfrac{\pi}{6} + 1 = 2\left(\dfrac{1}{2}\right)^2 + \left(\dfrac{1}{2}\right) + 1 = 2.$

Of course, you could always just find the values on your calculator. Just be sure that you are in **radian** mode. Also, **pay attention to order of operations** when you put the angles in your calculator. If you put in $\sin\pi\,/\,6$, the calculator will evaluate $\sin\pi$ first and then divide it by 6. You will get 0, which is NOT what you want. Therefore, you have to put in $\sin(\pi\,/\,6)$ so that the calculator gives you the correct value. Another typical error occurs when you have to square the sine (or any other trigonometric function). First you have to evaluate $\sin(\pi\,/\,6)$, *then* square the result. Otherwise the calculator will return an error. These are things that you should watch out for every time that you evaluate a trigonometric function or a logarithm.

PROBLEM 13: TRIGONOMETRIC EQUATIONS

Note that the trigonometric equation $2\cos\theta + 1 = 0$ has the same form as the equation $2x + 1 = 0$. Just as we can solve the latter equation, so too can we solve the former.

First, subtract 1 from both sides:

$$2\cos\theta = -1.$$

Next, divide both sides by 2:

$$\cos\theta = -\frac{1}{2}.$$

You should know the values of θ between $0°$ and $360°$ for which $\cos\theta = -\frac{1}{2}$ because they are special angles. They are $\theta = 240°$ and $\theta = 120°$. Because we are asked for the answer in the region $180° \leq \theta \leq 270°$, the answer is $\theta = 240°$.

PROBLEM 14: INVERSE VARIATION

When two variables, x, and y, are said to vary inversely, their product is always a constant. In other words, $xy = k$, where k is a constant.

Here, we are given that $x = 2$ when $y = 18$. Thus, $(2)(18) = k$ and $k = 36$. So we now know that $xy = 36$. If we plug in $y = 3$ and solve for m, we get: $m = 12$.

PROBLEM 15: CIRCLE RULES

Here's a good rule to know: Given a point exterior to a circle, *the square of the tangent segment to the circle is equal to the product of the lengths of the secant and its external segment.*

In other words, $(BC)(AC) = (DC)^2$.

Let's set the length of $\overline{BC} = x$. Then the length of $\overline{AC} = x + 6$. We are given that $DC = 4$, so we can plug into the equation and solve:

$$(x)(x + 6) = 4^2$$

$$x^2 + 6x = 16$$

$$x^2 + 6x - 16 = 0$$

$$(x - 2)(x + 8) = 0$$

$$x = 2 \text{ or } x = -8.$$

Since $\overline{BC}$ can't have a negative length, the answer is $\overline{BC} = 2$.

PROBLEM 16: TRANSFORMATIONS

Here we are asked to find the result of a composite transformation $(T_{-5,5} \circ T_{3,-1})$. *Whenever we have a composite transformation, always do the **right** one first.* The transformation $T_{3,-1}$ tells us to add 3 to the x-coordinate of a point and to add -1 to the y-coordinate of a point. The transformation $T_{-5,5}$ tells us to add -5 to the x-coordinate of a point and to add 5 to the y-coordinate of a point. Combined, we add -2 to the x-coordinate of a point and add 4 to the y-coordinate of a point. This is the same as the transformation $T_{-2,4}$.

PROBLEM 17: COMPLEX NUMBERS

To find the product of $4 - 3i$ and $2 + i$ we FOIL. This gives us:

$$(4)(2) + (4)(i) + (2)(-3i) + (-3i)(i) = 8 - 2i - 3i^2.$$

Next, because $i^2 = -1$ (by definition), we can rewrite the result as:

$$8 - 2i - 3i^2 = 8 - 2i - 3(-1) = 11 - 2i.$$

PROBLEM 18: LOGARITHMS

The definition of a logarithm tells us: $\log_B x = A$ means that $B^A = x$.

So, here, we can use this to rewrite the problem as:

$$x^{-2} = \frac{1}{4}.$$

Now, because $2^{-2} = \frac{1}{4}$, we can substitute for the expression on the right, which gives us $x^{-2} = 2^{-2}$, and thus, $x = 2$.

Multiple Choice Problems

For problems 19–35, you will find that you can sometimes *plug in* and *backsolve* to get the right answer. You will also find that a calculator will simplify the arithmetic. In addition, most of the problems only require knowing which formula to use.

PROBLEM 19: TRIGONOMETRY FORMULAS

Notice that this problem has the form $\sin A \cos B + \cos A \sin B$. Each Regents exam comes with a formula sheet that contains all of the trigonometric formulas that you will need to know. If you look under *Functions of the Sum of Two Angles*, you will find the formula $\sin(A + B) = \sin A \cos B + \cos A \sin B$. Thus we can rewrite the problem as $\sin 50° \cos 30° + \cos 50° \sin 30° = \sin(50° + 30°) = \sin 80°$.

The answer is (4).

Another way to answer this question was to use your calculator to find the value of $\sin 50° \cos 30° + \cos 50° \sin 30°$. **MAKE SURE THAT YOUR CALCULATOR IS IN DEGREE MODE.** You should get 0.9848. Now find the value of each of the answer choices with your calculator.

$\cos 80° = 0.1736$, so (1) is not the answer.

$\sin 20° = 0.3420$, so (2) is not the answer.

$\cos 20° = 0.9397$, so (3) is not the answer.

$\sin 80° = 0.9848$, so **(4) *is* the answer.**

PROBLEM 20: COMPOSITE FUNCTIONS

With composite functions, first we work with the function on the inside and then the function on the outside.

Here, we first plug 3 into $g(x)$. We obtain $g(3) = 3^3 = 27$.

Next, we plug 27 into $f(x)$. We obtain $f(27) = 27 - 3 = 24$.

The answer is (3).

PROBLEM 21: CONIC SECTIONS

A *line of symmetry* is a line that can be drawn through a graph that divides the graph into two identical halves that mirror each other. One half can then be "folded" over the line of symmetry and it will match the other half perfectly.

A circle has an infinite number of lines of symmetry.

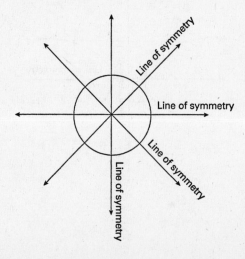

An ellipse has two lines of symmetry—one on the major axis, one on the minor axis.

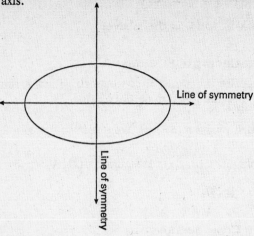

A hyperbola has two lines of symmetry—one on the transverse axis, one on the conjugate axis.

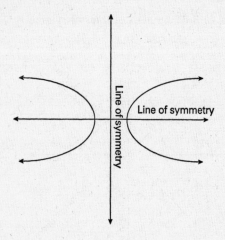

A parabola has only one line of symmetry—one that runs through the vertex and that splits the parabola into two equal halves.

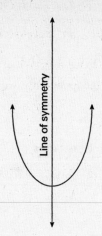

The answer is (3).

PROBLEM 22: TRIGONOMETRY GRAPHS

A graph of the form $y = a \sin bx$ or $y = a \cos bx$ has an amplitude of $|a|$ and a period of $\dfrac{2\pi}{b}$.

The amplitude of the graph is $\dfrac{1}{2}$. This eliminates choices (1) and (3).

One wavelength of the graph begins at 0 and ends at π, so the period is π. This means that b is 2. This eliminates choice (2).

Therefore, the answer is (4).

PROBLEM 23: ABSOLUTE VALUE EQUATIONS

When we have an absolute value equation we break it into two equations. First, rewrite the left side without the absolute value and leave the right side alone. Second, rewrite the left side without the absolute value, reverse the inequality, and make the right side negative.

Here, we have $|x - 3| > 5$, so we rewrite it as: $x - 3 > 5$ or $x - 3 < -5$. If we solve each of these independently, we get $x > 8$ or $x < -2$.

The answer is (4).

PROBLEM 24: EXPONENTIAL EQUATIONS

All exponential graphs of the form $y = b^x$, where $b > 0$, lie in quadrants I and II.

The answer is (1).

PROBLEM 25: FUNCTIONS

A graph of a function must pass the *vertical line test*. That is, no vertical line may be drawn that intersects the graph in more than one place. Here, the only graph that passes the vertical line test is (2).

The answer is (2).

PROBLEM 26: TRIGONOMETRY

$Arc\cos\dfrac{\sqrt{2}}{2}$ means "what angle θ has $\cos\theta = \dfrac{\sqrt{2}}{2}$?" Although there are an infinite number of answers, when the A in Arccosine is capitalized, you only use the principal angle. This means that, if the Arccosine is of a positive number, use the **first** quadrant answer; and, if the Arccosine is of a negative number, use the **fourth** quadrant answer. Here, $Arc\cos\dfrac{\sqrt{2}}{2} = 45°$.

Now, we just have to find $\tan 45°$. This is a special angle, so you should know that $\tan 45° = 1$.

The answer is (1).

Problem 27: Trigonometry

The tangent of an angle is negative in quadrants II *and* IV.

The sine of an angle is positive in quadrants I *and* II .

Because the cosecant of an angle is just the reciprocal of the sine of the angle, the sign of the cosecant of an angle is always the same as the sign of the sine of the same angle. Thus, *the cosecant of an angle is positive in quadrants* I *and* II   Therefore, the angle must be in quadrant II.

The answer is (2).

Problem 28: Trigonometry

In a unit circle, a terminal side with an angle of θ *has an x-coordinate of* cos θ *and a y-coordinate of* sin θ.

Because sine is negative in quadrant IV, the value of the y-coordinate will be $-\sin\dfrac{\pi}{3}$. You should know this value because it is a special angle. $-\sin\dfrac{\pi}{3} = -\dfrac{\sqrt{3}}{2}$

The answer is (1).

If you don't know that $-\sin\dfrac{\pi}{3} = -\dfrac{\sqrt{3}}{2}$, you can always find the value with your calculator. You should get -0.866

If you evaluate each of the answer choices with your calculator, you should find that answer choice (1) has the value -0.866.

Problem 29: Binomial Expansions

The binomial theorem says that if you expand $(a + b)^n$, you get the following terms:

$$_nC_0a^n + {}_nC_1a^{n-1}b^1 + {}_nC_2a^{n-2}b^2 + ... + {}_nC_{n-2}a^2b^{n-2} + {}_nC_{n-1}a^1b^{n-1} + {}_nC_nb^n$$

Therefore, if we expand $(x - 2y)^5$, we get:

$$_5C_0x^5 + {}_5C_1x^4(-2y)^1 + {}_5C_2x^3(-2y)^2 + {}_5C_3x^2(-2y)^3 + {}_5C_4x^1(-2y)^4 + {}_5C_5(-2y)^5.$$

Next, we use the rule that $_nC_r = \dfrac{n!}{(n-r)!\,r!}$. This gives us:

$$x^5 + 5x^4(-2y) + 10x^3(4y^2) + 10x^2(-8y^3) + 5x(16y^4) - 32y^5$$

which simplifies to $x^5 - 10x^4y + 40x^3y^2 - 80x^2y^3 + 80xy^4 - 32y^5$.

The fourth term is $-80x^2y^3$.

The answer is (3).

A shortcut is to know the rule that the *rth* term of the binomial expansion of $(a - b)^n$ is $_nC_{r-1}(a)^{n-r+1}(b)^{r-1}$. Thus the fourth term is:

$$_5C_{4-1}(x)^{5-4+1}(-2y)^{4-1},$$

which can be simplified to $_5C_3(x)^2(-2y)^3 = -80x^2y^3$

Problem 30: Statistics

You should know the following rule about normal distributions.

In a normal distribution, with a mean of $\bar{x}$ and a standard deviation of σ: approximately 68% of the outcomes will fall between $\bar{x} - \sigma$ and $\bar{x} + \sigma$; approximately 95% of the outcomes will fall between $\bar{x} - 2\sigma$ and $\bar{x} + 2\sigma$; approximately 99.5% of the outcomes will fall between $\bar{x} - 3\sigma$ and $\bar{x} + 3\sigma$.

Here, $\bar{x} - \sigma = 160$ and $\bar{x} + \sigma = 180$, so 68% of the women will be between 160 cm and 180 cm tall, which is 680 women.

Furthermore, because the distribution is normally distributed, half of these women will have heights between 160 cm and 170 cm, and half will have heights between 170 cm and 180 cm. Therefore, approximately 340 women will have heights between 170 cm and 180 cm.

The answer is (3).

PROBLEM 31: RATIONAL EXPRESSIONS

A function is undefined where it's denominator equals zero. If we set the denominator equal to zero we get:

$$\cos 2x = 1 \cdot$$

You should know the value of $2x$ for which $\cos 2x = 1$ because it is a special angle. It is $2x = 0$, and thus $x = 0$. Notice that $x = 0$ is not one of the answer choices. But, 2π has the same reference angle as 0. This means that $2x = 2\pi$ also works. Therefore, $x = \pi$ is also a solution.

The answer is (4).

Another way to get the right answer is to *backsolve*. Plug each of the answer choices into the problem and see which one gives you an error on your calculator.

Choice (1)· $\dfrac{1}{1 - \cos 2} \approx 0.706$ This is the wrong answer.

Choice (2)· $\dfrac{1}{1 - \cos 1} \approx 2.175$. This is the wrong answer.

Choice (3): $\dfrac{1}{1 - \cos \pi} = \dfrac{1}{2}$ This is the wrong answer, which leaves ..

Choice (4): $\dfrac{1}{1 - \cos 2\pi} = err$ This is the **right** answer.

Problem 32: Conic Sections

A graph of the form $xy = k$, where k is a constant, is a hyperbola.

The answer is (2).

Problem 33: Quadratic Inequalities

When we are given a quadratic inequality, the first thing that we do is factor it. We get: $x(x - 3) < 0$. The roots of the quadratic are $x = 0$, $x = 3$. If we plot the roots on a number line

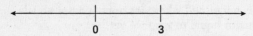

we can see that there are three regions: $x < 0$, $0 < x < 3$, and $x > 3$. Now we try a point in each region to see whether the point satisfies the inequality.

First, we try a number less than 0. Let's try -1. Is $(-1)(-1 - 3) < 0$? No.

Now, we try a number between 0 and 3. Let's try 1. Is $(1)(1 - 3) < 0$? Yes.

Last, we try a number greater than 3. Let's try 4. Is $(4)(4 - 3) < 0$? No.

Therefore, the region that satisfies the inequality is $0 < x < 3$.

The answer is (1).

PROBLEM 34: LAW OF SINES

First, let's draw a triangle and fill in the information. Any triangle will do for now

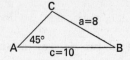

By asking for the possible number of triangles, the question is testing your knowledge of the *Ambiguous Case* of the Law of Sines. There is a simple test: Is $c \sin A < a < c$? If so, then the ambiguous case applies and there are two possible triangles.

$10 \sin 45° \approx 7.071$. Therefore, $10 \sin 45° < 8 < 10$, and there are two triangles.

The answer is (2).

Another way to get this right is to use the Law of Sines and see how many triangles can be made.

The Law of Sines says that, in any triangle, with angles A, B, and C, and opposite sides of a, b, and c, respectively, $\dfrac{a}{\sin A} = \dfrac{b}{\sin B} = \dfrac{c}{\sin C}$. (This is on the formula sheet.)

Here, $\dfrac{8}{\sin 45°} = \dfrac{10}{\sin C}$, so $\sin C = \dfrac{10 \sin 45°}{8} \approx 0.884$. (If we had gotten a number greater than 1 or less than -1 we would have NO triangles.) Therefore, $\angle C \approx 62°$. This would make $\angle B \approx 73°$. Now we have one triangle

But, there is a second angle for which $\sin C \approx 0.884$:

$$\angle C \approx 180° - 62° = 118°.$$

Can we make a triangle with angles of $45°, 118°$, and B? Yes, if $\angle B = 17°$. Therefore, there are two possible triangles.

PROBLEM 35: TRIGONOMETRIC AREA

The area of a triangle, if we are given the lengths of the two sides a and b, and their included angle θ, is $A = \frac{1}{2} ab \sin \theta$. In other words, if we are given SAS (side, angle, side) we find the area by multiplying $\frac{1}{2}$ by the product of the two sides by the sine of the included angle. Here, the two sides are a and b and the included angle is C.

Plugging in, we get:

$$A = \frac{1}{2}(8)(9) \sin 135°.$$

You should know that $\sin 135° = \frac{\sqrt{2}}{2}$ because it is a special angle. This give us:

$$36 \sin 135° = 36\left(\frac{\sqrt{2}}{2}\right) = 18\sqrt{2}.$$

The answer is (3).

Part II

Problem 36: Trigonometric Graphs

(a) A graph of the form $y = a \sin bx$ or $y = a \cos bx$ has an amplitude of $|a|$ and a period of $\frac{2\pi}{b}$. The first equation $y = \cos 2x$ has an amplitude of 1 and a period of π. The second equation $y = -2 \sin x$ has an amplitude of 2, is upside down (because of the minus sign), and has a period of 2π. The two graphs look like this:

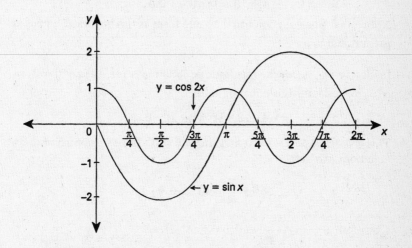

Remember, in graphing sines and cosines, the shape of the graph doesn't change, but it can be stretched or shrunk depending on the amplitude or period.

(b) We need to find when the vertical distance between the two graphs at any x-coordinate is 3. We can see that the distance is 3 only at

$$x = \frac{3\pi}{2}.$$

Problem 37: Trigonometric Equations

First, we use the trigonometric identity $\cos 2\theta = 1 - 2\sin^2\theta$ (it's on the formula sheet) and substitute it into the equation. This will make everything in the equations in terms of $\sin\theta$. We get:

$$2\left(1 - 2\sin^2\theta\right) - 3\sin\theta = 1$$

Next, distribute the 2:

$$2 - 4\sin^2\theta - 3\sin\theta = 1.$$

Now, put all terms on the same side of the equals sign and simplify:

$$4\sin^2\theta + 3\sin\theta - 1 = 0.$$

Notice that this equation has the same form as the quadratic equation $4x^2 + 3x - 1 = 0.$

Just as we can factor the quadratic equation into $(4x - 1)(x + 1) = 0$, so too can we factor this equation into:

$$(4\sin\theta - 1)(\sin\theta + 1) = 0.$$

This gives us $4\sin\theta - 1 = 0$ and $\sin\theta + 1 = 0$. If we solve each of these equations, we get:

$$\sin\theta = \frac{1}{4} \text{ and } \sin\theta = -1$$

Therefore $\theta = \sin^{-1}\left(\frac{1}{4}\right)$ and $\theta = \sin^{-1}(-1).$

We can find the solution to the first equation with the calculator. We get:

$$\theta \approx 14.478°.$$

This is not the only value for θ, however. Sine is also positive in quadrant II, so there is another angle that satisfies the first equation. It is:

$$\theta \approx 180° - 14.478° \approx 165.522°.$$

Rounding both of these to the nearest tenth of a degree, we get:

$$\theta = 14.5°, 165.5°$$

The solution to the second equation is easy because it is a special angle. You should know that $\sin 270° = -1$, and thus, $\theta = 270°$

Therefore, the solutions are $\theta = 14.5°, 165.5°, 270°$

PROBLEM 38: COMPLEX NUMBERS

(a) (1) A complex number of the form $a + bi$ is represented graphically by a point with the coordinates (a, b). In other words, the real part of the complex number is the x-coordinate, and the imaginary part is the y-coordinate.

Thus, the point $r = 2 - i$ has the coordinates $(2, -1)$ and the point $s = 4 + 3i$ has the coordinates $(4, 3)$.

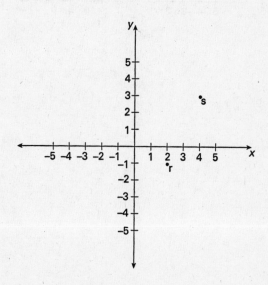

(2) Complex Numbers

In order to add two complex numbers, you simply add the real parts and the imaginary parts separately.

Thus, $(2-i)+(4+3i)=(2+4)+(-1+3)i=6+2i$ and has coordinates (6,2).

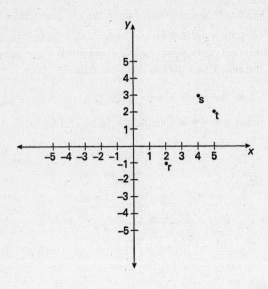

(3) Transformations

If a point $P(x,y)$ is rotated 90° counterclockwise about the origin, its image is $P'(-y,x)$. In other words, $P(x,y) \xrightarrow{Rot_{90°}} P'(-y,x)$. Here the image of t after a rotation of 90° counterclockwise about the origin is:

$$t(6,2) \xrightarrow{Rot_{90°}} t'(-2,6)$$

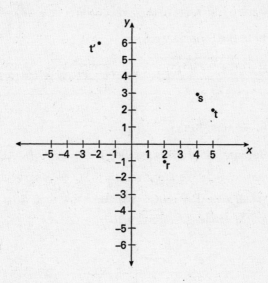

(b) Quadratic Equations

If we move all of the terms to the left side, we get.

$$2x + \frac{5}{x} - 2 = 0.$$

Next, we multiply through by x to get:
$$2x^2 - 2x + 5 = 0.$$
(Note that, because we had x in the denominator of the original expression, if we get the answer $x = 0$, we have to throw it out.)

We can now use the quadratic formula to solve for x.

The formula says that, given a quadratic equation of the form $ax^2 + bx + c = 0$, the roots of the equation are: $x = \dfrac{-b \pm \sqrt{b^2 - 4ac}}{2a}$

Plugging in to the formula we get:

$$x = \frac{-(-2) \pm \sqrt{(-2)^2 - 4(2)(5)}}{2(2)} = \frac{2 \pm \sqrt{4 - 40}}{4} = \frac{2 \pm 6i}{4} = \frac{1}{2} \pm \frac{3}{2}i$$

PROBLEM 39: CIRCLE RULES

(a) Because chord AB is parallel to diameter EC, $\overset{\frown}{AE}$ is congruent to $\overset{\frown}{BC}$. We are given that $\overset{\frown}{BC}$ is congruent to $\overset{\frown}{AB}$, so $m\overset{\frown}{AB} = m\overset{\frown}{AE} = m\overset{\frown}{BC}$. We also know that $\overline{EC}$ is a diameter of the circle, so $m\overset{\frown}{EAC} = 180$

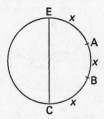

If we let $m\overset{\frown}{AB} = m\overset{\frown}{AE} = m\overset{\frown}{BC} = x$, then we have $3x = 180$, and thus, $x = 60$. Therefore, $m\overset{\frown}{AB} = m\overset{\frown}{AE} = m\overset{\frown}{BC} = 60$, but we are only asked for $m\overset{\frown}{AE} = 60$

(b) *The measure of an inscribed angle is half of the arc that it sub-tends.* Here, $\angle ABD$ subtends $\overparen{AED}$. We know that $\overline{EC}$ is a diameter of the circle, so $m\overparen{EDC} = 180$. We are also given that $m\overparen{CD} = 80$, so $m\overparen{ED} = 100$. Finally, we know from part (a) above that $m\overparen{AE} = 60$. Therefore, $m\overparen{AED} = 160$ and thus, $m\angle ABD = 80$.

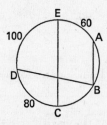

(c) *If two chords intersect within a circle, then the angle between the two chords is the average of their intercepted arcs.*

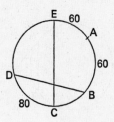

In other words, $m\angle DFC = \dfrac{m\overparen{CD} + m\overparen{EB}}{2}$. We know from part

(a) above that $m\overparen{AB} = m\overparen{AE} = 60$, so $m\overparen{EB} = 120$.

Substituting, we get:

$$m\angle DFC = \frac{80 + 120}{2} = 100.$$

(d) Here, we can use another rule: *The measure of an angle formed by a pair of secants or a secant and a tangent is equal to half of the difference between the larger and the smaller arcs that are formed by the secants or the secant and the tangent.*

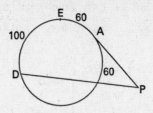

Here, the larger and smaller arcs formed by angle P are $\overset{\frown}{AED}$ and $\overset{\frown}{AB}$ respectively.

$$m\angle P = \frac{m\overset{\frown}{AED} - m\overset{\frown}{AB}}{2} = \frac{160 - 60}{2} = 50.$$

(e) *An angle formed by a tangent and a chord is equal to half of its intercepted arc.*

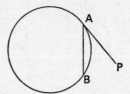

Therefore, $m\angle PAB = \frac{1}{2}\, m\overset{\frown}{AB} = \frac{1}{2}\, 60 = 30.$

PROBLEM 40:

(a) Rational Expressions

First, factor the denominator of the expression on the right side of the equals sign.

$$\frac{2x}{x+3} + \frac{3}{x-3} = \frac{8}{(x+3)(x-3)}.$$

Next, multiply through by $(x+3)(x-3)$ to clear the denominators.

$$2x(x-3) + 3(x+3) = 8.$$

Distribute the terms on the left side of the equals sign:

$$2x^2 - 6x + 3x + 9 = 8.$$

Subtract 8 from both sides and simplify:

$$2x^2 - 3x + 1 = 0.$$

Factor:

$$(2x-1)(x-1) = 0.$$

This means that $2x - 1 = 0$ or $x - 1 = 0$, and thus, $x = \frac{1}{2}$ or $x = 1$.

(b) Logarithms

(1) You should know the following log rules:

(i) $\log A + \log B = \log(AB)$

(ii) $\log A - \log B = \log\left(\frac{A}{B}\right)$

(iii) $\log A^B = B \log A$

Using rule (i), we can rewrite $\log_b R^2 S = \log_b R^2 + \log_b S$.

Next, using rule (iii), we can rewrite $\log_b R^2 S = 2\log_b R + \log_b S$.

Finally, we can substitute $\log_b R = 0.75$ and $\log_b S = 0.25$ and we get:

$$2\log_b R + \log_b S = 2(0.75) + 0.25 = 1.75.$$

(2) First, because $\sqrt[3]{R} = R^{\frac{1}{3}}$, we can rewrite $\log_b \dfrac{\sqrt[3]{R}}{RS} = \log_b \dfrac{R^{\frac{1}{3}}}{RS}$.

Next, using rule (ii), we can rewrite $\log_b \dfrac{R^{\frac{1}{3}}}{RS} = \log_b R^{\frac{1}{3}} - \log_b RS$.

Next, using rule (i), we can rewrite:

$$\log_b R^{\frac{1}{3}} - \log_b RS = \log_b R^{\frac{1}{3}} - \left(\log_b R + \log_b S\right).$$

Finally, using rule (iii), we can rewrite:

$$\log_b R^{\frac{1}{3}} - \left(\log_b R + \log_b S\right) = \frac{1}{3}\log_b R - \left(\log_b R + \log_b S\right).$$

Now, we can substitute $\log_b R = 0.75$ and $\log_b S = 0.25$ and we get:

$$\frac{1}{3}(0.75) - (0.75 + 0.25) = 0.25 - 1.00 = -0.75.$$

PROBLEM 41:

(a) Trigonometry

We can find the measure of the angle using the Law of Cosines.

The Law of Cosines says that, in any triangle, with angles A, B, and C, and opposite sides of a, b, and c, respectively, $a^2 = b^2 + c^2 - 2bc \cos A$. (This is on the formula sheet.) This time, we have SSS (side, side, side) and we are looking for the included angle. The angle we are looking for is A, and the side opposite the angle is a. Plugging in the information we are given, we get:

$$(6)^2 = (7)^2 + (10)^2 - 2(7)(10) \cos A$$

$$36 = 49 + 100 - 140 \cos A$$

$$140 \cos A = 49 + 100 - 36$$

$$140 \cos A = 113$$

$$\cos A = \frac{113}{140}$$

$$A \approx 36.18°.$$

Rounded to the nearest tenth of a degree, we get $A \approx 36.2°$.

(b) Trigonometric Area

The area of a triangle, if we are given the lengths of the two sides a and b, and their included angle θ, is $A = \frac{1}{2} ab \sin \theta$. In other words, if we are given SAS (side, angle, side) we find the area by multiplying $\frac{1}{2}$ by the product of the two sides by the sine of the included angle. Here, the two sides are b and c and the included angle is A. Plugging in, we get:

$$A = \frac{1}{2}(7)(10) \sin 36.2° \approx 20.7$$

PROBLEM 42:

(a) Statistics

The method for finding a standard deviation is simple, but time-consuming.

First, find the average of the scores. We do this by multiplying each score by the number of students that had that score, and adding them up.

$(95)(4) + (85)(13) + (75)(11) + (70)(6) + (65)(2) = 2860$. Next, we divide by the total number of students $(4 + 13 + 11 + 6 + 2) = 36$, to obtain the average $\bar{x} \approx 79.4$.

Second, subtract the average from each actual score.

$95 - 79.4 = 15.6$

$85 - 79.4 = 5.6$

$75 - 79.4 = -4.4$

$70 - 79.4 = -9.4$

$65 - 79.4 = -14.4$

Third, square each difference.

$(15.6)^2 = 243.36$

$(5.6)^2 = 31.36$

$(-4.4)^2 = 19.36$

$(-9.4)^2 = 88.36$

$(-14.4)^2 = 207.36$

Fourth, multiply the square of the difference by the corresponding number of students and sum.

$(243.36)(4) = 973.44$

$(31.36)(13) = 407.68$

$(19.36)(11) = 212.96$

$(88.36)(6) = 530.16$

$(207.36)(2) = 414.72$

$973.44 + 407.68 + 212.96 + 530.16 + 414.72 = 2538.96$.

Fifth, divide this sum by the total number of students.

$\dfrac{2538.96}{36} = 70.53$.

Last, take the square root of this number. This is the standard deviation.

$\sigma = \sqrt{70.53} \approx 8.4$.

(b) Statistics

You should know the following rule about normal distributions.

In a normal distribution, with a mean of $\bar{x}$ and a standard deviation of σ:

- approximately 68% of the outcomes will fall between $\bar{x} - \sigma$ and $\bar{x} + \sigma$.

- approximately 95% of the outcomes will fall between $\bar{x} - 2\sigma$ and $\bar{x} + 2\sigma$.

- approximately 99.5% of the outcomes will fall between $\bar{x} - 3\sigma$ and $\bar{x} + 3\sigma$.

Using this rule, 68% of 36 is 24.48, which rounds to 24 scores.

(c) Statistics

One standard deviation of the mean is $\bar{x} - \sigma$ and $\bar{x} + \sigma$. Here, we have $\bar{x} - \sigma = 79.4 - 8.4 = 71$ and $\bar{x} + \sigma = 79.4 + 8.4 = 87.8$.

The number of scores that fall between 71 and 87.8 is 24.

The percent of scores is thus:

$$\frac{24}{36} \times 100 = 66.7\%.$$

(d) Statistics

Two standard deviations of the mean is $\bar{x} - 2\sigma$ and $\bar{x} + 2\sigma$. Here, we have $\bar{x} - 2\sigma = 79.4 - 16.8 = 62.6$ and $\bar{x} + 2\sigma = 79.4 + 16.8 = 96.2$.

The number of scores that fall between 62.6 and 96.2 is 36; that is, all of them.

(e) Statistics

As we saw in part (d) above, all of the scores fall within two standarddeviations of the mean. Therefore, the answer is 100%.

Formulas

Pythagorean and Quotient Identities

$$\sin^2 A + \cos^2 A = 1 \qquad \tan A = \frac{\sin A}{\cos A}$$

$$\tan^2 A + 1 = \sec^2 A \qquad \cot A = \frac{\cos A}{\sin A}$$

$$\cot^2 A + 1 = \csc^2 A$$

Functions of the Sum of Two Angles

$$\sin (A + B) = \sin A \cos B + \cos A \sin B$$

$$\cos (A + B) = \cos A \cos B - \sin A \sin B$$

$$\tan (A + B) = \frac{\tan A + \tan B}{1 - \tan A \tan B}$$

Functions of the Difference of Two Angles

$$\sin (A - B) = \sin A \cos B - \cos A \sin B$$

$$\cos (A - B) = \cos A \cos B + \sin A \sin B$$

$$\tan (A - B) = \frac{\tan A - \tan B}{1 + \tan A \tan B}$$

Law of Sines

$$\frac{a}{\sin A} = \frac{b}{\sin B} = \frac{c}{\sin C}$$

Law of Cosines

$$a^2 = b^2 + c^2 - 2bc \cos A$$

Functions of the Double Angle

$$\sin 2A = 2 \sin A \cos A$$

$$\cos 2A = \cos^2 A - \sin^2 A$$

$$\cos 2A = 2 \cos^2 A - 1$$

$$\cos 2A = 1 - 2 \sin^2 A$$

$$\tan 2A = \frac{2 \tan A}{1 - \tan^2 A}$$

Functions of the Half Angle

$$\sin \frac{1}{2} A = \pm \sqrt{\frac{1 - \cos A}{2}}$$

$$\cos \frac{1}{2} A = \pm \sqrt{\frac{1 + \cos A}{2}}$$

$$\tan \frac{1}{2} A = \pm \sqrt{\frac{1 - \cos A}{1 + \cos A}}$$

Area of Triangle

$$K = \frac{1}{2} ab \sin C$$

Standard Deviation

$$\text{S.D.} = \sqrt{\frac{1}{n} \sum_{i=1}^{n} \left(x_i - \bar{x} \right)^2}$$

EXAMINATION
JUNE 1998

Part I

Answer 30 questions from this part. Each correct answer will receive 2 credits. No partial credit will be allowed. Write your answers in the spaces provided. Where applicable, answers may be left in terms of π or in radical form. [60]

1 If $f(x) = (2x)^2$, find $f(-4)$.

2 Solve for the positive value of x: $x^{\frac{2}{3}} = 9$

3 Solve for x: $\dfrac{x-3}{5} + \dfrac{4x}{3} = 4$

4 Solve for x: $\sqrt{2x - 8} - 1 = 5$

5 Evaluate: $\sum\limits_{n=1}^{5} n^2$

6 Find the image of $A(-3,2)$ under dilation with the center at the origin and a scale factor of -2.

7 In $\triangle ABC$, $\sin A : \sin B : \sin C = 4:5:6$. Find the value of c when $a = 10$

8 Find the value of the tan $\left(\text{Arc} \sin \dfrac{5}{6} \right)$

9 If a fair coin is flipped three times, what is the probability of obtaining exactly two heads?

10 In the accompanying diagram of circle O, secants $\overline{CBA}$ and $\overline{CED}$ intersect at C. If $AC = 12$, $BC = 3$, and $DC = 9$, find EC

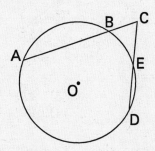

11 In a circle whose radius is 9 centimeters, what is the number of radians in a central angle if the length of the intercepted arc is 18 centimeters?

12 Find, in radical form, the area of $\triangle ABC$ if $a = 6$, $b = 6$, and m$\angle C = 45$.

13 Factor completely: $5x^2y^3 - 180y$

14 If P varies inversely as V and $P = 700$ when $V = 8$, find the value of V when $P = 350$.

[OVER]

15 In the accompanying diagram of circle *O*, m∠*AOB* = 80.

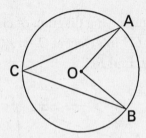

What is m∠*ACB*?

(1) 80   (3) 20

(2) 160   (4) 40

16 The value of $\cos 16° \cos 164° - \sin 16° \sin 164°$ is

(1) –1   (3) 0

(2) $-\dfrac{1}{2}$ (4) $\dfrac{\sqrt{3}}{2}$

17 If the graphs of the equations $xy = 12$ and $y = 2$ are drawn on the same set of axes, what is the total number of common points?

(1) 1   (3) 3

(2) 2   (4) 0

18 The expression $\dfrac{\dfrac{a}{b}-1}{\dfrac{a}{b}+1}$ is equivalent to

(1) $\dfrac{a+b}{a-b}$ (3) $\dfrac{1}{a-b}$

(2) $\dfrac{a-b}{a+b}$ (4) $\dfrac{1}{a+b}$

19 The value of $\sin \dfrac{3\pi}{2} + \cos \dfrac{2\pi}{3}$ is

(1) $\dfrac{1}{2}$ (3) $-1\dfrac{1}{2}$

(2) $1\dfrac{1}{2}$ (4) $-\dfrac{1}{2}$

20 The sum of $3\sqrt{-8}$ and $4\sqrt{-50}$ is

(1) $12\sqrt{-58}$ (3) $7i\sqrt{58}$

(2) $26i\sqrt{2}$ (4) $7i\sqrt{2}$

21 What is the solution set of the inequality $|3x+6| \le 30$?

(1) $-12 \le x \le 8$
(2) $-8 \le x \le 12$
(3) $x \le -12$ or $x \ge 8$
(4) $x \le -8$ or $x \ge 12$

[OVER]

22 The roots of the equation $x^2 + 6x + 11 = 0$ are

(1) real, rational, and unequal
(2) real, rational, and equal
(3) real, irrational, and unequal
(4) imaginary

23 If $\cos x = -\dfrac{\sqrt{2}}{2}$, in which quadrants could $\angle x$ terminate?

(1) I and IV   (3) II and IV
(2) I and III   (4) II and III

24 Which expression is equivalent to $\dfrac{\sin 2x}{\cos x}$?

(1) $2 \sin x$ (3) $\cos 2x$
(2) $\tan x$ (4) $2 \cos x$

25 If $\sin (x - 3)° = \cos (2x + 6)°$, then the value of x is

(1) −9   (3) 29
(2) 26   (4) 64

26 Which graph represents the solution set of
    $x^2 + 5x - 6 > 0$?

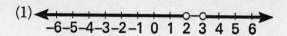

(1) number line with open circles at 2 and 3, segment between them

(2) number line with open circle at -6 and open circle at 2, rays extending outward

(3) number line with closed segment between 2 and 3 (open circles)

(4) number line with open circle at -6 and open circle at 2, segment between

27 What is the result of $T_{2,-1} \circ r_{y=-1}\ (2,0)$?
    (1) (2,0)   (3) (4,–3)
    (2) (2,–1)   (4) (–4,3)

28 In $\triangle ABC$, $a = 8$, $b = 2$, and $c = 7$. What is the value
    of $\cos C$?

    (1) $-\dfrac{19}{32}$ (3) $\dfrac{109}{112}$

    (2) $-\dfrac{11}{28}$ (4) $\dfrac{19}{32}$

29 What is the domain of $f(x) = \dfrac{1}{\sqrt{\left(4 - x^2\right)}}$?

    (1) $x < 2$ (3) $-2 < x < 2$
    (2) $|x| \le 2$ (4) all real numbers

[OVER]

30 In standard position, an angle of $\dfrac{7\pi}{3}$ radius has the same terminal side as an angle of

(1) 60°   (3) 240°
(2) 120°   (4) −420°

31 If the mean on a standardized test with a normal distribution is 54.3 and the standard deviation is 4.6, what is the best approximation of the percent of the scores that fall between 54.3 amd 63.5?

(1) 34   (3) 68
(2) 47.5   (4) 95

32 In the accompanying diagram of the unit circle, the ordered pair (x,y) represents the point where the terminal side of θ intersects the unit circle.

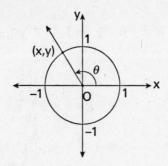

If $\theta = \dfrac{3\pi}{4}$, what is the value of x?

(1) 1   (3) $-\dfrac{\sqrt{2}}{2}$

(2) $-\dfrac{1}{2}$ (4) $\dfrac{\sqrt{3}}{2}$

33 What is the sum (S) and the product (P) of the roots of the equation $2x^2 - 4x + 1 = 0$?

(1) $S = \dfrac{1}{2}, P = 2$ (3) $S = -2, P = \dfrac{1}{2}$

(2) $S = 2, P = \dfrac{1}{2}$ (4) $S = -4, P = 1$

34 If $a = 5$, $c = 18$, and m$\angle A = 30$, what is total number of district triangles that can be constructed?

(1) 1 (3) 3
(2) 2 (4) 0

35 What is the middle term in the expansion $(x - 3y)^4$?

(1) $54x^2y^2$ (3) $9x^2y^2$
(2) $54xy^2$ (4) $9xy^2$

[OVER]

Part II

Answer four questions from this part. Clearly indicate the necessary steps, including appropriate formula substitutions, diagrams, graphs, charts, etc. Calculations that may be obtained by mental arithmetic or the calculator do not need to be shown. [40]

36 *a* On the same set of axes, sketch and label the graphs of the equations $y = \sin \frac{1}{2}x$ and $y = -2 \cos x$ in the interval $-\pi \le x \le \pi$. [8]

 b Using the graphs drawn in part a, determine the number of solutions to the equation $\sin \frac{1}{2}x = -2 \cos x$ in the interval $-\pi \le x \le \pi$. [2]

37 In the accompanying diagram of circle O. diameter $\overline{EOC}$ is extended through C to point P; diameter $\overline{AFOD}$, tangent $\overline{PD}$, and chords $\overline{AC}$, $\overline{CD}$. $\overline{BFE}$ are drawn; m$\angle COD = 60$; and m$\angle AFB = 100$.

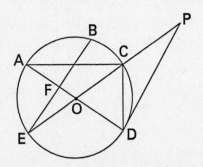

Find:

a m$\widehat{DE}$ [2]
b m$\angle P$ [2]
c m$\angle ACE$ [2]
d m$\widehat{AB}$ [2]
e m$\angle ACD$ [2]

38 a Given: $\log_b 3 = p$

                $\log_b 5 = q$

(1) Express $\log_b \dfrac{9}{5}$ in terms of p and q.   [2]

(2) Express $\log_b \sqrt[3]{15}$ in terms of p and q.   [2]

b Solve for x:

$$\log_4(x^2 + 3x) - \log_4(x + 5) = 1 \qquad [6]$$

39 a Only red cards and black cards are in a box.

The probability of drawing a black card is $\dfrac{3}{5}$.

A card is randomly drawn and replaced in the box after each draw. Five such draws are made. Find the probability that

(1) *exactly* two black cards will be drawn   [2]
(2) *at least* four black cards will be drawn   [4]

b For these measurements, find the standard deviation, to the *nearest hundredth*:

85, 88, 79 , 79, 80, 92, 94, 78, 80, 85   [4]

[OVER]

40 Find, to the *nearest ten minutes or nearest tenth of a degree*, all values of θ in the interval $0° \leq \theta < 360°$ that satisfy the equation $5 \sin^2 \theta - 7\cos \theta + 1 = 0$. [10]

41 *a* Solve for x: $\dfrac{x}{x+5} + \dfrac{9}{x-5} = \dfrac{50}{x^2-25}$ [5]

   b Solve for y and express the roots of the equation in simplest $a + bi$ form:

$$5y + \frac{5}{y} = 8 \quad [5]$$

42 *a* Given: $Z_1 = 4 - i$ and $Z_2 = -5 - 2i$.

   (1) Using the complex plane, graph and label Z_1 and Z_2. [2]

   (2) On the same plane, graph the sum of Z_1 and Z_2. [1]

   (3) Express the sum of Z_1 and Z_2 as a complex number. [1]

   b Forces of 40 pounds and 70 pounds act on a body at an angle measuring $60°$. Find the magnitude of the resultant of these forces to the *nearest hundredth of a pound*. [6]

ANSWER KEY

Part I

1. 64

2. 27

3. 3

4. 22

5. 55

6. (6,–4)

7. 15

8. $\dfrac{5}{\sqrt{11}}$

9. $\dfrac{3}{8}$

10. 4

11. 2

12. $9\sqrt{2}$

13. $5y(xy + 6)(xy - 6)$

14. 16

15. (4)

16. (1)

17. (1)

18. (2)

19. (3)

20. (2)

21. (1)

22. (4)

23. (4)

24. (1)

25. (3)

26. (2)

27. (3)

28. (4)

29. (3)

30. (1)

31. (2)

32. (3)

33. (2)

34. (4)

35. (1)

Part II

36. *b* 2   [2]

37. *a* 120 [2]

   b 30   [2]

   c 80   [2]

   d 80   [2]

   e 90   [2]

38. *a* (1) $2p - q$ [2]

   (2) $\frac{1}{3}(p + q)$[2]

   b –4,5 [6]

39. *a* (1) $\frac{720}{3125}$ [2]

   (2) $\frac{1053}{3125}$ [4]

   b 5.48   [4]

40. 53.1°, 306.9°
   or   [10]
   53°10′, 306°50′

41. *a* 1   [5]

   b $\frac{4}{5} \pm \frac{3}{5}i$ [5]

42. *a* (3) –1 – 3*i* [1]

   b 96.44 [6]

ANSWERS AND EXPLANATIONS
JUNE 1998

Part I

In problems 1–14, you must supply the answer. Because of this, you will not be able to **plug-in** or **backsolve**. But, you can use your calculator to simplify calculations. In addition, you will find that these problems tend to test your basic knowledge of Sequential Math III and are not tricky. Many can be solved just by knowing the right formula and by plugging in the numbers.

PROBLEM 1: FUNCTIONS

We plug in –4 for x: $f(-4) = (2 \cdot -4)^2 = (-8)^2 = 64$.

PROBLEM 2: EXPONENTIAL EQUATIONS

First, raise each side to the $\dfrac{3}{2}$ power, to get: $x = 9^{\frac{3}{2}}$.

The rule for raising a number to a fractional power is $x^{\frac{b}{a}} = \sqrt[a]{x^b}$.

This gives us: $9^{\frac{3}{2}} = \left(\sqrt{9}\right)^3 = \sqrt{729} = 27$.

PROBLEM 3: RATIONAL EXPRESSIONS

First, multiply through by 15 to clear the denominators.

$$(15)\frac{x-3}{5} + (15)\frac{4x}{3} = (15)4$$

$$3(x-3) + 5(4x) = 60$$

Distribute: $3x - 9 + 20x = 60$

Now, we can solve for x.

$23x - 9 = 60$

$23x = 69$

$x = 3$.

If there had been any variables in the denominator, we would have had to check the solution to make sure that the denominator isn't equal to zero at that value. We don't have that problem here.

PROBLEM 4: RADICAL EQUATIONS

First, we add 1 to both sides to isolate the radical. We get:

$$\sqrt{2x - 8} = 36.$$

Next, square both sides. $2x - 8 = 36$.

Now, solve for x.

$2x = 44$

$x = 22$.

BUT, whenever we have a radical in an equation, we have to check the root to see if it satisfies the original equation. Sometimes, in the process of squaring both sides, we will get answers that are invalid. So let's check the answer.

Does $\sqrt{44 - 8} - 1 = 6$? $6 - 1 = 5$; so, yes, it does. Therefore the answer is $x = 22$.

PROBLEM 5: SUMMATIONS

To evaluate a sum, we plug each of the consecutive values for n into the expression, starting at the bottom value and ending at the top value and sum the results. We get:

$$\sum_{n=1}^{5}\left(n^2\right) = 1^2 + 2^2 + 3^2 + 4^2 + 5^2 = 1 + 4 + 9 + 16 + 25 = 55.$$

Another way to do this is to use your calculator. Push **2nd STAT (LIST)** and under the **MATH** menu, choose **SUM**. Then, again push **2nd STAT** and under the **OPS** menu, choose **SEQ**.

We put in the following: *(expression, variable, start, finish, step)*

For *expression*, we put in the formula, using x as the variable

For *variable*, we ALWAYS put in x.

For *start*, we put in the bottom value.

For *finish*, we put in the top value.

For *step*, we ALWAYS put in 1.

Therefore, we put in the calculator: **SUM SEQ** $(x^2, x, 1, 5, 1)$. The calculator should return the value 55.

PROBLEM 6: TRANSFORMATIONS

A *dilation* means that we multiply the x- and y-coordinates by the scale factor. In this case, we get $(-3, 2) \xrightarrow{\;D_2\;} (6, -4)$.

PROBLEM 7: LAW OF SINES

The Law of Sines says that, in any triangle, with angles A, B, and C, and opposite sides of a, b, and c, respectively, $\dfrac{a}{\sin A} = \dfrac{c}{\sin C}$. (This is on the formula sheet.)

Here, we can rearrange the formula to:

$$\frac{\sin C}{\sin A} = \frac{c}{a}.$$

Plugging into the formula, we get:

$$\frac{6}{4} = \frac{c}{10}.$$

Multiplying both sides by 10, we get:

$$\frac{60}{4} = 15 = c.$$

PROBLEM 8: TRIGONOMETRY

$Arc \sin \frac{5}{6}$ means "what angle θ has $\sin \theta = \frac{5}{6}$?" Although there are an infinite number of answers, when the A in Arcsine is capitalized, you only use the principal angle. This means that, if the Arcsine is of a positive number, use the **first** quadrant answer; and, if the Arcsine is of a negative number, use the **second** quadrant answer. Now we can draw a triangle in the first quadrant, and using the fact that

$\sin \theta = \dfrac{opposite}{hypotenuse}$, label the sides.

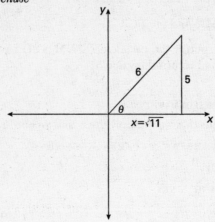

Now we can use the Pythagorean Theorem to find the third side, which we have labeled x.

$5^2 + x^2 = 6^2$

$x^2 = 11$

$x = \sqrt{11}$ (we use the positive value because we are in quadrant I.)

Now, using the definition of tangent as $\tan \theta = \dfrac{opposite}{adjacent}$, we get

$$\tan \theta = \frac{5}{\sqrt{11}}.$$

PROBLEM 9: PROBABILITY

*If the probability of a particular outcome is **p**, then the probability of that outcome occurring **r** times out of a possible **n** times is* $_nC_r(p)^r(1-p)^{n-r}$. This is known as *binomial probability*.

The probability of a fair coin coming up heads is $p = \dfrac{1}{2}$.

Using the formula, we find that the probability of 2 heads out of a possible 3 is $_3C_2\left(\dfrac{1}{2}\right)^2\left(1-\dfrac{1}{2}\right)^{3-2}$. This can be simplified to $_3C_2\left(\dfrac{1}{2}\right)^2\left(\dfrac{1}{2}\right)^1$.

We can then evaluate $_3C_2$ according to the formula $_nC_r = \dfrac{n!}{(n-r)!\,r!}$.
If we plug in, we get:

$$_3C_2 = \frac{3!}{(3-2)!\,2!} = \frac{3!}{1!\,2!} = \frac{3\cdot 2\cdot 1}{(1)(2\cdot 1)} = 3.$$

Therefore, the probability of 2 heads is $3\left(\dfrac{1}{2}\right)^2\left(\dfrac{1}{2}\right) = \dfrac{3}{8}$.

PROBLEM 10: CIRCLE RULES

Here's a good rule to know: Given a point exterior to a circle, *the product of the lengths of one secant and its external segment is equal to the product of the lengths of the other secant and its external segment.*

In other words, $(\overline{AC})(\overline{BC}) = (\overline{DC})(\overline{EC})$. We can simply plug into the equation and solve: $(12)(3) = (9)(\overline{EC})$

$\overline{EC} = 4$.

PROBLEM 11: ARC LENGTH

In a circle with a central angle, θ, measured in radians, and a radius, r; the arc length, s, is found by $s = r\theta$.

Here we have $r = 9$ and $s = 18$. Therefore, the central angle, θ, can be found by $18 = (9)(\theta)$.

$$\theta = 2$$

PROBLEM 12: TRIGONOMETRIC AREA

The area of a triangle, if we are given the lengths of the two sides a and b, and their included angle θ, is $A = \frac{1}{2} ab \sin \theta$. In other words, if we are given SAS (side, angle, side) we find the area by multiplying $\frac{1}{2}$ by the product of the two sides by the sine of the included angle. Here, the two sides are a and b and the included angle is C.

Plugging in, we get:

$$A = \frac{1}{2} (6)(6) \sin 45° .$$

You should know that $\sin 45° = \frac{\sqrt{2}}{2}$ because it is a special angle. This gives us:

$$A = \frac{1}{2} (6)(6) \frac{\sqrt{2}}{2} = 9\sqrt{2}$$

PROBLEM 13: FACTORING

We need to take out of both of the terms as many common factors as possible. Notice that both of the constant coefficients are divisible by 5, so we can take a 5 out of both terms, which will leave us with:
$5 \left(x^2 y^3 - 36y \right)$.

Next, notice that both terms have a y in common, so we can take a y out of both terms, which leaves us with an answer of: $5y\left(x^2 y^2 - 36 \right)$.

Finally, notice that the term in parentheses is the difference between two squares. A *difference of two squares* is an expression of the form $a^2 - b^2$, and can always be factored into $a^2 - b^2 = (a - b)(a + b)$. Here, the expression can be factored into: $5y(xy - 6)(xy + 6)$.

PROBLEM 14: INVERSE VARIATION

An *inverse variation* between V and P means that $PV = k$, where k is a constant.

If we put in the information for the first pair of values, we get that $(700)(8) = k$, and thus k = 5600. Now we have the equation PV = 5600. If we put in the information for the last pair of values, we can solve for V.

We get.

$$(350)V = 5600 \text{ and thus } V = 16.$$

Multiple Choice Problems

For problems 15–35, you will find that you can sometimes *plug-in* and *backsolve* to get the right answer. You will also find that a calculator will simplify the arithmetic. In addition, most of the problems again only require knowing which formula to use.

PROBLEM 15: CIRCLE RULES

We need to use two simple rules to solve this problem: (1) *A central angle has the same measure as the arc it subtends;* and (2) *The measure of an inscribed angle is always half of the measure of the arc that it subtends* (intercepts).

Using the first rule, we know that the measure of arc AB is the same as $m\angle AOB = 80$.

Now, using the second rule, we know that $m\angle ACB = \frac{1}{2} \cdot 80 = 40$.

The answer is (4).

PROBLEM 16: TRIGONOMETRY

Notice that this problem has the form $\cos A \cos B - \sin A \sin B$. Each Regents exam comes with a formula sheet that contains all of the trig formulas that you will need to know. If you look under *Functions of the Sum of Two Angles*, you will find the formula $\cos(A + B) = \cos A \cos B - \sin A \sin B$. Thus we can rewrite the problem as $\cos 16° \cos 164° - \sin 16° \sin 164° = \cos(16° + 164°) = \cos 180°$.

This is a special angle that you should know. $\cos 180° = -1$.

The answer is (1).

Another way to answer this question was to use your calculator to find the value of $\cos 16° \cos 164° - \sin 16° \sin 164°$. **MAKE SURE THAT YOUR CALCULATOR IS IN DEGREE MODE.** You should get -1.

PROBLEM 17: GRAPHING

First, we arrange the first equation to obtain $y = \dfrac{12}{x}$. Now, because both equations are solved for y, we can set them equal to each other and solve for x. The solution will be the x-coordinate of the intersection.

We get:

$$\frac{12}{x} = 2.$$

Therefore, $x = 6$ and there is only one solution.

The answer is (1).

PROBLEM 18: RATIONAL EXPRESSIONS

First, combine the fractions on the top and bottom using a common denominator b:

$$\frac{\dfrac{a}{b} - 1}{\dfrac{a}{b} + 1} = \frac{\dfrac{a}{b} - \dfrac{b}{b}}{\dfrac{a}{b} + \dfrac{b}{b}} = \frac{\dfrac{a-b}{b}}{\dfrac{a+b}{b}} .$$

Now, invert the bottom fraction and multiply it by the top one.

We get:

$$\left(\frac{a-b}{b}\right)\left(\frac{b}{a+b}\right) = \frac{a-b}{a+b} .$$

The answer is (2).

Another way to get the right answer is to *plug in*. Make up values for a and b in the problem and plug them in. For example, let's use $a = 3$, and $b = 5$.

We get:

$$\frac{\dfrac{3}{5} - 1}{\dfrac{3}{5} + 1} = -\frac{1}{4}$$

Now plug $a = 3$ $a = 3$ and $b = 5$ into the answer choices and pick the one that gives us $-\dfrac{1}{4}$.

Choice (1): $\dfrac{3+5}{3-5} = \dfrac{8}{-2} = -4$. Wrong.

Choice (2): $\dfrac{3-5}{3+5} = \dfrac{-2}{8} = -\dfrac{1}{4}$. Correct.

You should be able to tell by inspection that the other two can't give you $-\dfrac{1}{4}$, but if you aren't sure, test them for yourself.

PROBLEM 19: TRIGONOMETRY

You should know your trigonometric functions at the special angles but if you don't, put the calculator in radian mode and find the value of $\sin\left(\frac{3\pi}{2}\right) = -1$ and $\cos\left(\frac{2\pi}{3}\right) = -\frac{1}{2}$. This gives us:

$$\sin\left(\frac{3\pi}{2}\right) + \cos\left(\frac{2\pi}{3}\right) = -1 - \frac{1}{2} = -1\frac{1}{2}.$$

The answer is (3).

As mentioned above, you could always just find the values on your calculator. Just be sure that you are in **radian** mode. Also, **pay attention to order of operations** when you put the angles in your calculator. If you put in sin 3 π/2, the calculator will evaluate sin 3 π first and then divide it by 2. You will get 0, which is NOT what you want. Therefore, you have to put in sin(3 π/2) so that the calculator gives you the correct value of sin(3 π/2) = –1. This is something that you should watch out for every time that you evaluate a trigonometric function, a log, or a root.

PROBLEM 20: COMPLEX NUMBERS

We can rewrite $3\sqrt{-8}$ and $4\sqrt{-50}$ as $3\sqrt{8}\sqrt{-1}$ and $4\sqrt{50}\sqrt{-1}$, respectively.

Next, we use the definition that $i = \sqrt{-1}$ to get:

$$3\sqrt{8}\sqrt{-1} = 6\sqrt{2}\,i,$$

which we usually write as $6i\sqrt{2}$ to avoid thinking that the i is under the radical sign. Similarly, we get:

$$4\sqrt{50}\sqrt{-1} = 20i\sqrt{2}$$

Now we can add the two expressions to obtain $26i\sqrt{2}$

The answer is (2).

PROBLEM 21: ABSOLUTE VALUE INEQUALITIES

When we have an absolute value equation, we break it into two equations. First, *rewrite the left side without the absolute value and leave the right side alone. Second, rewrite the left side without the absolute value, negate the right side, and flip the inequality sign.*

Here, we get:

$$3x + 6 \leq 30 \text{ or } 3x + 6 \geq -30.$$

We can now solve each of these equations easily. For the first, we get:

$$x \leq 8.$$

For the second, we get:

$$x \geq -12.$$

The answer is (1).

PROBLEM 22: QUADRATIC EQUATIONS

We can determine the nature of the roots of a quadratic equation of the form $ax^2 + bx + c = 0$ by using the discriminant $b^2 - 4ac$.

You should know the following rule:

If $b^2 - 4ac < 0$, the equation has two imaginary roots.

If $b^2 - 4ac = 0$, the equation has one rational root

If $b^2 - 4ac > 0$, and $b^2 - 4ac$ is a perfect square, then the equation has two rational roots.

If $b^2 - 4ac > 0$, and $b^2 - 4ac$ is not a perfect square, then the equation has two irrational roots

Here, the discriminant is $b^2 - 4ac = (6)^2 - 4(1)(11) = 35 - 44 = -11$.

The answer is (4).

PROBLEM 23: TRIGONOMETRY

You should know the following quadrant rules about trig functions:

The sine of an angle is negative in quadrants III *and* IV.

The cosine of an angle is negative in quadrants II *and* III.

The tangent of an angle is negative in quadrants II *and* IV.

The answer is (4).

PROBLEM 24: TRIGONOMETRIC IDENTITIES

You can find the identities for double angles on the formula sheet under *Functions of the Double Angle*. It says that $\sin 2A = 2 \sin A \cos A$. Substituting this into the expression, we get:

$$\frac{2 \sin x \cos x}{\cos x}.$$

If we cancel like terms, we get:

$$\frac{2 \sin x \cos x}{\cos x} = 2 \sin x.$$

The answer is (1).

Another way to get the right answer is to *plug in*. For example, let $x = 30°$ and plug it into the problem. We get:

$$\frac{\sin 60°}{\cos 30°} = 1 \text{ (use your calculator if you have to).}$$

Now plug $x = 30°$ into each answer choice and pick the one that matches.

Choice (1): $2 \sin 30° = 1$. Correct! But, just in case, try the other choices.

Choice (2): $\tan 30° = \dfrac{\sqrt{3}}{3} \approx 0.577$. Wrong.

Choice (3): $\cos 60° = \dfrac{1}{2}$. Wrong.

Choice (4): $2 \cos 30° = \sqrt{3} \approx 1.732$. Wrong.

PROBLEM 25: TRIGONOMETRY

You should know that $\sin x = \cos(90° - x)$ and that $\cos x = \sin(90° - x)$. We could use either substitution, so if we use the latter substitution in the equation we get:

$$\sin(x - 3)° = \sin\left(90° - (2x + 6)°\right).$$

Therefore $(x - 3) = (90 - 2x - 1)$. If we solve for x, we get:

$$x = 29.$$

The answer is (3).

Another way to get the answer is to *backsolve*. Try each of the values for x in the problem and see which one works.

Choice (1): Does $\sin{-12°} = \cos{-12°}$? $-0.208 \neq 0.978$. Wrong answer.

Choice (2): Does $\sin 23° = \cos 58°$? $0.391 \neq 0.530$. Wrong answer.

Choice (3): Does $\sin 26° = \cos 64°$? $0.438 = 0.438$. Right!

Choice (4): Does $\sin 61° = \cos 134°$? $0.875 \neq -0.695$. Wrong answer

PROBLEM 26: QUADRATIC INEQUALITIES

When we are given a quadratic inequality, the first thing that we do is factor it. We get:

$$(x + 6)(x - 1) > 0.$$

The roots of the quadratic are $x = -6$, $x = 1$. If we plot the roots on a number line

we can see that there are three regions: $x < -6$, $-6 < x < 1$, and $x > 1$. Now we try a point in each region to see whether the point satisfies the inequality.

First, we try a number less than -6. Let's try -7. Is $(-7+6)(-7-1) > 0$? Yes.

Now, we try a number between -6 and 1. Let's try 0. Is $(0+6)(0-1) > 0$? No.

Last, we try a number greater than 1. Let's try 2. Is $(2+6)(2-1) > 0$? Yes.

Therefore, the regions that satisfy the inequality are:

$$x < -6 \text{ or } x > 1.$$

The answer is (2).

PROBLEM 27: TRANSFORMATIONS

Here we are asked to find the results of a composite transformation $(T_{2,-1} \circ r_y = -1)(2,0)$. *Whenever we have a composite transformation, always do the **right** one first.*

The reflection in the line $y = -1$ means that we take the distance from the point $(2,0)$ to the line and then move that distance in the other direction from the line. Here, the distance from $(2,0)$ to $y = -1$ is 1, so we move a distance of 1 in the other direction from $y = -1$, which gives us a new coordinate of $(2,-2)$

Next, the translation $T_{(x,y)}$ tells us to add x to the x-coordinate of the point being translated, and to add y to the y-coordinate of the point being translated. So, here, $T_{(2,-1)}(2,-2)$ means $2 + 2 = 4$ and $-2 + -1 = -3$. Therefore, the translation is $(2,-2) \xrightarrow{T_{(2,1)}} (4,-3)$.

The answer is (3).

PROBLEM 28: LAW OF COSINES

The Law of Cosines says that, in any triangle, with angles A, B, and C, and opposite sides of a, b, and c, respectively, $c^2 = a^2 + b^2 - 2ab \cos C$. (This is on the formula sheet.) If we have a triangle and we are given SSS (side, side, side), then the side opposite the angle we are looking for is c, and the angle is C.

Therefore, c = 7, and we are looking for $\cos C$.

If we substitute the values we get:

$$7^2 = 8^2 + 2^2 - 2(8)(2) \cos C.$$

We can solve this easily.

$49 = 64 + 4 - 32 \cos C$

$32 \cos C = 64 + 4 - 49$

$32 \cos C = 19$

$\cos C = \dfrac{19}{32}$.

The answer is (4).

PROBLEM 29: RATIONAL EXPRESSIONS

A function is undefined where it's denominator equals zero. If we set the denominator equal to zero we get:

$$\sqrt{4 - x^2} = 0.$$

Now notice that we have a second problem. Not only can the denominator not equal zero, but the radicand (the expression under the radical) must be positive. So we need to find where $4 - x^2 > 0$. This is an example of a quadratic inequality.

When we are given a quadratic inequality, the first thing that we do is factor it. We get:

$$(2-x)(2+x) > 0.$$

The roots of the quadratic are $x = 2$, $x = -2$. If we plot the roots on a number line

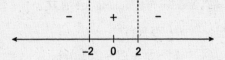

we can see that there are three regions: $x < -2$, $-2 < x < 2$, and $x > 2$. Now we try a point in each region to see whether the point satisfies the inequality.

First, we try a number less than -2. Let's try -3. Is $(2+3)(2-3) > 0$? No.

Now, we try a number between -2 and 2. Let's try 0. Is $(2-0)(2+0) > 0$? Yes.

Last, we try a number greater than 2. Let's try 3. Is $(2-3)(2+3) > 0$? No.

Therefore, the region that satisfies the inequality is $-2 < x < 2$.

The answer is (3).

PROBLEM 30: CONVERTING FROM RADIANS TO DEGREES

We simply multiply the angle by $\dfrac{180°}{\pi}$. We get:

$$\frac{7\pi}{3}\,\frac{180°}{\pi} = 420°.$$

To convert $420°$ to an angle in standard position, subtract $360°$ to get $60°$.

PROBLEM 31: STATISTICS

You should know the following rule about normal distributions.

> **In a normal distribution, with a mean of $\bar{x}$ and a standard deviation of σ:**
>
> - approximately 68% of the outcomes will fall between $\bar{x} - \sigma$ and $\bar{x} + \sigma$
>
> - approximately 95% of the outcomes will fall between $\bar{x} - 2\sigma$ and $\bar{x} + 2\sigma$
>
> - approximately 99.5% of the outcomes will fall between $\bar{x} - 3\sigma$ and $\bar{x} + 3\sigma$.

Here, $\bar{x} - 2\sigma = 45.1$ and $\bar{x} + 2\sigma = 63.5$, so we would expect approximately 95% of the outcomes will fall between 45.1 and 63.5. Because the distribution is symmetric about the mean, we would expect that one half of 95%, or 47.5% would fall between 54.3 and 63.5.

The answer is (2).

PROBLEM 32: TRIGONOMETRY

In a unit circle, a terminal side with an angle of θ has an x-coordinate of $\cos\theta$ and a y-coordinate of $\sin\theta$.

Therefore, the x-coordinate is $\cos\dfrac{3\pi}{4}$. You should know this value because it is a special angle. $\cos\dfrac{3\pi}{4} = -\dfrac{\sqrt{2}}{2}$.

The answer is (3).

PROBLEM 33: QUADRATIC EQUATIONS

If we are given the roots of a quadratic equation, we can figure out the equation by using the following rule:

Given an equation of the form $ax^2 + bx + c = 0$, the *sum of the roots is* $-\dfrac{b}{a}$ and the *product of the roots is* $\dfrac{c}{a}$.

Here, $S = -\dfrac{b}{a} = 2$ and $P = \dfrac{c}{a} = \dfrac{1}{2}$.

The answer is (1).

Another way to get the right answer is to use the quadratic formula to find the roots of each answer choice. Using the quadratic formula $x = \dfrac{-b \pm \sqrt{b^2 - 4ac}}{2a}$ we get:

$$x = \frac{-(-4) \pm \sqrt{(-4)^2 - 4(1)(2)}}{2(2)} = \frac{4 \pm \sqrt{8}}{4}.$$

If we now add the two roots, we get:

$$\frac{4 + \sqrt{8}}{4} + \frac{4 - \sqrt{8}}{4} = \frac{8}{4} = 2.$$

And, if we multiply the two roots, we get:

$$\frac{4 - \sqrt{8}}{4} = \frac{16 - 8}{16}$$

PROBLEM 34: LAW OF SINES

First, let's draw a triangle and fill in the information. Any triangle will do for now.

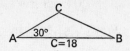

By asking for the number of distinct triangles, the question is testing your knowledge of the *Ambiguous Case* of the Law of Sines. There is a simple test: Is $c \sin A < a < c$? $18 \sin 30° = 9$. No. Therefore, there cannot be more than one possible triangle, *but* there could be *no* triangles. So now we can use the Law of Sines and see how many answers we get.

The Law of Sines says that in any triangle, with angles A, B, and C, and opposite sides of a, b, and c, respectively, $\dfrac{a}{\sin A} = \dfrac{b}{\sin B} = \dfrac{c}{\sin C}$. (This is on the formula sheet.)

Here, $\dfrac{5}{\sin 30°} = \dfrac{18}{\sin C}$, so $\sin C = \dfrac{18 \sin 30°}{5} = 1.8$. Because the sine of an angle cannot be greater than 1 or less than –1, there is no value of C that satisfies the equation and thus there are NO triangles.

The answer is (4).

PROBLEM 35: BINOMIAL EXPANSIONS

The binomial theorem says that if you expand $(a + b)^n$, you get the following terms:

$$_nC_0 a^n + {}_nC_1 a^{n-1}b^1 + {}_nC_2 a^{n-2}b^2 + ... + {}_nC_{n-2}a^2b^{n-2} + {}_nC_{n-1}a^1b^{n-1} + {}_nC_n b^n.$$

Therefore, if we expand $(x - 3y)^4$, we get:

$$_4C_0(x)^4 + {}_4C_1(x)^3(-3y)^1 + {}_4C_2(x)^2(-3y)^2 + {}_4C_3(x)^1(-3y)^3 + {}_4C_4(-3y)^4$$

Next, we use the rule that $_nC_r = \dfrac{n!}{(n-r)!\,r!}$. This gives us:

$$(x)^4 + 4(x)^3(-3y)^1 + 6(x)^2(-3y)^2 + 4(x)^1(-3y)^3 + (-3y)^4$$

which simplifies to $x^4 - 12x^3y + 54x^2y^2 - 108xy^3 + 81y^4$.

The middle term is $54x^2y^2$.

The answer is (1).

A shortcut is to know the rule that the *rth* term of the binomial expansion of $(a-b)^n$ is $_nC_{r-1}(a)^{n-r+1}(b)^{r-1}$. Thus the middle term is: $_4C_{3-1}(x)^{4-3+1}(-3y)^{3-1}$, which can be simplified to $_4C_2(x)^2(-3y)^2 = 54x^2y^2$.

Part II

Problem 36: Trigonometric Graphs

(a) A graph of the form $y = a \sin bx$ or $y = a \cos bx$ has an amplitude of $|a|$ and a period of $\frac{2\pi}{b}$. The first equation $y = \sin \frac{1}{2}x$ has an amplitude of 1 and a period of 4π. The second equation $y = -2 \cos x$ has an amplitude of 2 and a period of 2π, and is upside down because of the minus sign. The two graphs look like this:

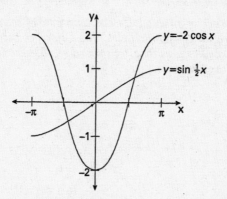

Remember, in graphing sines and cosines, the shape of the graph doesn't change, but it can be stretched or shrunk depending on the amplitude or period.

(b) We can find the intersections two ways: First, we can look at the number of intersections that we graphed on the paper. **The graphs intersect twice.**

The second way is to solve the equation algebraically. However, because you are only asked for the number of intersections, not the values of the intersections, the easiest thing to do is to look at the graphs.

PROBLEM 37: CIRCLE RULES

(a) *A central angle has the same measure as the arc it subtends.*

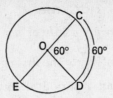

Here, we are given that $m\angle COD = 60$. We also know that $\overline{EOC}$ is a diameter of the circle, so $m\angle COD + m\angle DOE = 180$. This means that $m\angle DOE = 180 - 60 = 120$ and therefore $m\widehat{DE} = 120$.

(b) *The measure of an angle formed by a pair of secants or a secant and a tangent is equal to half of the difference between the larger and the smaller arcs that are formed by the secants or the secant and the tangent.*

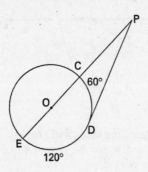

Here, the larger and smaller arcs formed by angle P are $\widehat{DE}$ and $\widehat{CD}$ respectively.

$$m\angle P = \frac{m\widehat{DE} - m\widehat{CD}}{2} = \frac{120 - 60}{2} = 30$$

(c) *The measure of an inscribed angle is half of the arc it subtends (intercepts)*

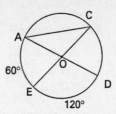

We found in part (a) above that $m\overset{\frown}{DE} = 120$, and we are given that $\overline{AFOD}$ is a diameter, so we know that $m\overset{\frown}{AE} = 180 - 120 = 60$

Therefore, $m\angle ACE = \dfrac{60}{2} = 30$

(d) *If two chords intersect within a circle, then the angle between the two chords is the average of their intercepted arcs.*

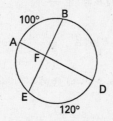

In other words, $m\angle AFB = \dfrac{m\overset{\frown}{AB} + m\overset{\frown}{DE}}{2}$ We know from part

(a) above that $m\overset{\frown}{DE} = 120$, and we were given that $m\angle AFB = 100$,

so: $100 = \dfrac{m\overset{\frown}{AB} + 120}{2}$ and we can easily find that $m\overset{\frown}{AB} = 80$.

(e) *The measure of an inscribed angle is half of the arc it subtends (in-tercepts).*

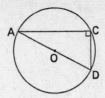

We are given that $\overline{AFOD}$ is a diameter, so we know that $m\widehat{AED} = 180$. Therefore, $m\angle ACD = \dfrac{180}{2} = 90$.

PROBLEM 38: LOGARITHMS

(a) **(1)** You should know the following log rules:

(i) $\log A + \log B = \log(AB)$

(ii) $\log A - \log B = \log\left(\dfrac{A}{B}\right)$

(iii) $\log A^B = B \log A$

Using rule (ii), we can rewrite $\log_b \dfrac{9}{5} = \log_b 9 - \log_b 5$.

Next, because $9 = 3^2$, we can rewrite $\log_b 9 - \log_b 5 = \log_b 3^2 - \log_b 5$.

Now, using rule (iii), we can rewrite $\log_b 9 - \log_b 5 = 2 \log_b 3 - \log_b 5$.

Finally, we can substitute $\log_b 3 = p$ and $\log_b 5 = q$ and we get: $2p - q$.

(2) First, because $\sqrt[3]{15} = 15^{\frac{1}{3}}$, we can rewrite $\log_b \sqrt[3]{15} = \frac{1}{3} \log_b 15$.

Next, we know that $15 = 5 \cdot 3$, so we can rewrite $\frac{1}{3} \log_b (5 \cdot 3)$. Now, using rule (i), we get:

$$\frac{1}{3} \log_b (5 \cdot 3) = \frac{1}{3} \left(\log_b 5 + \log_b 3 \right).$$

Finally, we can substitute $\log_b 3 = p$ and $\log_b 5 = q$ and we get:

$$\frac{1}{3}(q + p).$$

(b) Using rule (i), we get $\log_4 \left(\dfrac{x^2 + 3x}{x + 5} \right) = 1$. Then, by the definition of a log, we get:

$$\frac{x^2 + 3x}{x + 5} = y.$$

Cross-multiply: $x^2 - 3x = 4x + 20$

Simplify: $x^2 - x - 20 = 0$

Factor: $(x - 5)(x + 4) = 0$

So $x = 5$ or -4.

PROBLEM 39:

(a) Probability

(1) This problem requires that you know something called *binomial probability*. The rule is: *If the probability of a particular outcome is* **p**, *then the probability of that outcome occurring* **r** *times out of a possible* **n** *times is* ${}_nC_r(p)^r(1-p)^{n-r}$.

The probability of drawing a black card is $p = \dfrac{3}{5}$.

The probability of drawing exactly two black cards out of a possible five is:

$$\,_5C_2\left(\frac{3}{5}\right)^2\left(1-\frac{3}{5}\right)^{5-2} = \,_5C_2\left(\frac{3}{5}\right)^2\left(\frac{2}{5}\right)^3.$$

The rule for finding $\,_nC_r$ is $\,_nC_r = \dfrac{n!}{(n-r)!\,r!}$, so,

$\,_5C_2 = \dfrac{5!}{3!\,2!} = \dfrac{5\cdot4\cdot3\cdot2\cdot1}{(3\cdot2\cdot1)(2\cdot1)} = 10$. Thus, the probability of

drawing a black card exactly two out of four times is:

$$10\left(\frac{3}{5}\right)^2\left(\frac{2}{5}\right)^3 = \frac{720}{3125}.$$

(2) The probability of drawing *at least* four black cards out of a possible five is the probability of drawing four black cards *plus* the probability of drawing five.

The probability of drawing four black cards is:

$$\,_5C_4\left(\frac{3}{5}\right)^4\left(\frac{2}{5}\right)^{5-4} = \,_5C_4\left(\frac{3}{5}\right)^4\left(\frac{2}{5}\right).$$

The probability of drawing five black cards is:

$$\,_5C_5\left(\frac{3}{5}\right)^5\left(\frac{2}{5}\right)^{5-5} = \,_5C_5\left(\frac{3}{5}\right)^5\left(\frac{2}{5}\right)^0.$$

We can evaluate $\,_5C_4 = \dfrac{5!}{1!\,4!} = \dfrac{5\cdot4\cdot3\cdot2\cdot1}{(1)(4\cdot3\cdot2\cdot1)} = 5$ and

$\,_5C_5 = \dfrac{5!}{0!\,5!} = \dfrac{5\cdot4\cdot3\cdot2\cdot1}{(1)(5\cdot4\cdot3\cdot2\cdot1)} = 1$ (By the way, $0! = 1$). Or

we could use the rule that $\,_nC_{n-1} = n$ and that $\,_nC_n = 1$.

Therefore, the probability of drawing four black cards is:

$$5\left(\frac{3}{5}\right)^4\left(\frac{2}{5}\right) = \frac{810}{3125},$$

and the probability of drawing five black cards is:

$$\left(\frac{3}{5}\right)^5 = \frac{243}{3125}.$$

The sum of the two probabilities is:

$$\frac{810}{3125} + \frac{243}{3125} = \frac{1053}{3125}$$

(b) Statistics

The method for finding a standard deviation is simple, but time-consuming.

To simplify matters, let's put the scores in order: 78, 79, 79, 80, 80, 85, 85, 88, 92, 94.

First, find the average of the scores.

$$\bar{x} = \frac{78 + 79 + 79 + 80 + 80 + 85 + 85 + 88 + 92 + 94}{10} = 84$$

Second, subtract the average from each actual score.

$78 - 84 = -6$

$79 - 84 = -5$

$79 - 84 = -5$

$80 - 84 = -4$

$80 - 84 = -4$

$85 - 84 = 1$

$85 - 84 = 1$

$88 - 84 = 4$

$92 - 84 = 8$

$94 - 84 = 10$

Third, square each difference.

$(-6)^2 = 36$

$(-5)^2 = 25$

$(-5)^2 = 25$

$(-4)^2 = 16$

$(-4)^2 = 16$

$(1)^2 = 1$

$(1)^2 = 1$

$(4)^2 = 16$

$(8)^2 = 64$

$(10)^2 = 100$

Fourth, find the sum of these squares.

$36 + 25 + 25 + 16 + 16 + 1 + 1 + 16 + 64 + 100 = 300$

Fifth, divide this sum by the total number of games.

$$\frac{300}{10} = 30.$$

Last, take the square root of this number. This is the standard deviation.

$\sigma = \sqrt{30} \approx 5.48$ (Rounded to the nearest hundredth)

First, we use the trigonometric identity $\sin^2 \theta = 1 - \cos^2 \theta$ (it comes from the fact that $\sin^2 \theta + \cos^2 \theta = 1$, which is on the formula sheet) and substitute it into the equation. We get:

$$5 \left(1 - \cos^2 \theta\right) - 7 \cos \theta + 1 = 0$$

We then simplify and get:

$$5 - 5\cos^2 \theta - 7\cos \theta + 1 = 0$$
$$5\cos^2 \theta + 7\cos \theta - 6 = 0$$

This is a quadratic equation of the form $5x^2 + 7x - 6 = 0$, which factors into $\left(5\cos \theta - 3\right)\left(\cos \theta + 2\right) = 0$.

This gives us $5\cos \theta - 3 = 0$ and $\cos \theta + 2 = 0$. If we solve each of these equations, we get:

$$\cos \theta = \frac{3}{5} \text{ and } \cos \theta = -2.$$

Therefore $\theta = \cos^{-1}\left(\frac{3}{5}\right)$ or $\theta = \cos^{-1}(-2)$.

Or we could use the quadratic formula $x = \dfrac{-b \pm \sqrt{b^2 - 4ac}}{2a}$ to get:

$$\cos \theta = \frac{-7 \pm \sqrt{7^2 - 4(5)(-6)}}{10},$$

which simplifies to $\cos \theta = \dfrac{-7 \pm \sqrt{169}}{10} = \dfrac{3}{5}$ or -2.

We can find the solution to the first equation with the calculator. You should get $\theta \approx 53.1°$. This is not the only value for θ however. Cosine is also positive in quadrant IV, so there is another angle that satisfies the first equation. It is: $\theta \approx 360° - 53.1° \approx 306.9°$.

Because $-1 \leq \cos \theta \leq 1$, there is no value of θ that satisfies the second equation. Therefore, the solutions are $\theta \approx 53.1°, 306.9°$.

PROBLEM 41: RATIONAL EXPRESSIONS

(a) First, factor the numerator and denominator of the expression on the right of the equals sign:

$$\frac{x}{x+5} + \frac{9}{x-5} = \frac{50}{(x+5)(x-5)}$$

Next, multiply through by $(x+5)(x-5)$ to clear the denominators of all of the expressions: $x(x-5) + 9(x+5) = 50$

Distributing on the left side, we get: $x^2 - 5x + 9x + 45 = 50$

This can be simplified to: $x^2 + 4x - 5 = 0$.

This can be factored into: $(x+5)(x-1) = 0$

Setting each of the factors equal to zero, we get

$$x = -5 \text{ or } x = 1.$$

HOWEVER, because the original equation is undefined at $x = -5$, we have to throw out that answer Thus, the only answer is $x = 1$

(b) Quadratic Equations

If we move all of the terms to the left side, we get $5y + \dfrac{5}{y} - 8 = 0$

Next, we multiply through by y to get.

$$5y^2 - 8y + 5 = 0$$

(Note that, because we had y in the denominator of the original expression, if we get the answer $y = 0$, we have to throw it out.)

We can now use the quadratic formula to solve for y.

The formula says that, given a quadratic equation of the form $ax^2 + bx + c = 0$, the roots of the equation are

$$x = \frac{-b \pm \sqrt{b^2 - 4ac}}{2a}$$

Plugging in to the formula we get:

$$y = \frac{-(-8) \pm \sqrt{(-8)^2 - 4(5)(5)}}{2(5)} = \frac{8 \pm \sqrt{-36}}{10} = \frac{8 \pm 6i}{10} = \frac{8}{10} \pm \frac{6}{10}i = \frac{4}{5} \pm \frac{3}{5}i$$

PROBLEM 42:

(a) (1) Complex Numbers

A complex number of the form $a + bi$ is represented graphically by a point with the coordinates (a, b). In other words, the real part of the complex number is the x-coordinate, and the imaginary part is the y-coordinate.

Thus, the point $Z_1 = 4 - i$ has the coordinates $(4, -1)$ and the point $Z_2 = -5 - 2i$ has the coordinates $(-5, -2)$.

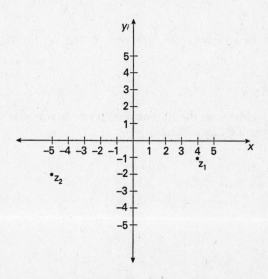

(a)   (2) In order to add two complex numbers, you simply add the real parts and the imaginary parts separately. Thus, $(4 - i) + (-5 - 2i) = (4 - 5) + (-1 - 2)i = -1 - 3i$ and has coordinates $(-1, -3)$.

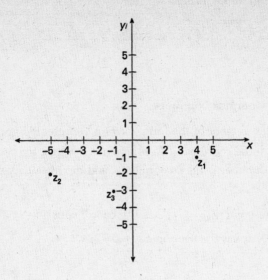

(a) (3) As we saw in part (2) above, $Z_1 + Z_2 = -1 - 3i$.

(b) Vectors

Vector problems on the Regents exams can be solved with the Law of Cosines or the Law of Sines.

First, let's draw a picture of the situation.

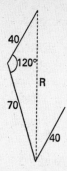

We find the resultant force by making a parallelogram and finding the length of the segment marked R. Because adjacent angles in a parallelogram are supplementary, we know that the angle opposite R is $180° - 60° = 120°$. So now we have SAS and we can use the Law of Cosines.

The Law of Cosines says that, in any triangle, with angles A, B, and C, and opposite sides of a, b, and c, respectively, $c^2 = a^2 + b^2 - 2ab \cos C$. (This is on the formula sheet.) If we have a triangle and we are given SAS (side, angle, side), then the side opposite the included angle is c, and the included angle is C. In this case, the included angle is $120°$, and its opposite side is R.

$$R^2 = 70^2 + 40^2 - 2(70)(40) \cos 120°$$

$$R^2 = 4900 + 1600 - 5600 \cos 120°$$

$$R^2 = 9300$$

$$R \approx 96.44 \text{ pounds.}$$

Formulas

Pythagorean and Quotient Identities

$$\sin^2 A + \cos^2 A = 1 \qquad \tan A = \frac{\sin A}{\cos A}$$

$$\tan^2 A + 1 = \sec^2 A \qquad \cot A = \frac{\cos A}{\sin A}$$

$$\cot^2 A + 1 = \csc^2 A$$

Functions of the Sum of Two Angles

$$\sin (A + B) = \sin A \cos B + \cos A \sin B$$

$$\cos (A + B) = \cos A \cos B - \sin A \sin B$$

$$\tan (A + B) = \frac{\tan A + \tan B}{1 - \tan A \tan B}$$

Functions of the Difference of Two Angles

$$\sin (A - B) = \sin A \cos B - \cos A \sin B$$

$$\cos (A - B) = \cos A \cos B + \sin A \sin B$$

$$\tan (A - B) = \frac{\tan A - \tan B}{1 + \tan A \tan B}$$

Law of Sines

$$\frac{a}{\sin A} = \frac{b}{\sin B} = \frac{c}{\sin C}$$

Law of Cosines

$$a^2 = b^2 + c^2 - 2bc \cos A$$

Functions of the Double Angle

$$\sin 2A = 2 \sin A \cos A$$

$$\cos 2A = \cos^2 A - \sin^2 A$$

$$\cos 2A = 2 \cos^2 A - 1$$

$$\cos 2A = 1 - 2 \sin^2 A$$

$$\tan 2A = \frac{2 \tan A}{1 - \tan^2 A}$$

Functions of the Half Angle

$$\sin \frac{1}{2} A = \pm\sqrt{\frac{1 - \cos A}{2}}$$

$$\cos \frac{1}{2} A = \pm\sqrt{\frac{1 + \cos A}{2}}$$

$$\tan \frac{1}{2} A = \pm\sqrt{\frac{1 - \cos A}{1 + \cos A}}$$

Area of Triangle

$$K = \frac{1}{2} ab \sin C$$

Standard Deviation

$$\text{S.D.} = \sqrt{\frac{1}{n} \sum_{i=1}^{n} \left(x_i - \bar{x}\right)^2}$$

EXAMINATION
AUGUST 1998

Part I

Answer 30 questions from this part. Each correct answer will receive 2 credits. No partial credit will be allowed. Write your answers in the spaces provided on the separate answer sheet. Where applicable, answers may be left in terms of π or in radical form.   [60]

1   Express $\dfrac{7\pi}{5}$ radians in degrees.

2   If $f(x) = x-2$ and $g(x) = x^2$, find $f(g(3))$.

3   If $\sin A = -1$, and $0° \le A < 360°$, find $m\angle A$.

4   In $\triangle ABC$, $a = 15$, $c = 10$, and $\sin A = 0.45$. Find $\sin C$.

5   Find the coordinates of P', the image of $P(3, -1)$ under the transformation $(x, y) \rightarrow (-y, -x)$.

6   Find the value of $\sin 135°$ in radical form.

7   Evaluate: $\displaystyle\sum_{x=1}^{4} \left(x^2 - 3\right)$

[OVER]

8　When the graphs of $2 + 4i$ and $3 - 7i$ are drawn on the same set of axes, in which quadrant will the sum of these expressions lie?

9　Solve for the positive value of x: $\dfrac{x}{3} - \dfrac{4}{x} = \dfrac{4}{3}$

10　If $a = 4$, evaluate $a^{\frac{1}{2}} + a^0 + a^{-2}$

11　In the accompanying diagram, chords $\overline{AB}$ and $\overline{CD}$ intersect at E. If $m\,\widehat{AC} = 75$ and $m\,\widehat{DB} = 45$, find $m\angle AED$.

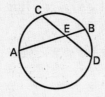

12　What is the image of $(6, 5)$ under a counterclockwise rotation of $180°$?

13　If $\log_n 8 = 3$, find the value of n.

14 In the accompanying diagram, p and q are lines of symmetry in regular pentagon $ABCDE$. find $r_p \circ r_q(B)$.

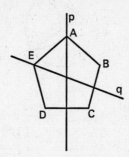

15 If x varies inversely as y and $x = 12$ when $y = 4$, what is the value of y when $x = 16$?

16 For which value of θ is $\dfrac{\cos\theta}{\sin\theta}$ undefined in the interval $-\pi < \theta < \pi$?

Directions (17–35): For *each* question chosen, write on the separate answer sheet the *numeral* preceding the word or expression that best completes the statement or answers the question.

17 What is the value of y if $y = \sin\left(\text{Arc}\tan\dfrac{5}{12}\right)$?

(1) $\dfrac{5}{13}$ (3) $\dfrac{13}{12}$

(2) $\dfrac{12}{13}$ (4) $\dfrac{13}{5}$

[OVER]

18 In the accompanying diagram of circle O, m$\angle AOC = 108$.

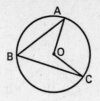

What is m$\angle ABC$?

    (1) 27   (3) 108

    (2) 54   (4) 216

19 What is the value of x in the equation $3^{x-3} = 1$?

    (1)   1   (3) 3

    (2) $\dfrac{1}{3}$ (4) 0

20 Which equation is represented by the graph in the accompanying diagram?

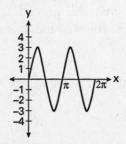

    (1) $y = 3 \sin 2x$ (3) $y = 3 \sin x$

    (2) $y = 2 \sin 3x$ (4) $y = 2 \sin 4x$

21 The expression $\cos^2 40 - \sin^2 40$ has the same value as

    (1) sin 20   (3) cos 80

    (2) sin 80   (4) cos 20

22 If a fair die is tossed five times, what is the probability of getting exactly three 6's?

    (1) $\dfrac{125}{7776}$ (3) $_5C_3\left(\dfrac{1}{6}\right)^2\left(\dfrac{5}{6}\right)^3$

    (2) $_5C_3\left(\dfrac{1}{6}\right)^3\left(\dfrac{5}{6}\right)^2$ (4) $\dfrac{25}{7776}$

23 What is the solution set for the inequality $x^2 - 2x < 8$?

    (1) $-2 < x < 4$ (3) $x < -2$ or $x > 4$

    (2) $-4 < x < 2$ (4) $x < -4$ or $x > 2$

24 What is the solution set of the equation $|2x - 5| = 3$?

    (1) $\{\ \}$ (3) $\{4\}$

    (2) $\{-4, 4\}$ (4) $\{1, 4\}$

25 The expression $\dfrac{\sin 2x}{\sin(-x)}$ is equivalent to

    (1) $-2 \sin x$ (3) $-2 \cos x$

    (2) $2 \sin x$ (4) $2 \cos x$

[OVER]

26 Which equation has both 3 and 6 as roots?

(1) $\sqrt{x-2} = x - 4$　　(3) $\sqrt{x-2} = \dfrac{3}{x}$

(2) $\sqrt{x-2} = 4 - x$　　(4) $\sqrt{x-2} = \dfrac{x}{3}$

27 In $\triangle ABC$, $a = 4$, $b = 3$, and $\cos C = -\dfrac{1}{2}$. What is the length of c?

(1) 7　　　　　　　　(3) $\sqrt{37}$

(2) $\sqrt{13}$　　　　　　(4) $\sqrt{19}$

28 Which equation has rational roots?

(1) $x^2 + 8x - 8 = 0$　　(3) $2x^2 + 4x + 5 = 0$

(2) $x^2 + 8x + 9 = 0$　　(4) $3x^2 + 8x + 4 = 0$

29 When drawn on a set of axes, which equation is an ellipse?

(1) $2x^2 + y = 12$　　　(3) $2x^2 + 2y^2 = 12$

(2) $2x^2 + y^2 = 12$　　(4) $2x^2 - y^2 = 12$

30 When simplified, i^{99} is equivalent to

(1) 1　　　　　　　　(3) i

(2) -1　　　　　　　(4) $-i$

31 On a standardized test with a normal distribution of scores, the mean score is 82 and the standard deviation is 6. Which interval contains 95% of the scores?

(1) 70–82   (3) 76–88
(2) 70–94   (4) 76–94

32 In the interval $0 \le \theta < 2\pi$, the number of solutions of the equation $\sin \theta = \cos \theta$ is

(1) 1   (3) 3
(2) 2   (4) 4

33 In isosceles triangle ABC, $\overline{AB} \cong \overline{BC}$, $m\angle B = 45$, and $AB = 3\sqrt{2}$. The area of the triangle is

(1) $\dfrac{9}{2}$ (3) $\dfrac{9\sqrt{2}}{2}$

(2) $9\sqrt{2}$ (4) $\dfrac{3\sqrt{2}}{2}$

34 What is the fourth term in the expansion $(a + b)^5$?

(1) $10a^2b^3$ (3) $5a^2b^3$
(2) $10a^3b^2$ (4) $5a^3b^2$

[OVER]

35 Which diagram could represent the graph of an equation with imaginary roots?

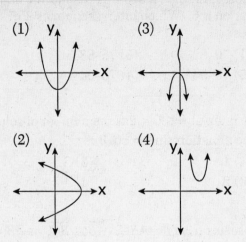

(1) Y

(3) Y

(2) Y

(4) Y

Answers to the following questions are to be written on paper provided by the school.

Part II

Answer four questions from this part. Clearly indicate the necessary steps, including appropriate formula substitutions, diagrams, graphs, charts, etc. Calculations that may be obtained by mental arithmetic or the calculator do not need to be shown.   [40]

36 *a* On the same set of axes, sketch and label the graphs of the equations $y = 3\cos\frac{1}{2}x$ and $y = -2\sin x$ in the interval $0 \le x \le 2\pi$ [8]

   b In the interval $0 \le x \le 2\pi$, which value of x satisfies the equation $3\cos\frac{1}{2}x = -2\sin x$? [2]

37 Find all values of x in the interval $0° \le x < 360°$ that satisfy the equation $3\cos 2x + 2\sin x = -1$. Express your answer to the *nearest ten minutes or nearest tenth of a degree*   [10]

38 *a* Simplify. $\dfrac{\dfrac{x}{x-3} + \dfrac{4}{x}}{1 - \dfrac{1}{3-x}}$ [6]

   b Solve the equation for x and express the roots in simplest $a + bi$ form·

   $$4x^2 - 12x + 25 = 0 \quad [4]$$ [OVER]

39 *a* (1) On graph paper, sketch and label the graph of the equation $y = 2^x$ in the interval $-2 \le x \le 2$.
    [2]

    (2) On the same set of axes, reflect the graph drawn in part *a* (1) in the *y*-axis and label it *r*.
    [2]

    (3) Write an equation of the graph drawn in part *a* (2).
    [2]

    b Solve for *x* to the *nearest hundredth*:

    $$2^x = \frac{3}{16} \qquad [4]$$

40 In parallelogram *ABCD*, *AB* = 14, *BC* = 20, and m$\angle B$ = 54.

    a Find, to the *nearest tenth*, the length of diagonal $\overline{BD}$. [6]

    b Find m $\angle DBC$ to the *nearest degree*. [4]

41 The table below shows the heights of a group of 20 students.

Height (inches)	Frequency
72	3
71	2
70	1
69	2
68	4
67	2
66	4
65	2

a Find the mean and the standard deviation to the *nearest tenth*. [4]

b If one student's height is chosen at random, what is the probability that the height falls within one standard deviation of the mean? [2]

c If three students' heights are chosen at random, what is the probability that *at most* one of them falls within one standard deviation of the mean? [4]

[OVER]

42 In accompanying diagram of circle 0, m $\overparen{AB}$:$\overparen{BC}$ = 1:2 diameter $\overline{CA}$ and chord $\overline{AE}$ are drawn; chord $\overline{EC}$ is parallel to chord $\overline{AB}$; chord $\overline{BC}$ is extended through C to D: and tangent $\overline{DE}$ is drawn.

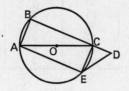

Find:

a   m $\overparen{BC}$ [2]

b   m $\overparen{CE}$ [2]

c   m ∠AEC   [2]

d   m ∠CED   [2]

e   m ∠BDE   [2]

ANSWER KEY

Part I

1. 252
2. 7
3. 270
4. 0.3
5. (1,−3)
6. $\dfrac{1}{\sqrt{2}}$
7. 18
8. IV
9. 6
10. $3\dfrac{1}{16}$
11. 120
12. (−6,−5)
13. 2
14. D
15. 3
16. 0
17. (1)

18. (2)
19. (3)
20. (1)
21. (3)
22. (2)
23. (1)
24. (4)
25. (3)
26. (4)
27. (3)
28. (4)
29. (2)
30. (4)
31. (2)
32. (2)
33. (3)
34. (1)
35. (4)

Part II

36. *a* see explanations section   42. *a* 120

    b π   *b* 60

37. $$90°, 221.8°, 318.2°$$ *c* 90

              or   *d* 30

    $90°, 221°50', 318°10'$

38. *a* $\dfrac{x+6}{x}$ *e* 60

    b $\dfrac{3}{2} + 2i$

39. *a* (1) see explanations section.

       (2) see explanations section.

       (3) $y = 2^{-x}$

    b –2.42

40. *a* 30.4

    b 22

41. *a* $\bar{x} = 68.3$

    b $\dfrac{13}{20}$

    c $\dfrac{2254}{8000}$

ANSWERS AND EXPLANATIONS
AUGUST 1998

Part I

In problems 1–16, you must supply the answer. Because of this, you will not be able to *plug in* or *backsolve*. But, you can use your calculator to simplify calculations. In addition, you will find that these problems tend to test your basic knowledge of Sequential Math III and are not tricky. Many can be solved just by knowing the right formula and by plugging in the numbers.

PROBLEM 1: CONVERTING FROM RADIANS TO DEGREES

All we have to do is multiply $\dfrac{7\pi}{5}$ by $\dfrac{180°}{\pi}$

$$\frac{7\pi}{5}\left(\frac{180°}{\pi}\right) = 252°$$

PROBLEM 2: COMPOSITE FUNCTIONS

When evaluating a composite function, start with the one on the inside.

Here, we first plug 3 into $g(x)$. We obtain $g(3) = 3^2 = 9$.

Next, we plug 9 (the value that we found when we evaluated $g(3)$) into $f(x)$. We obtain $f(9) = 9 - 2 = 7$

PROBLEM 3: TRIGONOMETRY

You should know your special angles and that $\sin 270° = -1$

Therefore, $m\angle A = 270$.

PROBLEM 4: LAW OF SINES

The Law of Sines says that, in any triangle, with angles A, B, and C, and opposite sides of a, b, and c, respectively, $\dfrac{a}{\sin A} = \dfrac{c}{\sin C}$ (This is on the formula sheet).

Here, $\dfrac{15}{0.45} = \dfrac{10}{\sin C}$.

Next, cross-multiply to get:

$$15 \sin C = 4.5.$$

Finally, divide both sides by 15:

$$\sin C = 0.3$$

PROBLEM 5: TRANSFORMATIONS

Here we are asked to find the results of a transformation $(x, y) \rightarrow (-y, -x)$. In other words, switch the x- and y-coordinates and multiply each coordinate by -1.

Thus we get·

$$(3, -1) \rightarrow (1, -3)$$

PROBLEM 6: TRIGONOMETRY

This angle is in the second quadrant. In order to find the reference angle, subtract it from $180°$.

$$180° - 135° = 45°$$

Next, sine is positive in quadrants I and II. Therefore, $\sin 135° = \sin 45°$.

You should know the value of $\sin 45°$ because it is a special angle. The value is $\sin 45° = \dfrac{1}{\sqrt{2}}$

PROBLEM 7: SUMMATIONS

To evaluate a sum, we plug each of the consecutive values for x into the expression, starting at the bottom value and ending at the top value and sum the results. We get:

$$\sum_{x=1}^{4}\left(x^2 - 3\right) = \left[1^2 - 3\right] + \left[2^2 - 3\right] + \left[3^2 - 3\right] + \left[4^2 - 3\right] = -2 + 1 + 6 + 13 = 18.$$

Another way to do this is to use your calculator. Push **2nd STAT** (**LIST**) and under the **MATH** menu, choose **SUM.** Then, again push **2nd STAT** and under the **OPS** menu, choose **SEQ.**

We put in the following: *(expression, variable, start, finish, step).*

For *expression*, we put in the formula, using x as the variable.

For *variable*, we ALWAYS put in x.

For *start*, we put in the bottom value.

For *finish*, we put in the top value.

For *step*, we ALWAYS put in 1.

Therefore, we put in the calculator: **SUM SEQ** $(x^\wedge 2 - 3, x, 1, 4, 1)$. The calculator should return the value 18.

PROBLEM 8: COMPLEX NUMBERS

To find the sum of $2 + 4i$ and $3 - 7i$ we add the real components and the imaginary components separately. This gives us:

$$(2 + 3) + \left(4 + (-7)\right)i = 5 - 3i.$$

Next, to determine the quadrant, we look at the signs of the components. *The real component corresponds to the x-coordinate of the graph, and the imaginary component corresponds to the y-coordinate of the graph.* Because the real component is positive, so is the x-coordinate. Because the imaginary component is negative, so is the y-coordinate. Therefore, the sum of $2 + 4i$ and $3 - 7i$ lies in quadrant IV.

PROBLEM 9: RATIONAL EXPRESSIONS

First, multiply through by $3x$ to clear the denominators of the expressions. This gives us $3x\dfrac{x}{3} - 3x\dfrac{4}{x} = 3x\dfrac{4}{3}$, which simplifies to $x^2 - 12 = 4x$.

Now we have a quadratic expression.

Subtract $4x$ from both sides to get:

$$x^2 - 4x - 12 = 0 \cdot$$

This can be factored into $(x-6)(x+2) = 0$, which yields roots of $x = 6$ or $x = -2$.

Because we are asked for the positive value of x, the answer is $x = 6$.

PROBLEM 10: EXPONENTS

A number raised to a negative power is the same as the reciprocal of that number raised to the corresponding positive power. Here $a^{-2} = \dfrac{1}{a^2}$.

A number raised to $\dfrac{1}{n}$ is the same as the nth root of that number. Here $a^{\frac{1}{2}} = \sqrt{a}$.

Any number raised to the power 0 is 1. Here $a^0 = 1$.

Now we can substitute 4 for a into the expression:

$$\sqrt{4} + 1 + \frac{1}{4^2} = 2 + 1 + \frac{1}{16} = 3\frac{1}{16} \text{ or } \frac{49}{16}.$$

PROBLEM 11: CIRCLE RULES

If two chords intersect in a circle, then the angle formed by their intersection is the average of the arcs formed by those angles.

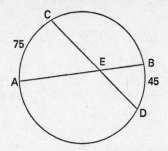

This means that $m\angle AEC = \dfrac{m\overset{\frown}{AC} + m\overset{\frown}{DB}}{2} = \dfrac{75 + 45}{2} = 60$. Now, we can find $m\angle AED$ because it is the supplement of $\angle AEC$.

Therefore, $m\angle AED = 180 - 60 = 120$.

PROBLEM 12: TRANSFORMATIONS

You should know the following rotation rules:

$$R_{90°}(x, y) = (-y, x)$$

$$R_{180°}(x, y) = (-x, -y)$$

$$R_{270°}(x, y) = (y, -x)$$

Therefore, $(6, 5) \xrightarrow{\ R_{180°}\ } (-6, -5)$.

PROBLEM 13: LOGARITHMS

The definition of a logarithm is that $\log_b x = a$ means $b^a = x$.

Here, $\log_n 8 = 3$ means that $n^3 = 8$ and thus $n = 2$.

PROBLEM 14: TRANSFORMATIONS

Here we are asked to find the results of a composite transformation $\left(r_{p'} \circ r_q\right)(B)$. *Whenever we have a composite transformation, always do the **right** one first.*

r_q means a reflection in the line q. This means that the point B is reflected onto the point C.

Next, we do the other transformation. r_p means a reflection in the line p. This means that the point B is now reflected onto the point D.

Therefore, $\left(r_{p'} \circ r_q\right)(B) \to (D)$.

PROBLEM 15: INVERSE VARIATION

An *inverse variation* between x and y means that $xy = k$, where k is a constant.

If we look at the first pair of points, we see that $(12)(4) = 48$. Therefore, k must equal 48.

Now we have the equation $xy = 48$. If we plug in the second pair of points, we get $(16)(x) = 48$, and thus $x = 3$.

PROBLEM 16: TRIGONOMETRY

Where the denominator of a fraction is zero, the fraction is undefined. Here we need to know where $\sin\theta = 0$. Sine is zero at 0, π, and $-\pi$ radians, among other points. So, in the given interval, the function is undefined at 0.

Multiple Choice Problems

For problems 17–35, you will find that you can sometimes *plug in* and *backsolve* to get the right answer. You will also find that a calculator will simplify the arithmetic. In addition, most of the problems again only require knowing which formula to use.

PROBLEM 17: TRIGONOMETRY

$Arc\tan\dfrac{5}{12}$ means "what angle θ has $\tan\theta = \dfrac{5}{12}$?" Although there are an infinite number of answers, when the A in Arctangent is capitalized, you only use the principal angle. This means that, if the Arctangent is of a positive number, use the **first** quadrant answer; and, if the Arctangent is of a negative number, use the **second** quadrant answer. Now we can draw a triangle in the first quadrant, and using the fact that $\tan\theta = \dfrac{opposite}{adjacent}$, label the sides.

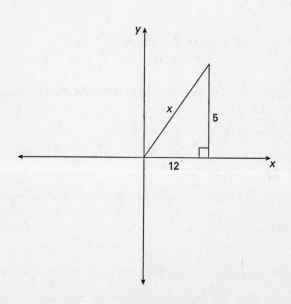

Now we can use the Pythagorean theorem to find the third side, which we have labeled x.

$$5^2 + 12^2 = x^2$$

$$x^2 = 169$$

$x = 13$ (We use the positive value because we are in quadrant I.)

Now, using the definition of sine as $\sin \theta = \dfrac{opposite}{hypotenuse}$, we get $\sin \theta = \dfrac{5}{13}$.

The answer is (1).

PROBLEM 18: CIRCLE RULES

A central angle has the same measure as the arc that it subtends (intercepts). Thus, the measure of arc AC is 108. Next, *the measure of an inscribed angle is always half of the measure of the arc that it subtends*. Therefore, $m\angle ABC = 54$.

The answer is (2).

PROBLEM 19: EXPONENTIAL EQUATIONS

$3^{(x-3)} = 1$.

Any number raised to the power zero is 1, so in this equation $x - 3 = 0$ and thus, $x = 3$.

The answer is (3).

PROBLEM 20: TRIGONOMETRY GRAPHS

A graph of the form $y = a \sin bx$ *or* $y = a \cos bx$ *has an amplitude of* $|a|$ *and a period of* $\dfrac{2\pi}{b}$.

Here, the amplitude is 3 and the period is π. Thus, $\dfrac{2\pi}{b} = \pi$, so $b = 2$. Therefore, the equation of the graph is $y = 3\sin 2x$.

The answer is (1).

PROBLEM 21: TRIGONOMETRY FORMULAS

Notice that this problem has the form $\cos^2 A - \sin^2 A$. If you look on the formula sheet, under *Functions of the Double Angle*, you will find the formula $\cos 2A = \cos^2 A - \sin^2 A$. Thus we can rewrite the problem as $\cos^2 40 - \sin^2 40 = \cos 80$.

The answer is (3).

Another way to get this right is to evaluate $\cos^2 40 - \sin^2 40$ on a calculator. You should get 0.174. Now evaluate each of the answer choices to see which one of them matches.

Choice (1): $\sin 20 = 0.342$. Wrong.

Choice (2): $\sin 80 = 0.985$. Wrong.

Choice (3): $\cos 80 = 0.174$. Correct! But, just in case, check choice (4).

Choice (4): $\cos 20 = 0.940$. Wrong.

PROBLEM 22: PROBABILITY

If the probability of a particular outcome is ***p***, *then the probability of that outcome occurring* ***r*** *times out of a possible* ***n*** *times is* ${}_nC_r(p)^r(1-p)^{n-r}$ This is known as *binomial probability*.

The probability of a fair die coming up 6 is $p = \dfrac{1}{6}$ because there are six sides to a die and it is equally likely to get any of the six numbers

Thus, the probability of getting exactly three 6's out of a possible 5 is ${}_5C_3\left(\dfrac{1}{6}\right)^2\left(1-\dfrac{1}{6}\right)^{5-3}$. This can be simplified to ${}_5C_3\left(\dfrac{1}{6}\right)^3\left(\dfrac{5}{6}\right)^2$.

The answer is (2).

PROBLEM 23: QUADRATIC INEQUALITIES

When we are given a quadratic inequality, the first thing that we do is factor it. Rewrite this as $x^2 - 2x - 8 < 0$ and we get:

$$(x - 4)(x + 2) < 0.$$

The roots of the quadratic are $x = 4$, $x = -2$. If we plot the roots on a number line

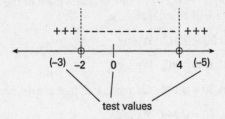

we can see that there are three regions: $x < -2$, $-2 < x < 4$, and $x > 4$. Now we try a point in each region to see whether the point satisfies the inequality.

First, we try a number less than -2. Let's try -3. Is $(-3 - 4)(-3 + 2) < 0$? No.

Now, we try a number between -2 and 4. Let's try 0. Is $(0 - 4)(0 + 2) < 0$? Yes.

Last, we try a number greater than 4. Let's try 5. Is $(5 - 4)(5 + 2) < 0$? No.

These test values -3, 0, and 5 are shown in parentheses in the diagram. Therefore, the region that satisfies the inequality is $-2 < x < 4$

The answer is (1).

PROBLEM 24: ABSOLUTE VALUE EQUATIONS

When we have an absolute value equation we break it into two equations. First, rewrite the left side without the absolute value and leave the right side alone. Second, rewrite the left side without the absolute value, and change the sign of the right side.

Here, we have $|2x - 5| = 3$, so we rewrite it as:

$$2x - 5 = 3 \text{ or } 2x - 5 = -3.$$

If we solve each of these independently, we get $x = 4$ or $x = 1$. Thus, the solution set is $\{1, 4\}$.

The answer is (4).

PROBLEM 25: TRIGONOMETRIC IDENTITIES

You can find the identities for double angles on the formula sheet under *Functions of the Double Angle*. It says $\sin 2A = 2 \sin A \cos A$ Substituting this into the expression, we get:

$$\frac{\sin 2x}{\sin(-x)} = \frac{2 \sin x \cos x}{\sin(-x)}.$$

Next, you should know that $\sin(-x) = -\sin x \, \sin(-x) = -\sin x$, which gives us.

$$\frac{2 \sin x \cos x}{\sin(-x)} = \frac{2 \sin x \cos x}{-\sin x}$$

If we cancel like terms, we get:

$$\frac{2 \sin x \cos x}{-\sin x} = -2 \cos x.$$

The answer is (3).

Another way to get the problem right is to *plug in*. Make up a value for x and plug it into the problem. For example, let $x = 30°$. Using a calculator, we get.

$$\frac{\sin 60°}{\sin(-30°)} \approx -1.732$$

Now, plug $x = 30°$ into the answer choices and pick the one that matches; or that is closest, because sometimes rounding will throw you off by a little bit.

Choice (1): $-2 \sin 30° = -1$. Wrong!

Choice (2): $2 \sin 30° = 1$. Wrong!

Choice (3): $-2 \cos 30° \approx -1.732$. Correct!

We don't need to check Choice (4).

PROBLEM 26: RADICAL EQUATIONS

The simplest thing to do is to try 3 in each answer choice and see if it works. If it doesn't, throw out the answer. If it does, try 6 and see if it works.

Choice (1): Does $\sqrt{3-2} = 3 - 4$? $\sqrt{1} \neq -1$, so throw out this answer. (Remember, when evaluating a radical, we only use the positive answer.)

Choice (2): Does $\sqrt{3-2} = 4 - 3$? $\sqrt{1} = 1$, so now we try 6 in the equation. Does $\sqrt{6-2} = 4 - 6$? $\sqrt{4} \neq -2$, so throw out this answer.

Choice (3): Does $\sqrt{3-2} = \dfrac{3}{3}$? $\sqrt{1} = 1$, so now we try 6 in the equation.

Does $\sqrt{6-2} = \dfrac{3}{6}$? $\sqrt{4} \neq \dfrac{1}{2}$, so throw out this answer.

This leaves us with choice (4) which we hope is right. Otherwise, we made a mistake.

Does $\sqrt{3-2} = \dfrac{3}{3}$? $\sqrt{1} = 1$, so now we try 6 in the equation. Does

$\sqrt{6-2} = \dfrac{6}{3}$? $\sqrt{4} = 2$, so we have a winner!

The answer is (4).

PROBLEM 27: LAW OF COSINES

The Law of Cosines says that, in any triangle, with angles A, B, and C and opposite sides of a, b, and c, respectively, $c^2 = a^2 + b^2 - 2ab \cos C$. (This is on the formula sheet.) If we have a triangle and we are given SAS (side, angle, side), then the side opposite the included angle is c and the included angle is C.

If we substitute the values we get:

$$c^2 = 3^2 + 4^2 - 2(3)(4)\left(-\frac{1}{2}\right).$$

We can solve this for c.

$$c^2 = 9 + 16 - 24\left(-\frac{1}{2}\right)$$

$$c^2 = 9 + 16 + 12$$

$$c^2 = 37$$

$$c = \sqrt{37}$$

The answer is (3).

PROBLEM 28: QUADRATIC EQUATIONS

We can determine the nature of the roots of a quadratic equation of the form $ax^2 + bx + c = 0$ by using the discriminant $b^2 - 4ac$.

You should know the following rule:

If $b^2 - 4ac < 0$, the equation has two imaginary roots.

If $b^2 - 4ac = 0$, the equation has one rational root.

If $b^2 - 4ac > 0$, and $b^2 - 4ac$ is a perfect square, then the equation has two rational roots.

If $b^2 - 4ac > 0$, and $b^2 - 4ac$ is not a perfect square, then the equation has two irrational roots.

So, if we plug the coefficients of each choice into the discriminant, we can determine the nature of the roots.

Choice (1): We get: $b^2 - 4ac = 64 - 4(1)(-8) = 96$. This is not a perfect square, so the roots are irrational.

Choice (2): We get: $b^2 - 4ac = 64 - 4(1)(9) = 28$. This is not a perfect square, so the roots are irrational.

Choice (3): We get: $b^2 - 4ac = 16 - 4(2)(5) = -24$ This is negative, so the roots are complex.

Choice (4): We get: $b^2 - 4ac = 64 - 4(3)(4) = 16$. This *is* a perfect square, so the roots are rational (which are unequal).

The answer is (4).

PROBLEM 29: CONIC SECTIONS

An ellipse is an equation of the form $ax^2 + by^2 = c$, where a and b are positive and not equal to each other. Choice (2) is the only one where both x and y are squared and where both of the coefficients are positive and unequal.

The answer is (2).

PROBLEM 30: COMPLEX NUMBERS

In order to find a power of i, you divide the exponent by 4 and just use the remainder as the power. Then you use the following rule (which you should know):

$i^0 = 1$

$i^1 = i$

$i^2 = -1$

$i^3 = -1$

If we divide 4 into 99, we get 24 with a remainder of 3 . Checking against the rule above, we get $i^{99} = i^3 = -i$

The answer is (4).

PROBLEM 31: STATISTICS

You should know the following rule about normal distributions:

> In a normal distribution, with a mean of $\bar{x}$ and a standard deviation of σ:
>
> - approximately 68% of the outcomes will fall between $\bar{x} - \sigma$ and $\bar{x} + \sigma$
>
> - approximately 95% of the outcomes will fall between $\bar{x} - 2\sigma$ and $\bar{x} + 2\sigma$
>
> - approximately 99.5% of the outcomes will fall between $\bar{x} - 3\sigma$ and $\bar{x} + 3\sigma$.

Here, $\bar{x} - 2\sigma = 82 - 12 = 70$ and $\bar{x} + 2\sigma = 82 + 12 = 94$, so 95% of the scores will be between 70 and 94

The answer is (2).

PROBLEM 32: TRIGONOMETRIC EQUATIONS

One way to solve this is to divide both sides by $\cos\theta$. This gives us

$$\frac{\sin\theta}{\cos\theta} = 1$$

Next, using the identity $\dfrac{\sin\theta}{\cos\theta} = \tan\theta$, we can rewrite the equation as:

$$\tan\theta = 1$$

You should know the values of θ between 0 and 2π for which this is true because they are special angles. They are $\theta = \dfrac{\pi}{4}$ and $\theta = \dfrac{5\pi}{4}$

Thus, there are two values for which this is true.

Another way to get this right was to know from your special angles, or from the basic graphs of sine and cosine (shown below), that there are two angles where the functions are equal—one in quadrant I, and one in quadrant III.

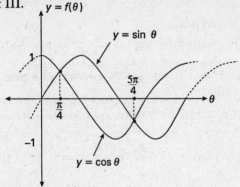

$y = f(\theta)$

$y = \sin \theta$

$\dfrac{5\pi}{4}$

$\dfrac{\pi}{4}$

θ

$y = \cos \theta$

The answer is (2).

PROBLEM 33: TRIGONOMETRIC AREA

The area of a triangle, if we are given the lengths of the two sides a and b, and their included angle θ, is $A = \dfrac{1}{2}ab\sin\theta$. In other words, if we are given SAS (side, angle, side) we find the area by multiplying $\dfrac{1}{2}$ by the product of the two sides by the sine of the included angle.

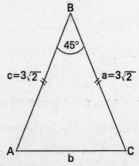

B

45°

$c=3\sqrt{2}$

$a=3\sqrt{2}$

A

b

C

Here, we are given that side AB is $3\sqrt{2}$. Because the triangle is isosceles, we know that side BC is also $3\sqrt{2}$. The included angle is $45°$. Plugging in, we get:

$$A = \frac{1}{2}\left(3\sqrt{2}\right)\left(3\sqrt{2}\right)\frac{1}{\sqrt{2}} = \frac{9\sqrt{2}}{2}.$$

(You should know that $\sin 45° = \frac{1}{\sqrt{2}}$, but if you don't, use your calculator).

The answer is (3).

PROBLEM 34: BINOMIAL EXPANSIONS

The binomial theorem says that if you expand $(a+b)^n$, you get the following terms:

$$_nC_0a^n + \ _nC_1a^{n-1}b^1 + \ _nC_2a^{n-2}b^2 + .. + \ _nC_{n-2}a^2b^{n-2} + \ _nC_{n-1}a^1b^{n-1} + \ _nC_nb^n$$

Therefore, if we expand $(a+b)^5$, we get:

$$_5C_0(a)^5 + \ _5C_1(a)^4(b)^1 + \ _5C_2(a)^3(b)^2 + \ _5C_3(a)^2(b)^3 + \ _5C_4(a)^1(b)^4 + \ _5C_5(b)^5.$$

Because we are asked to evaluate the fourth term, we only need to find $_5C_3(a)^2(b)^3$.

Next, we use the rule that $_nC_r = \dfrac{n!}{(n-r)!\,r!}$ This gives us.

$$_5C_3 = \frac{5!}{3!\,2!} = 10.$$

Therefore, the fourth term is $10a^2b^3$.

The answer is (1).

A shortcut is to know the rule that the *rth* term of the binomial expansion of $(a-b)^n$ is $_nC_{r-1}(a)^{n-r+1}(b)^{r-1}$. Thus the fourth term is.

$-3,0$, $_5C_{4-1}(a)^{5-4+1}(b)^{4-1}$, which can be simplified to $_5C_3(a)^2(b)^3 = 10a^2b^3$

PROBLEM 35: QUADRATIC EQUATIONS

If a graph does not intercept the x-axis, it does not have any real roots. Thus, because graph (4) does not cross the x-axis, it does not have any real roots and could have imaginary roots.

Part II

PROBLEM 36: TRIGONOMETRIC GRAPHS

(a) A graph of the form $y = a \sin bx$ or $y = a \cos bx$ has an amplitude of $|a|$ and a period of $\frac{2\pi}{b}$. Therefore, the graph of the equation $y = 3\cos\frac{1}{2}x$ has an amplitude of 3 and a period of 4π. The second equation, $y = -2 \sin x$, has an amplitude of 2, is upside down (because of the minus sign), and has a period of 2π. The two graphs look like this:

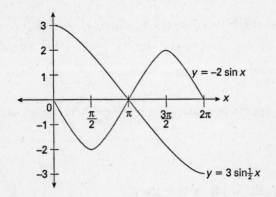

Remember, in graphing sines and cosines, the shape of the graph doesn't change, but it can be stretched or shrunk depending on the amplitude or period.

(b) As we can see by the graph, the answer is $x = \pi$.

First, add 1 to both sides of the equation to get $3\cos 2x + 2\sin x + 1 = 0$.
Next, use the trigonometric identity $\cos 2x = 1 - 2\sin^2 x$ to get
$3(1 - 2\sin^2 x) + 2\sin x + 1 = 0$.

This can be simplified to $3 - 6\sin^2 x + 2\sin x + 1 = 0$, then to
$6\sin^2 x - 2\sin x - 4 = 0$, and finally, to $3\sin^2 x - \sin x - 2 = 0$.

Note that the trigonometric equation $3\sin^2 x - \sin x - 2 = 0$ has the same
form as a quadratic equation $3x^2 - x - 2 = 0$. Just as we could factor
the quadratic equation, so too can we factor the trigonometric equation. We get:

$$(3\sin x + 2)(\sin x - 1) = 0.$$

This gives us $3\sin x + 2 = 0$ and $\sin x - 1 = 0$. If we solve each of these
equations, we get:

$$\sin x = -\frac{2}{3} \text{ and } \sin x = 1.$$

Therefore $x = \sin^{-1}\left(-\frac{2}{3}\right)$ and $x = \sin^{-1}(1)$. The solution to the second equation is easy because it is a special angle, $x = 90°$.

We need to use a calculator to find the solution to the other equation.
Make sure that you are in **Degree** mode. You should get:

$$x = -41.8°.$$

In order to find the equivalent angles that are between $0°$ and $360°$
we draw a little picture and put in the answer to find the reference
angle.

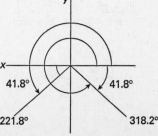

We can see that the reference angle is $41.8°$, thus the third quadrant angle is $180° + 41.8° = 221.8°$ and the fourth quadrant angle is $360° - 41.8° = 318.2°$.

PROBLEM 38: RATIONAL EXPRESSIONS

(a) The first thing that we do is get a common denominator. In this case, we want all terms to have a denominator of $x(x-3)$. If we multiply the top and bottom of the first term in the numerator by x, we get

$\dfrac{x^2}{x(x-3)}$. If we multiply the top and bottom of the second term in

the numerator by $x-3$, we get $\dfrac{4(x-3)}{x(x-3)}$, which simplifies to $\dfrac{4x-12}{x(x-3)}$.

If we multiply the top and bottom of the first term in the denominator

by $x(x-3)$, we get $\dfrac{x(x-3)}{x(x-3)}$, which simplifies to $\dfrac{x^2-3x}{x(x-3)}$. Finally, if

we multiply the top and bottom of the second term in the denomina-

tor by $-x$, we get $\dfrac{x}{x(x-3)}$.

Then our expression becomes:

$$\frac{\dfrac{x^2+4x-12}{x(x-3)}}{\dfrac{x^2-2x}{x(x-3)}}.$$

If we invert the bottom fraction and multiply it by the top one, we get:

$$\frac{x^2+4x-12}{x(x-3)} \cdot \frac{x(x-3)}{x^2-2x}$$

This simplifies to:

$$\frac{x^2+4x-12}{x^2-2x}.$$

Now we can factor the numerator and denominator to obtain:

$$\frac{(x+6)(x-2)}{x(x-2)}.$$

Finally, we can cancel like terms to get:

$$\frac{x+6}{x}$$

(b) Quadratic Equations

We can use the quadratic formula to solve for x.

The formula says that, given a quadratic equation of the form $ax^2 + bx + c = 0$, the roots of the equation are: $x = \dfrac{-b \pm \sqrt{b^2 - 4ac}}{2a}$.

Plugging in to the formula we get:

$$x = \frac{-(-12) \pm \sqrt{(-12)^2 - 4(4)(25)}}{2(4)} = \frac{12 \pm \sqrt{-256}}{8} = \frac{12 \pm 16i}{8} = \frac{3}{2} \pm 2i$$

(a) (1) The graph of an equation of the form $y = a^x$ always has the same general shape. Here, the graph goes through the points $\left(-2, \frac{1}{4}\right)$, $\left(-1, \frac{1}{2}\right)$, $(0, 1)$, $(1, 2)$, and $(2, 4)$.

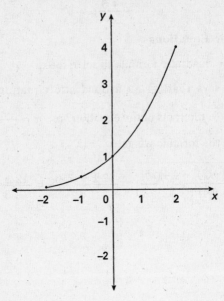

(a) (2) The transformation $r_{y\text{-}axis}$ means that we reflect the graph in the y-axis; that is, we switch the left and right sides. It looks like this:

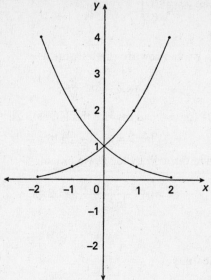

Don't forget to label the graph r.

(a) (3) $r_{y\text{-}axis}$ means a reflection in the y-axis. This means that we multiply each x-coordinate by -1, and leave the y-coordinates alone. The equation becomes: $y = 2^{-x}$.

(b) Logarithms

Whenever we are given an exponential equation where we need to solve for an exponent, we can use logarithms to get the answer. Here, we first take the *log* of both sides.

$$\log 2^x = \log \frac{3}{16}$$

Now we can use the rules of logarithms to solve the equation.

You should know the following log rules:

(i) $\log A + \log B = \log(AB)$

(ii) $\log A - \log B = \log\left(\dfrac{A}{B}\right)$

(iii) $\log A^B = B \log A$

Using rule (iii), we can rewrite the left side as:

$$x \log 2 = \log \frac{3}{16}.$$

Next, using rule (ii), we can rewrite the left side as:

$$x \log 2 = \log 3 - \log 16.$$

Now, if we divide through by $\log 2$, we obtain:

$$x = \frac{\log 3 - \log 16}{\log 2} \approx -2.42$$

PROBLEM 40:

(a) Law of Cosines

First, let's draw a picture of the situation.

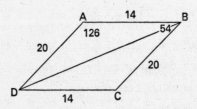

We need to find the length of the diagonal $\overline{BD}$. Because adjacent angles in a parallelogram are supplementary, we find that $m\angle A = 126$. Now we have SAS and we can use the Law of Cosines to solve for the desired side.

The Law of Cosines says that, in any triangle, with angles A, B, and C, and opposite sides of a, b, and c, respectively, $c^2 = a^2 + b^2 - 2ab \cos C$ (This is on the formula sheet). If we have a triangle and we are given SAS (side, angle, side), then the side opposite the included angle is c, and the included angle is C. In this case, the included angle is $126°$, and its opposite side is $\overline{BD}$.

$$\overline{BD}^2 = 14^2 + 20^2 - 2(14)(20)\cos 126°$$

$$\overline{BD}^2 = 196 + 400 - 560\cos 126° = 596 + 329.16$$

$$\overline{BD}^2 \approx 925.16$$

$$\overline{BD} \approx 30.4$$

(b) Law of Sines

Now we want to find angle DBC, which we have labeled x. We could use the Law of Cosines again, but just for the fun of it, let's use the Law of Sines.

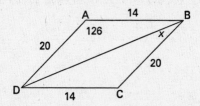

The Law of Sines says that, given a triangle with angles A, B, and C and opposite sides of a, b, and c:

$$\frac{a}{\sin A} = \frac{b}{\sin B} \quad \text{(This is on the formula sheet)}.$$

Substituting, we have:

$$\frac{30.4}{\sin 126°} = \frac{14}{\sin x}$$

First, we cross-multiply:

$$30.4 \sin x = 14 \sin 126°.$$

Next, we divide through by 30.4:

$$\sin x = \frac{14 \sin 126°}{30.4} \approx 0.3726$$

Finally, use inverse sine to find the angle:

$$x = \sin^{-1}(0.3726) \approx 22°.$$

PROBLEM 41: STATISTICS

(a) The method for finding a standard deviation is simple, but time-consuming.

First, find the average of the heights. We do this by multiplying each height by the number of readings that had that height, and adding them up.

$$(72)(3) + (71)(2) + (70)(1) + (69)(2) + (68)(4) + (67)(2) + (66)(4) + (65)(2) = 1366.$$

Next, we divide by the total number of readings $(3 + 2 + 1 + 2 + 4 + 2 + 4 + 2) = 20$, to obtain the average $\bar{x} = 68.3$.

Second, subtract the average from each actual height.

$72 - 68.3 = 3.7$

$71 - 68.3 = 2.7$

$70 - 68.3 = 1.7$

$69 - 68.3 = 0.7$

$68 - 68.3 = -0.3$

$67 - 68.3 = -1.3$

$66 - 68.3 = -2.3$

$65 - 68.3 = -3.3$

Third, square each difference.

$(3.7)^2 = 13.69$

$(2.7)^2 = 7.29$

$(1.7)^2 = 2.89$

$$(0.7)^2 = 0.49$$

$$(-0.3)^2 = 0.09$$

$$(-1.3)^2 = 1.69$$

$$(-2.3)^2 = 5.29$$

$$(-3.3)^2 = 10.89$$

Fourth, multiply the square of the difference by the corresponding number of heights and sum.

$$(13.69)(3) = 41.07$$

$$(7.29)(2) = 14.58$$

$$(2.89)(1) = 2.89$$

$$(0.49)(2) = 0.98$$

$$(0.09)(4) = 0.36$$

$$(1.69)(2) = 3.38$$

$$(5.29)(4) = 21.16$$

$$(10.89)(2) = 21.78$$

$$41.07 + 14.58 + 2.89 + 0.98 + 0.36 + 3.38 + 21.16 + 21.78 = 106.20$$

Fifth, divide this sum by the total number of heights.

$$\frac{106.2}{20} = 5.31.$$

Last, take the square root of this number. This is the standard deviation.

$$\sigma = \sqrt{5.31} \approx 2.3.$$

(b) Probability

Here, $\bar{x} - \sigma = 68.3 - 2.3 = 66$ and $\bar{x} + \sigma = 68.3 + 2.3 = 70.6$. The number of heights that lie between 66 and 70.6 is 13. Therefore, the probability that a student's height is within one standard deviation of the mean is $\dfrac{13}{20}$.

(c) Probability

This problem requires that you know something called *binomial probability*. The rule is: *If the probability of a particular outcome is **p**, then the probability of that outcome occurring **r** times out of a possible **n** times is* $_nC_r(p)^r(1-p)^{n-r}$

The probability that at most one student's height will fall within one standard deviation of the mean is the sum of the probability that there will be one student and the probability that there will be no students.

The probability that a student's height will fall within one standard deviation of the mean is $\dfrac{13}{20}$.

Therefore, the probability that one student's height fall within one standard deviation of the mean is: $_3C_1\left(\dfrac{13}{20}\right)^1\left(1-\dfrac{13}{20}\right)^2$ and the probability of there being no students is $_3C_0\left(\dfrac{13}{20}\right)^0\left(1-\dfrac{13}{20}\right)^3$.

The rule for finding $_nC_r$ is $_nC_r = \dfrac{n!}{(n-r)!\,r!}$. Therefore,

$_3C_1 = \dfrac{3!}{1!\,2!} = \dfrac{3\cdot2\cdot1}{(1)(2\cdot1)} = 3$ and $_3C_0 = \dfrac{3!}{0!\,3!} = \dfrac{3\cdot2\cdot1}{(1)(3\cdot2\cdot1)} = 1$ (By the way, $0! = 1$). Or you could use the rule that $_nC_1 = n$ and that $_nC_0 = 1$.

Thus, the probability that at most one student's height will fall within one standard deviation of the mean is:

$$3\left(\dfrac{13}{20}\right)^1\left(\dfrac{7}{20}\right)^2 + 1\left(\dfrac{13}{20}\right)^0\left(\dfrac{7}{20}\right)^3 = \dfrac{1911}{8000} + \dfrac{343}{8000} = \dfrac{2254}{8000}$$

(a) We are given that AC is a diameter, so $m\overgroup{ABC} = 180$. We are given that $m\overgroup{AB} : m\overgroup{BC} = 1 : 2$, so if we let $m\overgroup{AB} = x$, then $m\overgroup{BC} = 2x$. Now we have $x + 2x = 3x = 180$, so $x = 60$. Thus, $m\overgroup{AB} = 60$ and $m\overgroup{BC} = 120$.

(b) *The measure of an inscribed angle is half of the arc it subtends (intercepts).*

Angle ABC is inscribed in arc AEC, and angle AEC is inscribed in arc ABC. Therefore, $m\angle ABC = m\angle AEC = 90$. In fact, *any angle that subtends a semicircle has a measure of 90.* Now we have quadrilateral $ABCE$ that has a pair of opposite right angles and a pair of parallel sides, so it is a rectangle. This means that $m\overline{AB} = m\overline{CE}$.

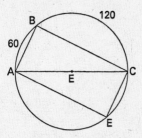

In a circle, chords of equal length cut off equal arcs. This means that $m\overgroup{AB} = m\overgroup{CE} = 60$.

(c) As we showed in part (b) above, $m\angle ABC = m\angle AEC = 90$.

(d) *The measure of an angle formed by a chord and a tangent is equal to half of the arc that the angle subtends (intercepts).*

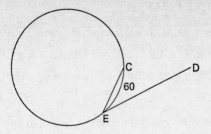

Here, because $m\overset{\frown}{CE} = 60$ (see part (b) above), we find that $m\angle CED = 30$.

(e) *The measure of an angle formed by a pair of secants or a secant and a tangent is equal to half of the difference between the larger and the smaller arcs that are formed by the secants or the secant and the tangent.*

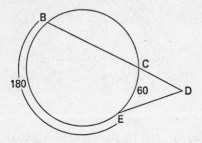

Here, the larger and smaller arcs formed by angle BDE are $\overset{\frown}{ABC}$ and $\overset{\frown}{CE}$, respectively. We found the measures of these arc in parts (a) and (b) above. Thus, $m\angle BDE = \dfrac{m\overset{\frown}{ABC} - m\overset{\frown}{CE}}{2} = \dfrac{180 - 60}{2} = 60$.

Formulas

Pythagorean and Quotient Identities

$$\sin^2 A + \cos^2 A = 1 \qquad \tan A = \frac{\sin A}{\cos A}$$

$$\tan^2 A + 1 = \sec^2 A \qquad \cot A = \frac{\cos A}{\sin A}$$

$$\cot^2 A + 1 = \csc^2 A$$

Functions of the Sum of Two Angles

$$\sin (A + B) = \sin A \cos B + \cos A \sin B$$

$$\cos (A + B) = \cos A \cos B - \sin A \sin B$$

$$\tan (A + B) = \frac{\tan A + \tan B}{1 - \tan A \tan B}$$

Functions of the Difference of Two Angles

$$\sin (A - B) = \sin A \cos B - \cos A \sin B$$

$$\cos (A - B) = \cos A \cos B + \sin A \sin B$$

$$\tan (A - B) = \frac{\tan A - \tan B}{1 + \tan A \tan B}$$

Law of Sines

$$\frac{a}{\sin A} = \frac{b}{\sin B} = \frac{c}{\sin C}$$

Law of Cosines

$$a^2 = b^2 + c^2 - 2bc \cos A$$

Functions of the Double Angle

$$\sin 2A = 2 \sin A \cos A$$

$$\cos 2A = \cos^2 A - \sin^2 A$$

$$\cos 2A = 2 \cos^2 A - 1$$

$$\cos 2A = 1 - 2 \sin^2 A$$

$$\tan 2A = \frac{2 \tan A}{1 - \tan^2 A}$$

Functions of the Half Angle

$$\sin \frac{1}{2} A = \pm\sqrt{\frac{1 - \cos A}{2}}$$

$$\cos \frac{1}{2} A = \pm\sqrt{\frac{1 + \cos A}{2}}$$

$$\tan \frac{1}{2} A = \pm\sqrt{\frac{1 - \cos A}{1 + \cos A}}$$

Area of Triangle

$$K = \frac{1}{2} ab \sin C$$

Standard Deviation

$$\text{S.D.} = \sqrt{\frac{1}{n} \sum_{i=1}^{n} \left(x_i - \bar{x}\right)^2}$$

EXAMINATION
JANUARY 1999

Part I

Answer 30 questions from this part. Each correct answer will receive 2 credits. No partial credit will be allowed. Write your answers in the spaces provided on the separate answer sheet. Where applicable, answers may be left in terms of π or in radical form. [60]

1   In the accompanying diagram of circle O, $\overrightarrow{XA}$ and $\overrightarrow{XB}$ are tangents and $m\angle XAB = 75$. Find $m\angle X$.

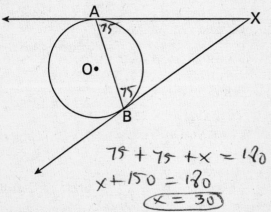

$$75 + 75 + x = 180$$
$$x + 150 = 180$$
$$\boxed{x = 30}$$

2   Translation T maps point $(2,6)$ to point $(4, -1)$. What is the image of point $(-1, 3)$ under translation T?

$(1, -4)$

3   Express the sum of $2\sqrt{-49}$ and $-3\sqrt{-16}$ as a monomial in terms of i.

$$2 \cdot 7i + (-3 \cdot 4i)$$
$$14i - 12i$$
$$\boxed{2i}$$

4   If $f(x) = x^2 - 3x$, find $f(-1.8)$.

$F(-1.8) = (-1.8)^2 - 3(-1.8)$
$3.24 + 5.4 = 8.64$

5   If $f(x) = \sin 3x + \cos x$, what is $f\left(\dfrac{\pi}{2}\right)$?

$F(\pi/2) = \sin 3 \cdot \pi/2 + \cos \pi/2$
$= \sin 270 + \cos 90/2 = -1 + 0 = -1$

6   Evaluate: $\displaystyle\sum_{k=1}^{3}(k+1)$

$(1+1)^2 + (2+1)^2 + (3+1)^2$
$2^2 + 3^2 + 4^2 = 4 + 9 + 16 = 29$

7   In the accompanying diagram of circle O, diameter $\overline{AB}$ is perpendicular to chord $\overline{CD}$ at E, $CD = 8$, and $EB = 2$. What is the length of the diameter of circle O?

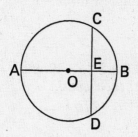

8   Express 75° in radian measure.

$\dfrac{75 \cdot 180}{2}$

$75 \cdot \dfrac{\pi}{180} = \dfrac{5\pi}{12}$

9   In the accompanying diagram, tangent $\overline{AB}$ and secant $\overline{ACD}$ are drawn to circle O from point A. If $AC = 4$ and $CD = 12$, find AB.

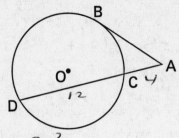

A) · $CA = BA^2$
$16 \cdot 4 = x^2$ $BA = 8$
$\sqrt{64} = \sqrt{x^2}$

10 Express in simplest form: $\dfrac{\frac{3}{4} + \frac{3}{x}}{\frac{1}{x} + \frac{1}{4}}$ $= 3$

11 In $\triangle ABC$, $m\angle A = 33$, $a = 12$, and $b = 15$. Find $\sin B$ to the *nearest thousandth*.

42·906

12 Factor completely: $x^4 - 16$

$(x^2 - 4)(x^2 + 4)$
$(x+2)(x-2)(x^2+4)$

13 In $\triangle ABC$, $a = 2$, $c = 6$, and $\cos B = \dfrac{1}{6}$. Find b.

$b = 6$

14 The width of a rectangle with constant area varies inversely as its length. If the width is 4 when the length is 12, find the width when the length is 16.

3

Directions (15–35): For *each* question chosen, write on the separate answer sheet the *numeral* preceding the word or expression that best completes the statement or answers the question.

15 For which value(s) of x is the function $f(x) = \dfrac{x^2 - 9}{x - 7}$ undefined?

    (1) 9   (3) 3, only

    (2) 3 and –3   (4) 7

16 The solution of set $2^{x+1} = 8$ is

    (1) { }   (3) $\{3\}$

    (2) $\{2\}$ (4) $\{4\}$

17 The expression cos 70° cos 10° + sin 70° sin 10° is equivalent to

    (1) cos 60°   (3) sin 60°

    (2) cos 80°   (4) sin 80°

18 If the image of A after a dilation of –2 is $A'(-8,6)$, what are the coordinates of A?

    (1) (4, –3)   (3) (16, –12)

    (2) (–4, 3)   (4) (–16,12)

19   If θ is an angle in Quadrant I and $\tan^2 \theta - 4 = 0$,
     what is the value of θ to *the nearest degree*?

       (1) 1   (3) 63

       (2) 2   (4) 75

20   If $\log_4 x = 3$, then x is equal to

       (1) 7   (3) 64

       (2) 12   (4) 81

21   The value of sin (Arc cos 1) is

       (1) 1   (3) $\frac{1}{2}\sqrt{3}$

       (2) $\frac{1}{2}$ (4) 0

22   If cos $A > 0$ and csc $A < 0$, in which quadrant does
     the terminal side of $\angle A$ lie?

       (1) I   (3) III

       (2) II   (4) IV

23   If -1 and 7 are the roots of the quadratic equa-
     tion $x^2 + kx - 7 = 0$, then k must be

       (1) -7 (3) 6

       (2) -6 (4) 8

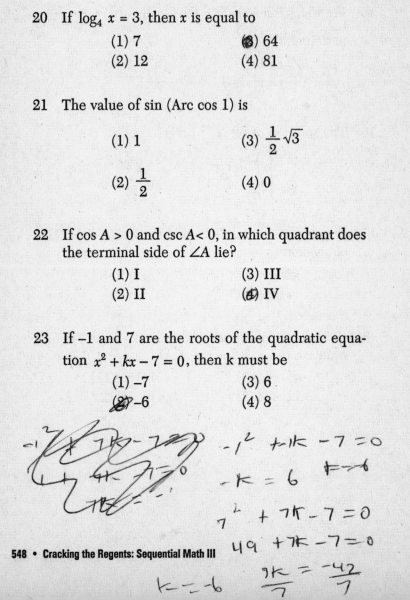

24 The expression sec x sin $2x$ is equivalent to

    (1) $\dfrac{1}{2}$ (3) 2 cos x

    (2) 2   (4) 2 sin x

25 A fair die is tossed five times. What is the probability of obtaining exactly three 4's?

    (1) $\dfrac{250}{7776}$ (3) $\dfrac{1250}{7776}$

    (2) $\dfrac{10}{7776}$ (4) $\dfrac{90}{1024}$

26 If $|2x + 3| < 1$, then the solution set contains

    (1) only negative real numbers
    (2) only positive real numbers
    (3) both positive and negative real numbers
    (4) no real numbers

27 Which relation is a function?

    (1) $y = \cos x$ (3) $x = y^2$

    (2) $x = 4$ (4) $x^2 + y^2 = 16$

28 The roots of the equation $x^2 + 4x + 2 = 0$ are

    (1) real, rational, and equal
    (2) real, rational, and unequal
    (3) real, irrational, and unequal
    (4) imaginary

29 In the accompanying diagram of a unit circle, $\overline{BA}$ is tangent to circle O at A, $\overline{CD}$ is perpendicular to the x-axis, and $\overline{OC}$ is a radius.

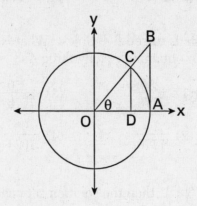

Which distance represents sin θ?

(1) *OD*   (3) *BA*

(2) *CD*   (4) *OB*

30 A standardized test with a normal distribution of scores has a mean score of 43 and a standard deviation of 6.3. Which range would contain the score of a student in the 90th percentile?

(1) 30.4 – 36.7   (3) 43.0 – 49.3

(2) 36.7 – 43.0   (4) 49.3 – 55.6

31 The graph of the equation $xy = 5$ forms

(1) an ellipse   (3) a line

(2) a hyperbola   (4) a parabola

32 The value of $(1 - i)^2$ is

      (1) 0   (3) $-2i$

      (2) 2   (4) $2 - 2i$

33 The solution set of the equation $\sqrt{2x + 15} = x$ is

      (1) $\{5, -3\}$ (3) $\{-3\}$

      (2) $\{5\}$ (4) $\{\ \}$ $x^2 - 2x - 15 = 0$

34 What are the coordinates of the image of $P(-2,5)$ after a clockwise rotation of 90° about the origin?

      (1) $(-5, -2)$ (3) $(2,5)$

      (2) $(-2, -5)$ (4) $(5,2)$

35 What is the best approximation of the standard deviation of the measures $-4, -3, 0, 8,$ and 9?

      (1) 1   (3) 5

      (2) 2   (4) 10

Answers to the following questions are to be written on paper provided by the school.

Part II

Answer four questions from this part. Clearly indicate the necessary steps, including appropriate formula substitutions, diagrams, graphs, charts, etc. Calculations that may be obtained by mental arithmetic or the calculator do not need to be shown. [40]

36 a On the same set of axes, sketch and label the graphs of the equations $y = -\sin x$ and $y = 2\cos x$ in the interval $0 \leq x \leq 2\pi$. [8]

   b Using the graphs sketched in part a, determine the number of solutions to the equation $2\cos x = -\sin x$ in the interval $0 \leq x \leq 2\pi$. [2]

37 In the accompanying diagram of cirlce O, $m\widehat{AC} = 140$, $m\widehat{AE} = 130$, $m\widehat{AB} : m\widehat{BC} = 6{:}4$, $\overline{PD}$ is a tangent, secant $\overline{PCE}$ intersects diameter $\overline{AD}$ at F, and secant $\overline{PBA}$ is drawn.

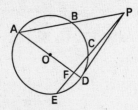

Find:

a $m\widehat{ED}$ [2]

b $m\widehat{AB}$ [2]

c $m\angle BAD$ [2]

d $m\angle APE$ [2]

e $m\angle EFD$ [2]

38 *a* Express in simplest form:

$$\frac{2x-8}{x^2+x-12} \div \frac{20-5x}{2x^2-5x-3} \qquad [4]$$

b Prove the following identity:

$$\frac{\sin\theta}{1+\cos\theta} + \frac{1+\cos\theta}{\sin\theta} = 2\cot\theta\sec\theta \qquad [6]$$

39 *a* On graph paper, sketch the graph of the equation $y = 3^x$ in the interval $-2 \le x \le 2$.   [2]

  b On the same set of axes, sketch the graph of the equation $y = 6$.   [1]

  c Based on the graphs sketched in parts *a* and *b*, between which two consecutive integers does the solution of $3^x = 6$ lie? Explain your answer. [1, 2]

  d Find x to the *nearest hundredth*: $3^x = 6$ [4]

40 In $\triangle ABC$, $AC = 8$, $BC = 17$, and $AB = 20$.

  a Find the measure of the largest angle to the nearest degree.   [6]

  b Find the area of $\triangle ABC$ to the *nearest integer*. [4]

41 *a* A mathematics quiz has five multiple- choice questions. There are four possible responses for each question. Jennifer selects her responses at random on every question.

>   (1) What is the probability she will select the correct response for at most one question?   [3]

>   (2) What is the probability she will select the correct response to at least three questions?   [4]

b Find, in simplest form, the middle term in the expansion of $\left(x^2 + \dfrac{1}{x}\right)^6$.   [3]

42 Find, to the *nearest degree*, all values of θ in the interval $0° \le \theta < 360°$ that satisfy the equation $2 \sin^2 \theta + 2 \cos \theta - 1 = 0$.   [10]

ANSWER KEY

Part I

(1) 30

(2) (1,–4)

(3) $2i$

(4) 8.64

(5) –1

(6) 29

(7) 10

(8) $\dfrac{5\pi}{12}$

(9) 8

(10) 3

(11) 0.681

(12) $\left(x^2 + 4\right)(x + 2)(x - 2)$

(13) 6

(14) 3

(15) 4

(16) 2

(17) 1

(18) 1

(19) 3

(20) 3

(21) 4

(22) 4

(23) 2

(24) 4

(25) 1

(26) 1

(27) 1

(28) 3

(29) 2

(30) 4

(31) 2

(32) 3

(33) 2

(34) 4

(35) 3

Part II

(36) *b* 2   [2]

(37) *a* 50   [2]

     b 84   [2]

     c 48   [2]

     d 37   [2]

     e 95   [2]

(40) *a* 100   [6]

     b 67   [4]

(41) *a* (1) $\dfrac{648}{1024}$ [3]

     (2) $\dfrac{106}{1024}$ [4]

     b $20x^3$ [3]

(42) 111 and 249 [10]

(38) $a - \dfrac{2(2x+1)}{5(x+4)}$ [4]

(39) *c* 1 and 2   [1]

The graph of $y = 6$ intersects the graph of $y = 3^x$ for some value of x between 1 and 2.   [2]

d 1.63   [4]

The University of the State of New York

REGENTS HIGH SCHOOL EXAMINATION

SEQUENTIAL MATH – COURSE III

Part I Score ....
Part II Score ....
Total Score ....
Rater's Initials: ....

ANSWER SHEET

Pupil .... Sex: ☐ Male ☐ Female   Grade ....

Teacher ....   School ....

Your answers to Part I should be recorded on this answer sheet.

Part I

Answer 30 questions from this part.

1	11	21	31 ....
2	12	22	32 ....
3	13	23	33 ....
4	14	24	34 ....
5	15	25	35 ....
6	16	26	
7	17	27	
8	18	28	
9	19	29	
10	20	30	

Your answers for Part II should be placed on paper provided by the school.

The declaration below should be signed when you have completed the examination.

I do hereby affirm, at the close of this examination, that I had no unlawful knowledge of the questions or answers prior to the examination, and that I have neither given nor received assistance in answering any of the questions during the examination.

_____

Signature

Part I Score	....
Part II Score	....
Total Score	....
Rater's Initials:	....

_____

ANSWER SHEET

Pupil .... Sex: ☐ Male ☐ Female   Grade ....

Teacher .... School ....

Your answers to Part I should be recorded on this answer sheet.

Part I

Answer 30 questions from this part.

1	11	21	31 ....
2	12	22	32 ....
3	13	23	33 ....
4	14	24	34 ....
5	15	25	35 ....
6	16	26	
7	17	27	
8	18	28	
9	19	29	
10	20	30	

Your answers for Part II should be placed on paper provided by the school.

The declaration below should be signed when you have completed the examination.

I do hereby affirm, at the close of this examination, that I had no unlawful knowledge of the questions or answers prior to the examination, and that I have neither given nor received assistance in answering any of the questions during the examination.

_____

Signature

The University of the State of New York

REGENTS HIGH SCHOOL EXAMINATION

SEQUENTIAL MATH – COURSE III

ANSWER SHEET

Pupil .... Sex: ☐ Male ☐ Female Grade ....

Teacher .... School ....

Your answers to Part I should be recorded on this answer sheet.

Part I

Answer 30 questions from this part.

1	11	21	31 ....
2	12	22	32 ....
3	13	23	33 ....
4	14	24	34 ....
5	15	25	35 ....
6	16	26	
7	17	27	
8	18	28	
9	19	29	
10	20	30	

Your answers for Part II should be placed on paper provided by the school.

The declaration below should be signed when you have completed the examination.

I do hereby affirm, at the close of this examination, that I had no unlawful knowledge of the questions or answers prior to the examination, and that I have neither given nor received assistance in answering any of the questions during the examination.

_____

Signature

The University of the State of New York

REGENTS HIGH SCHOOL EXAMINATION

SEQUENTIAL MATH – COURSE III

ANSWER SHEET

Pupil .... Sex: ☐ Male ☐ Female Grade ....

Teacher .... School ....

Your answers to Part I should be recorded on this answer sheet.

Part I

Answer 30 questions from this part.

1	11	21	31 ....
2	12	22	32 ....
3	13	23	33 ....
4	14	24	34 ....
5	15	25	35 ....
6	16	26	
7	17	27	
8	18	28	
9	19	29	
10	20	30	

Your answers for Part II should be placed on paper provided by the school.

The declaration below should be signed when you have completed the examination.

I do hereby affirm, at the close of this examination, that I had no unlawful knowledge of the questions or answers prior to the examination, and that I have neither given nor received assistance in answering any of the questions during the examination.

_____

Signature

The University of the State of New York

REGENTS HIGH SCHOOL EXAMINATION

SEQUENTIAL MATH – COURSE III

Part I Score	. . . .
Part II Score	. . . .
Total Score	. . . .
Rater's Initials:	. . . .

ANSWER SHEET

Pupil . . . .   Sex: ☐ Male ☐ Female   Grade . . . .

Teacher . . . .   School . . . .

Your answers to Part I should be recorded on this answer sheet.

Part I

Answer 30 questions from this part.

1	11	21	31 . . . .
2	12	22	32 . . . .
3	13	23	33 . . . .
4	14	24	34 . . . .
5	15	25	35 . . . .
6	16	26	
7	17	27	
8	18	28	
9	19	29	
10	20	30	

Your answers for Part II should be placed on paper provided by the school.

The declaration below should be signed when you have completed the examination.

I do hereby affirm, at the close of this examination, that I had no unlawful knowledge of the questions or answers prior to the examination, and that I have neither given nor received assistance in answering any of the questions during the examination.

_____

Signature

The University of the State of New York

REGENTS HIGH SCHOOL EXAMINATION

SEQUENTIAL MATH – COURSE III

_____

Part I Score . . . .
Part II Score . . . .
Total Score . . . .
Rater's Initials: . . . .

ANSWER SHEET

Pupil . . . .   Sex: ☐ Male ☐ Female   Grade . . . .

Teacher . . . .   School . . . .

Your answers to Part I should be recorded on this answer sheet.

Part I

Answer 30 questions from this part.

1	11	21	31 . . . .
2	12	22	32 . . . .
3	13	23	33 . . . .
4	14	24	34 . . . .
5	15	25	35 . . . .
6	16	26	
7	17	27	
8	18	28	
9	19	29	
10	20	30	

Your answers for Part II should be placed on paper provided by the school.

The declaration below should be signed when you have completed the examination.

I do hereby affirm, at the close of this examination, that I had no unlawful knowledge of the questions or answers prior to the examination, and that I have neither given nor received assistance in answering any of the questions during the examination.

_____

Signature

The University of the State of New York

REGENTS HIGH SCHOOL EXAMINATION

SEQUENTIAL MATH – COURSE III

Part I Score ....
Part II Score ....
Total Score ....
Rater's Initials: ....

ANSWER SHEET

Pupil .... Sex: ☐ Male ☐ Female   Grade ....

Teacher .... School ....

Your answers to Part I should be recorded on this answer sheet.

Part I

Answer 30 questions from this part.

1	11	21	31 ....
2	12	22	32 ....
3	13	23	33 ....
4	14	24	34 ....
5	15	25	35 ....
6	16	26	
7	17	27	
8	18	28	
9	19	29	
10	20	30	

Your answers for Part II should be placed on paper provided by the school.

The declaration below should be signed when you have completed the examination.

I do hereby affirm, at the close of this examination, that I had no unlawful knowledge of the questions or answers prior to the examination, and that I have neither given nor received assistance in answering any of the questions during the examination.

_____

Signature

The University of the State of New York

REGENTS HIGH SCHOOL EXAMINATION

SEQUENTIAL MATH – COURSE III

Part I Score ....
Part II Score ....
Total Score ....
Rater's Initials: ....

ANSWER SHEET

Pupil ... . '. ....   Sex: ☐ Male ☐ Female   Grade ... ...

Teacher ... . School ....

Your answers to Part I should be recorded on this answer sheet.

Part I

Answer 30 questions from this part.

1	11	21	31 ....
2	12	22	32 ....
3	13	23	33 ....
4	14	24	34 ....
5	15	25	35 ....
6	16	26	
7	17	27	
8	18	28	
9	19	29	
10	20	30	

Your answers for Part II should be placed on paper provided by the school.

The declaration below should be signed when you have completed the examination.

I do hereby affirm, at the close of this examination, that I had no unlawful knowledge of the questions or answers prior to the examination, and that I have neither given nor received assistance in answering any of the questions during the examination.

_____

Signature

The University of the State of New York

REGENTS HIGH SCHOOL EXAMINATION

SEQUENTIAL MATH – COURSE III

ANSWER SHEET

Pupil .... Sex: ☐ Male ☐ Female Grade ....

Teacher .... School ....

Your answers to Part I should be recorded on this answer sheet.

Part I

Answer 30 questions from this part.

1	11	21	31 ....
2	12	22	32 ....
3	13	23	33 ....
4	14	24	34 ....
5	15	25	35 ....
6	16	26	
7	17	27	
8 ... . , ...	18	28	
9	19	29	
10	20	30	

Your answers for Part II should be placed on paper provided by the school.

The declaration below should be signed when you have completed the examination.

I do hereby affirm, at the close of this examination, that I had no unlawful knowledge of the questions or answers prior to the examination, and that I have neither given nor received assistance in answering any of the questions during the examination.

_____

Signature

SEQUENTIAL MATH – COURSE III

_____

Part I Score	....
Part II Score	....
Total Score	....
Rater's Initials:	....

ANSWER SHEET

Pupil .... Sex: ☐ Male ☐ Female   Grade ....

Teacher .... School ....

Your answers to Part I should be recorded on this answer sheet.

Part I

Answer 30 questions from this part.

1	11	21	31 ....
2	12	22	32 ....
3	13	23	33 ....
4	14	24	34 ....
5	15	25	35 ....
6	16	26	
7	17	27	
8	18	28	
9	19	29	
10	20	30	

Your answers for Part II should be placed on paper provided by the school.

The declaration below should be signed when you have completed the examination.

I do hereby affirm, at the close of this examination, that I had no unlawful knowledge of the questions or answers prior to the examination, and that I have neither given nor received assistance in answering any of the questions during the examination.

_____

Signature

The University of the State of New York

REGENTS HIGH SCHOOL EXAMINATION

SEQUENTIAL MATH – COURSE III

Part I Score	. . . .
Part II Score	. . . .
Total Score	. . . .
Rater's Initials:	. . . .

ANSWER SHEET

Pupil . . . . Sex: ☐ Male ☐ Female   Grade . . . .

Teacher . . . . School . . . .

Your answers to Part I should be recorded on this answer sheet.

Part I

Answer 30 questions from this part.

1	11	21	31 . . . .
2	12	22	32 . . . .
3	13	23	33 . . . .
4	14	24	34 . . . .
5	15	25	35 . . . .
6	16	26	
7	17	27	
8	18	28	
9	19	29	
10	20	30	

Your answers for Part II should be placed on paper provided by the school.

The declaration below should be signed when you have completed the examination.

I do hereby affirm, at the close of this examination, that I had no unlawful knowledge of the questions or answers prior to the examination, and that I have neither given nor received assistance in answering any of the questions during the examination.

_____

Signature

The University of the State of New York

REGENTS HIGH SCHOOL EXAMINATION

SEQUENTIAL MATH – COURSE III

Part I Score	. . . .
Part II Score	. . . .
Total Score	. . . .
Rater's Initials:	. . . .

ANSWER SHEET

Pupil . . . . Sex: ☐ Male ☐ Female Grade . . . .

Teacher . . . . School . . . .

Your answers to Part I should be recorded on this answer sheet.

Part I

Answer 30 questions from this part.

1 . . . .   11 . . . .   21 . . . .   31 . . . .

2 . . . .   12 . . . .   22 . . . .   32 . . . .

3 . . . .   13 . . . .   23 . . . .   33 . . . .

4 . . . .   14 . . . .   24 . . . .   34 . . . .

5 . . . .   15 . . . .   25 . . . .   35 . . . .

6 . . . .   16 . . . .   26 . . . .

7 . . . .   17 . . . .   27 . . . .

8 . . . .   18 . . . .   28 . . . .

9 . . . .   19 . . . .   29 . . . .

10 . . . .   20 . . . .   30 . . . .

Your answers for Part II should be placed on paper provided by the school.

The declaration below should be signed when you have completed the examination.

I do hereby affirm, at the close of this examination, that I had no unlawful knowledge of the questions or answers prior to the examination, and that I have neither given nor received assistance in answering any of the questions during the examination.

_____

Signature

The University of the State of New York

REGENTS HIGH SCHOOL EXAMINATION

SEQUENTIAL MATH – COURSE III

———

Part I Score	. . . .
Part II Score	. . . .
Total Score	. . . .
Rater's Initials:	. . . .

ANSWER SHEET

Pupil . . . .   Sex: ☐ Male ☐ Female   Grade . . . .

Teacher . . . .   School . . . .

Your answers to Part I should be recorded on this answer sheet.

Part I

Answer 30 questions from this part.

1	11	21	31 . . . .
2	12	22	32 . . . .
3	13	23	33 . . . .
4	14	24	34 . . . .
5	15	25	35 . . . .
6	16	26	
7	17	27	
8	18	28	
9	19	29	
10	20	30	

Your answers for Part II should be placed on paper provided by the school.

The declaration below should be signed when you have completed the examination.

I do hereby affirm, at the close of this examination, that I had no unlawful knowledge of the questions or answers prior to the examination, and that I have neither given nor received assistance in answering any of the questions during the examination.

_____

Signature

About the Author

David Kahn studied Applied Mathematics and Physics at the University of Wisconsin and has taught courses in calculus, pre-calculus, algebra, trigonometry, and geometry at the college and high school level. He has taught Princeton Review courses for the SAT I, SAT II, GRE, GMAT, and the LSAT, as well as trained other teachers in the same. He has been an educational consultant for many years and has tutored more students in mathematics than he can count! (Actually, he can count them. Take a look at the acknowledgments.)

NOTES

NOTES

NOTES

NOTES

NOTES

www.review.com

Expert Advice

Counselor-O-Matic

Pop Surveys

Paying for it

www.review.com

THE PRINCETON REVIEW

Getting In

Word du Jour

www.review.com

www.review.com

College Talk

Find-O-Rama College Search

www.review.com

Best Schools

SAT Survival

www.review.com

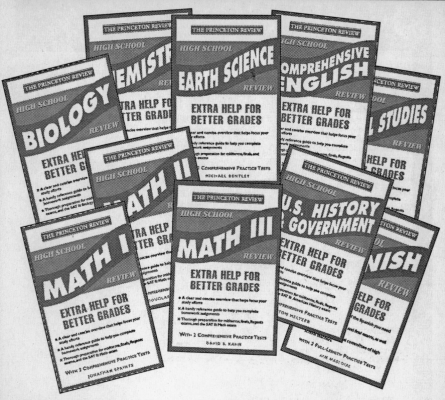

FIND US...

International

Hong Kong
4/F Sun Hung Kai Centre
30 Harbour Road, Wan Chai,
Hong Kong
Tel: (011)85-2-517-3016

Japan
Fuji Building 40, 15-14
Sakuragaokacho, Shibuya Ku,
Tokyo 150, Japan
Tel: (011)81-3-3463-1343

Korea
Tae Young Bldg, 944-24,
Daechi- Dong, Kangnam-Ku
The Princeton Review- ANC
Seoul, Korea 135-280,
South Korea
Tel: (011)82-2-554-7763

Mexico City
PR Mex S De RL De Cv
Guanajuato 228 Col. Roma
06700 Mexico D.F., Mexico
Tel: 525-564-9468

Montreal
666 Sherbrooke St.
West, Suite 202
Montreal, QC H3A 1E7 Canada
Tel: (514) 499-0870

Pakistan
1 Bawa Park - 90 Upper Mall
Lahore, Pakistan
Tel: (011)92-42-571-2315

Spain
Pza. Castilla, 3 - 5° A, 28046
Madrid, Spain
Tel: (011)341-323-4212

Taiwan
155 Chung Hsiao East Road
Section 4 - 4th Floor,
Taipei R.O.C., Taiwan
Tel: (011)886-2-751-1243

Thailand
Building One, 99 Wireless Road
Bangkok, Thailand 10330
Tel: (662) 256-7080

Toronto
1240 Bay Street, Suite 300
Toronto M5R 2A7 Canada
Tel: (800) 495-7737
Tel: (716) 839-4391

Vancouver
4212 University Way NE,
Suite 204
Seattle, WA 98105
Tel: (206) 548-1100

National (U.S.)

We have over 60 offices around the U.S. and
run courses in over 400 sites. For courses and locations
within the U.S. call 1 (800) 2/Review and you will be
routed to the nearest office.